'The perspective on space Bowen brings is meant to—and does—challenge the reader. There is much of value here to chew on, and any serious space power student or practitioner needs to pay careful attention to his arguments.'

Everett Dolman, Professor of Comparative Military Studies and Strategy, US Air Force's Air Command and Staff College, and author of *Astropolitik*

'This book sets out to describe how the militarisation of space is not a new policy issue or trend in world politics, but a long-established historical fact and present reality. Original and compelling, this will be of interest to anyone seeking a history of technology's role in space strategy.'

John J. Klein, author of *Understanding Space Strategy*

'A fresh, timely and detailed account of the military-political origins of spacepower and the future of warfare in space.'

Deganit Paikowsky, author of *The Power of the Space Club*

'Bowen persuasively demonstrates that space technology is the outcome of war, and driven by military applications. By exposing the hidden face of space, this book makes a critical intervention towards a peaceful world.'

Alice Gorman, author of *Dr Space Junk vs the Universe*

'*Original Sin* makes the valid argument that, far from us existing in a new "space race", the history of space exploration has always been about power, politics, political economy and the pursuit of national interests on the global stage. A satisfying and insightful read from a deeply knowledgeable expert.'

Tim Stevens, Senior Lecturer in Global Security and head of the King's Cyber Security Research Group, King's College London

'Just as space security is becoming more urgent, Bleddyn Bowen has delivered a remarkable account of the seventy-year-long militarised Global Space Age. *Original Sin*'s compelling narrative will be of great interest to scholars of military strategy, Cold War historians, and policymakers.'

Aaron Bateman, Assistant Professor of History and International Affairs, George Washington University

ORIGINAL SIN

BLEDDYN E. BOWEN

Original Sin

Power, Technology and War in Outer Space

OXFORD UNIVERSITY PRESS

Oxford University Press is a department of the
University of Oxford. It furthers the University's objective
of excellence in research, scholarship, and education
by publishing worldwide.

Oxford New York
Auckland Cape Town Dar es Salaam Hong Kong Karachi
Kuala Lumpur Madrid Melbourne Mexico City Nairobi
New Delhi Shanghai Taipei Toronto

With offices in
Argentina Austria Brazil Chile Czech Republic France Greece
Guatemala Hungary Italy Japan Poland Portugal Singapore
South Korea Switzerland Thailand Turkey Ukraine Vietnam

Published in the United States of America by
Oxford University Press
198 Madison Avenue, New York, NY 10016

Library of Congress Cataloging-in-Publication Data is available
Bleddyn E. Bowen.
Original Sin: Power, Technology and War in Outer Space.
ISBN: 9780197677315

Printed in the United Kingdom on acid-free paper
by Bell and Bain Ltd, Glasgow

I Dad, Mary, Vanessa, ac Alwyn

‘If we say that we have no sin, we deceive ourselves,
and the truth is not in us.’

1 John, 1:8, World English Bible

CONTENTS

ACKNOWLEDGEMENTS

No book exists in a social and intellectual vacuum, despite the isolating effects of a pandemic. I have drawn upon the help, support, and knowledge of many people throughout the research and writing of this book. In particular, a massive thanks to Aaron Bateman at the Space Policy Institute, George Washington University, who reviewed most of the book and provided excellent insights, reading suggestions, and information across several areas. Special thanks to Cameron Hunter and Andrew Futter, both friends and colleagues at the School of History, Politics, and International Relations (HyPIR) at the University of Leicester who reviewed parts of the book and also served as reliable touchstones for ideas, arguments, insights from their areas of expertise, and general support in other ways, involving the pub when pandemic restrictions allowed! Thanks also to HyPIR for institutional support with production requirements at the end of the publishing process. *Diolch* to Catrin Edwards and David Morgan-Owen who humoured conversations on the book's general arguments and helped me make sense to audiences outside of my own area. Thanks also to those I have the privilege of being able to talk to about goings-on in space: Martin Barstow, Mia Brown, Ralph 'Dinz' Dinsley, Theresa Hitchens, Andrew Jones, Andy Netherwood, Christopher Newman, Victoria Samson, and Brian Weeden. Any errors or shortcomings in the book are of course my own.

I also want to thank the many undergraduate and postgraduate students at HyPIR who took my astropolitics classes. Writing my lectures and talking through these subjects in seminars with you has helped me improve my arguments and narratives. It's always interesting to see how astropolitics is seen by those only just discovering it and who have not been that interested in space—prior to sitting in my classes!

Thanks as well to Michael Dwyer for getting this book started when we had that conversation in London, and his colleagues at Hurst for all the support and flexibility as the final phases of the process came about. Thanks to Tony Lyons for creating the lovely design for the book cover and the diagram. I would also like to thank the anonymous peer reviewers for their comments that helped improve the manuscript. Thanks are due as well to the numerous journalists, media producers, panel and guest lecture organisers, and podcast hosts who have invited me to speak or interview with you as you have allowed me to further develop and refine my thoughts on the subject of this book, as well as improve my ability to convey complex and obscure academic matter to broader audiences.

Fel yr arfer rwy'n hynod ddiolchgar i'r teulu, yn enwedig dros Nadolig 2021 wrth geisio cwrdd â'r amserlen dynn i gwblhau'r llyfr yng nghanol dyletswyddau dysgu a gweinyddu yng Nghaerlŷr. Rwy'n gobeithio bydd y llyfr hwn yn haws i'w ddarllen i gymharu â'r un cyntaf—ond cawn weld!

COMMON ABBREVIATIONS

A2/AD	Anti-Access/Area Denial
ABM	Anti-Ballistic Missile
ASAT	Anti-satellite
ASLV	Advanced Space Launch Vehicle
BMD	Ballistic Missile Defence
C4ISR	Command, Control, Communications, Computers, Intelligence, Surveillance, Reconnaissance
CA	Coarse/Acquisition (civil GPS signal)
CBERS	China-Brazil Earth Resources Satellite
CHEOS	China High Resolution Earth Observation System
CIA	Central Intelligence Agency
CNES	*Centre national d'études spatiales*/National Centre for Space Studies
COMINT	Communications Intelligence
DoD	Department of Defense
DRDO	Defence Research and Development Organisation
DSP	Defense Support Program
ELDO	European Launcher Development Organisation. One of two precursor agencies to ESA.
ELINT	Electronic Intelligence
eLORAN	Enhanced Long-Range Navigation
EORSAT	Electronic Intelligence Ocean Surveillance Satellite
ESA	European Space Agency
ESRO	European Space Research Organisation. One of two precursor agencies to ESA.

EU	European Union
FBI	Federal Bureau of Investigation
FOBS	Fractional Orbital Bombardment System
GCHQ	Government Communications Headquarters
GEO	Geostationary/Geosynchronous Orbit
GLONASS	Globalnaya Navigazionnaya Sputnikovaya Sistema/ Global Navigation Satellite System
GNSS	Global Navigation Satellite System
GPS	Global Positioning System
GSLV	Geosynchronous Space Launch Vehicle
GSSAP	Geosynchronous Space Situational Awareness Program
HAND	High Altitude Nuclear Detonation
HBTSS	Hypersonic and Ballistic Tracking Space Sensor
HEO	Highly Elliptical Orbit
HGV	Hypersonic Glide Vehicle
HUMINT	Human Intelligence
ICBM	Intercontinental Ballistic Missile
IGY	International Geophysical Year
IMINT	Imagery Intelligence
INCOSPAR	Indian National Committee on Space Research
IR	International Relations
IRBM	Intermediate Range Ballistic Missile
IS	Istrebitel Sputnikov/Satellite Destroyer
ISR	Intelligence, Surveillance, Reconnaissance
ISRO	Indian Space Research Organisation
ISS	International Space Station
IT	Information Technology
ITU	International Telecommunications Union
LEO	Low-Earth Orbit
MANPADS	Man-Portable Air Defence System

MaRV	Manoeuvrable Re-entry Vehicle
MEO	Medium-Earth Orbit
MIDAS	Missile Defense Alarm System
MoD	Ministry of Defence
MRBM	Medium-Range Ballistic Missile
MTCR	Missile Technology Control Regime
NACA	National Advisory Committee for Aeronautics
NASA	National Aeronautics and Space Administration
NATO	North Atlantic Treaty Organization
NRO	National Reconnaissance Office
OODA	Observe, Orient, Decide, Act
OST	Outer Space Treaty
PGM	Precision Guided Munition
PLA	People's Liberation Army
PLARF	People's Liberation Army Rocket Force
PLASSF	People's Liberation Army Strategic Support Force
PNT	Position, Navigation, Timing
PSLV	Polar Space Launch Vehicle
RAF	Royal Air Force
RMA	Revolution in Military Affairs
RORSAT	Radar Ocean Reconnaissance Satellite
SA	Selective Availability (ability to switch off all non-military GPS signals)
SAR	Synthetic Aperture Radar
SATCOM	Satellite Communications
SBI	Space-based Interceptor
SBIRS	Space-based Infrared System
SDI	Strategic Defense Initiative
SIGINT	Signals Intelligence
SITE	Satellite Instructional Television Experiment
SLBM	Submarine-Launched Ballistic Missile

SLV	Space Launch Vehicle
SPACEINT	Space Intelligence
SPOT	Satellite Pour l'Observation de la Terre/Satellite for Observation of Earth
SRBM	Short-Range Ballistic Missile
SSA	Space Situational Awareness
SSBN	Nuclear-powered ballistic missile-carrying submarine
STM	Space Traffic Management
STS	Space Transportation System
TELINT	Telemetry Intelligence
TENCAP	Tactical Exploitation of National Capabilities Program
UFWD	United Front Work Department
UN	United Nations
UNCOPUOS	United Nations Committee on the Peaceful Uses of Outer Space
USAF	United States Air Force
USSF	United States Space Force

INTRODUCTION

Gorau chwedl, gwirionedd

The best tale is truth

Founding the U.S. Space Force (USSF) in 2019, U.S. President Donald Trump proclaimed that space had become a 'war-fighting domain' and that the new branch of the U.S. military would need to dominate 'the ultimate high ground' of outer space in the wars of the future.[1] The new military service beamed the military uses of outer space and the prospect of space warfare into the mainstream of mass media and expert military, security, and policy communities. In the wake of the USSF's establishment, many of Washington's allies responded with their own military space reorganisations. The North Atlantic Treaty Organization (NATO) declared space an 'operational domain' and stated that significant attacks on satellite systems could constitute an Article V-triggering event—the alliance's collective self-defence clause responding to acts of war on any member of the alliance.[2] The South Korean Chief of Staff of the Air Force commented that 'space is no longer a mere area of curiosity; rather, it has now become a key domain for our national security and only rigorous preparation will ensure our survival in the future space environment'.[3] The French Minister for the Armed Forces commented in the foreword to the 2019 Defence Space Strategy that space 'is now a "new front" we have to defend.'[4] The Air Vice-Marshal of the Royal Australian Air Force equivocated, saying that 'space is a war-fighting domain but we're not going to militarise space'.[5] Australia, Britain, and France have each set up new 'Space Command' structures in their defence ministries as the United States re-established the one it shut down in 2002. In the Biden administration there has allegedly been a 'dawning realisation' that U.S. satellite systems that underpin its military and economic power are vulnerable to many methods of attack.[6]

Casual observers or those new to space can be forgiven for thinking there is a new scramble over the domination of outer space, or that the militarisation or weaponisation of space are novel, unprecedented issues. According to more dramatic and alarmist opinions, it is a reflection of a so-called 'great power competition' that may possibly extend to the Moon as U.S.-Russian-Chinese relations sour.[7] For many, 'space' conjures majestic images of the planets, stars, and nebulae, the Apollo Moon landings, Martian rovers, or guitar-wielding astronauts on the International Space Station. These encapsulate space as a realm where humans work together in the pursuit of knowledge for the common benefit of all humankind. Yet these feats of science and exploration in space, which inspire many, are only the tip of the iceberg of a seventy-year-long militarised Global Space Age. Robotic adventures on the Moon or astronauts on rival space stations are high-profile scientific or exploration activities that have little bearing on terrestrial military power today.

High-technology economic, industrial, and military power rely on thousands of satellites orbiting Earth, owned or operated by over eighty countries or companies registered within them. This is not something that happened overnight. Military and intelligence space programmes, such as spy satellites and military satellite communications systems, have been deliberately hidden for most of our Global Space Age. Seventy years of techno-industrial competition in the global space sector has been known to only those inside it and to liberal arts and humanities academics who cared to look for it. Publicly-released satellite imagery of the Russian invasion of Ukraine in 2022 demonstrated spy satellite technologies to mass audiences that governments and militaries have enjoyed for decades. In this book I try to expose what has long been a secretive and niche area of expertise to a much wider audience, and show what the growth of technological infrastructures, military power, and economic competition in space means for the practice and study of International Relations (IR) today. Space is not some new environment that transforms our social and political world, or somewhere that we cannot apply our terrestrial political concepts and practices. As the Welsh proverbs at the top of each chapter imply, our terrestrially-bound social experiences, conventions, and concepts apply to our understandings of outer space. Despite the technical wizardry involved, the Global Space Age is ultimately about what people do with machines, why they do these things, and how they affect other humans, their freedoms, restrictions, and environment.

Space is not a new or emerging military and security issue. Earth orbit has always been a place of military exploitation and economic competition since satellites began flying above the atmosphere.[8] The 'militarisation of space' is not a new trend in world politics in the twenty-first century—it is a long-established historical fact and present reality. The question of setting up an independent Space Force in the Pentagon is almost as old as its parent service, the U.S. Air Force (USAF), but it took presidential enthusiasm to finally make it happen and push it through the bureaucracy. It mostly amounted to a 'rearranging of the deckchairs', a bureaucratic reorganisation, of the USAF's existing duties and capabilities in Earth orbit. It was not a massive break in direction in U.S. space policy or strategy, or the herald of an arms race in outer space. It follows China's standing up of the 'Space Troops' in the People's Liberation Army Strategic Support Force (PLASSF) in 2015, which manages China's military satellite infrastructure, which itself dates back several decades. In the same year, Russia re-established its Space Forces, a branch within the Russian Aerospace Forces.

Military and spying activities in space are not just as old as humanity's Global Space Age, in fact they *are* the Global Space Age. Creating missiles to carry nuclear bombs and spy satellites to find targets for those city-destroying bombs drove the earliest rocket programmes that put the first satellites and people into space. The fundamental space technologies created by the military-industrial complexes of the Second World War and the Cold War committed space technology's original sin as a tool for warfare, intelligence gathering, and self-interested political-economic power. Space technologies were created by and reinforced the trends of the 'replacement of the market by command, public-sector investment in private-sector technological development, the role of technocratic elites in the policy process.'[9] The Space Age was born from these military-industrial complexes as they expanded into nuclear-missile-space technologies, creating new national and transnational space industries that the major powers are jockeying for influence over today, showing no easing of the original sin of space technology.

Military infrastructure and spy satellites in orbit are now among the more important techno-geographic, physical, material aspects of international relations. 'Spacepower' is now an important provider and instrument of state and military power, like seapower and airpower. Spacepower is the use and denial of thousands of machines in Earth orbit that provide important data gathering and communications services for governments, militaries, and economic infrastructure on Earth itself. Scott Pace, the former Executive Director of the U.S. National Space Council in the Trump administration, commented that 'heavy-lift rockets are strategic national assets, like aircraft carriers.'[10] The power and influence that space technology provides is like any other multi-generational, multi-billion dollar, and globe-spanning technological infrastructure on Earth. This creates power which can be exercised as influence upon others, or used to do things in the face of opposition. For some political theorists, politics is simply the exercise of influence.[11] Spacepower is pursued by states for the purposes of war, development, and prestige, meeting specific needs and goals because the expense and difficulty of space activities require huge resources marshalled by a central authority to do them. The Global Space Age has seen the creation of multiple technological infrastructures that many states have spent decades building to make their military forces more effective and high-technology economies more competitive today.

Countries or states that control space technologies, orbital infrastructures, and space industries influence the use and rules of using outer space. They are 'space powers', in a similar way to how we may understand 'sea powers' controlling ships, ports and shipping industries, influencing the uses of the seas for military, political, and economic purposes. Spacepower is relevant to any state that wishes to develop its capacities for war, development, and prestige. Spacepower provides useful tools in each of those terrestrial goals, and often manifests in what can be called 'space policy'. Space policy usually means the activities and goals of a state and an agency that in some way is related to space. Space policy does not restrict itself to space science, space commerce, or military space activities either. Space is a place, and not an abstract or single policy issue.

'Space policy' is a misnomer because one policy could not govern an entire environment. We do not speak in terms of a state's 'sea policy' or 'air policy', but we do have maritime strategy, marine science projects, aerial bombing campaigns, and aviation industries. Britain has maritime doctrine and defence policy to explain and direct the use and development of the Royal Navy, but they have little to say about the deployment of the scientific vessel RRS *Sir David Attenborough* and its uncrewed submersible *Boaty McBoatface* in their marine science missions. Space is just as diverse—military space strategies and space data commerce strategies would have little to say about space science and vice versa. Yet 'space policy' is often taken to mean all or any of them. They are very different kinds of activity governed by different parts of the state that happen in the same physical place, like the military, commercial, and scientific uses of the seas and the air. A Chinese rover wielding nothing more dangerous than a potato on the far side of the Moon has very little to do with China's military satellite infrastructure in orbit above Earth. This book is about the military, intelligence, and security aspects of spacepower. Space exploration and scientific projects play a minor role in terms of practical military and economic impact.

This political reality of outer space is at odds with the spirit of the 1967 Outer Space Treaty, which declares in Article I that 'the exploration and use of outer space... shall be carried out for the benefit and in the interests of all countries, irrespective of their degree of economic or scientific development, and shall be the province of all mankind'.[12] The reality is that space technologies have been developed by specific people, countries, and organisations for a mixture of reasons, including self-serving ones, and some people have benefitted whilst others have suffered from or been threatened by them. This book is a journey through the bloody origins of space technologies and how their continued development around the globe reflects the self-interested and self-serving nature of the political and economic coalitions and networks of interests that build them. Firmly grounded in political and strategic reality, this book is a jarring rejoinder to idealist, egalitarian, and utopian perceptions and visions of space as exclusively a place of unique science and international cooperation. Space is just as politically and socially complicated and tragic as Earth due to the original sin of space technology and its militaristic heritage.

The era of astropolitics as a specialist field of study and activity has long since arrived. Astropolitics refers to the political aspects of any activity in or to do with outer space. No activity escapes politics, not even international cooperation in space science. Who cooperates with whom, why, and on what terms is political and not necessarily something that happens for purely selfless reasons. Though, looking at the scholarly field of International Relations (IR), it does not feel like we are living in a space age. Humanities and social science research on space remains a very niche field, even though IR scholar Michael Sheehan explained in 2007 that 'space and politics are, and always have been, inseparably interlinked... Politics has always been at the heart of mankind's exploration and utilisation of space, and the space programmes themselves have never been able to transcend terrestrial international politics'.[13] Questioning why some countries develop their own space systems rather than relying on another's is an astropolitical matter. To study astropolitics is to investigate behaviour and its consequences in outer space according to political and social concepts, and not simply recounting the partisan politics of a legislature that factors into space policy-making. A space policy of any kind is political. Recognising the impacts of space on people, societies, and polities, good and bad, is political analysis. Astropolitics therefore looks at the 'why', 'should', and 'so what' kind of questions that space activities pose us with, not the relatively simple technical questions of whether something 'can' be done.

Space is not just for the scientists—we need to discuss the political, strategic, and social causes and consequences of our Global Space Age. This book looks at these political aspects of space technologies through the international relations and power politics of the states that built them. In other words, it is a technopolitical and strategic story of the Global Space Age, which connects large technological systems with political institutions, practices, and interests which in turn create new forms of power, influence, and possibilities for political action, or constrain them.[14] In the process, the key technologies of the Global Space Age are also explained to a non-technical audience, hopefully producing a book that is as enlightening to politically-versed readers and researchers who are entirely new to space activities as well as space scientists and engineers who are interested in the politics of their crafts.

Argument

There are three main arguments in this book. First, space technologies have not been developed for the benefit of all humankind. As Michael Sheehan points out, the 'central driving force for all space programmes has been political objectives.'[15] Space technologies were and are developed to meet military-political objectives. In Christian theology, original sin refers to 'both the first historical sin of humankind and the bondage to sin that afflicts all of humankind thereafter'.[16] Space systems—rockets, satellite constellations, their ground infrastructures, and peripherals on Earth—have not escaped their militarised and competitive origins, or original sin. Yet we need not accept a fatalist, doomed, or cynical future only redeemable by a divine entity as some Augustinian Christian doctrines might suggest. For those wishing to transform things or prevent them from getting any worse, a realistic assessment of the political and military origins and interests in space technologies must be the starting point. Political reform cannot happen without understanding past and present political reality in space.

The biggest stain on benign images of space technology and the Space Age is that it helped bring about the prospect of nuclear Armageddon. Further, breakthrough technical advances in rocketry came at the cost of tens of thousands of slave labourers worked to death in the Mittelwerk factory in Nazi Germany. The progenitor of much missile and rocket technologies—the German V-2—was not only a missile strike vehicle that killed 3,000 people by the end of the Second World War but killed 20,000 through the use of genocidal labour practices. In Mittelwerk were 'slaves like living skeletons… thousands of corpses stacked here and there as garbage awaiting pickup.'[17] Historians are settled on the fact that chief designer Wernher von Braun knew of the horrifying conditions at Mittelwerk and the concentration camps that served it. Concerns about the Nazi sympathies and guilt of the German rocket experts were brushed aside in the post-war United States lest their expertise fall into Soviet hands.[18]

Were it not for its military potential, the handful of early space powers in the 1950s and 1960s would not have poured huge sums of money into what were once exotic, temperamental, and unproven technologies. As the political theorist Daniel Deudney argues, 'the full extent of the involvement of space in the apparatus for nuclear annihilation remains dramatically underappreciated.'[19] Using rockets to hurl city-busting nuclear bombs around Earth and developing satellites to better aim them at each other, the first space powers had opened up Earth orbit for a smorgasbord of military and intelligence-gathering applications with satellites. Military and political patrons have long funded scientific development, and space science is no different.[20]

Space technology's original sin goes much further than missile and nuclear technology. The satellite and Moon races of the 1950s and 1960s were in part deliberate covers for developing essential military space technologies such as more capable missiles, nuclear warheads, and spy satellites. Professional communities refer to the civil and military uses of such technologies as 'dual use', and the multiple uses of base technologies are common in space as they are on Earth. Yet much of the original technological drives were military first, with civil and commercial applications developed afterwards.

The U.S. Space Shuttle was in part originally designed and funded to meet the Pentagon's and the Intelligence Community's needs. The Hubble Space Telescope is an adapted KH-11 spy satellite that simply looks to the cosmos rather than down on Earth. The International Space Station was a product of intense political bargaining between the U.S., Europe, and Japan, and included Russian deal-making on missile technology exports controls which tore up a prior Soviet-Indian rocket engine sales agreement. Civilian and military space agencies often cooperate on major projects and activities and personnel move from one side to another. On the ground, many space powers benefitted from previous imperial and colonial conquests. Imperial territories became home to satellite control systems and launch sites.

Space technologies support a wide gamut of military capabilities at the cutting edge of the state's ability to kill, destroy, and control populations, having moved on from its original use in nuclear warfighting to now helping tip the scales in non-nuclear or conventional warfare and counter-insurgency campaigns. Behind international cooperation and science in space is a global drive building military and economic infrastructure made up of thousands of satellites and other technological systems that we now cannot easily do without.

Recognising that some of the most iconic achievements in space science and exploration are products of, or highly influenced by, military space technologies and interests dampens the alarmism some want to generate in media narrative and policy discussion by simply pointing out the mere existence of 'militarised' aspects of certain space activities or management structures.[21]

The anarchy of the international system of space-faring states that compete militarily among themselves exists in space as the tethers of modernity have reached into Earth orbit; it is not some distant dystopian vision of the future. Without a supreme authority above them, the major states and most powerful governments of Earth use space to meet their own goals, check the interests of others, and impose their own influence on others where possible. As Geoffrey Herrera succinctly put it, 'the technologies we have (and don't have) are not inevitable. They are not separate from social, economic, or political forces.'[22] Global space infrastructures are the accumulation of generations of decisions and choices made by people and organisations for practical purposes and ambitions, and often reflect the interests and preoccupations of the major powers of the day and the technopolitical and economic elites that live within them.

Second, the Space Age we live in is a global one. Whilst the Space Age may be global, our knowledge of it is not. Space in the Cold War was not simply a story of two superpowers, and neither is it a story of a looming U.S.-China rivalry today. Today, over 80 states, wealthy and poor, large and small, are using space technologies, building their own space industries, or have their own ambitions of space technological development. Space technology is not a luxury for the superpowers of the day. It is part of high-technology industries and infrastructure. Activities, technologies, and bureaucracies in space need to be looked at in a global context involving many countries and societies. At least in English-language academic books, most knowledge of space policy and politics, past and present, focuses on the United States,[23] and to a lesser extent, the Soviet Union and Europe.[24] There are very few books that present a global view of the international politics of space.[25] This book tries to do just that and make a contribution to such ranks, bringing in the many specialist research perspectives needed for a global, materialist view of astropolitics.

Space technology has diffused through transnational networks and political connections. Rather than imitating the two superpowers, other states charted their own waypoints in space-technological development. Much of space history focuses on the perspectives of the distant metropoles of the major space powers, and not the marginalised, far-flung imperial holdings on Earth that were forced to host terrestrial space infrastructure, such as spaceports and ground stations which directly impacted marginalised peoples. Spacepower impacts everyone and presents opportunities and dangers for every polity on Earth. It is as global a subject as the study and practice of international relations can physically get, whilst also imposing a multitude of local effects on Earth. U.S.-crewed spaceflights were supported from Diego Garcia, where a facility was built after the indigenous Chagos Islanders had been forcibly removed by the British. The Soviet Union's major spaceport was located in Kazakhstan, the site of Moscow's many imperial grand designs for generations in 'modernising' the Steppe.

Global satellite networks have pushed globalisation and global interconnections further than what the industrial empires of the nineteenth century had already imposed with telegraph cables and twentieth-century telecoms corporations achieved with telephone and fibre optic cables. Beyond the superpowers, the other major powers of the Cold War invested in and developed what they saw as the new essential technologies and industries that ensured they would not be totally dependent on the superpowers for economic development, intelligence, political power, and military potential. Approaching things from a global and connected mindset, Brieg Powel argues, 'reveals the ongoing formation of substances through relations and, importantly, exposes the breadth of interactions that are essential to the everyday functioning of international relations.'[26] The transnational patterns of competition, cooperation, coordination, and controlling space technological development alter between allies and rivals as each want to meet their own interests.

The technopolitics of space infrastructure is an important feature of international relations and global politics. Technopolitics refers to the political and social values that the creators of technology intentionally or unintentionally put into the machines they create. Technical devices are built by some people for specific reasons and to benefit specific people and groups. These are the political and social 'values' of technologies. Cynthia Enloe tells us that 'asking how something has been made implies it has been made by someone. Suddenly there are clues to trace; there is also blame, credit and responsibility to apportion, not just at the start but at each point along the way.'[27] Things that have been made for some humans' needs that then become useful start to alter the behaviour of others around them in turn. Geoffrey Herrera argues that:

> certain, significant technologies are politically malleable in their development and diffusion phases, yet grow increasingly harder to change, and have a greater impact on politics, once mature... Technology is a very important determinant of the overall capability for interaction in any social system—including the international. We can theorise a class of internationally significant technologies by investigating those that affect the interaction capacity of the international system. Technologies of communications, transportation, and violence feature prominently in this class.[28]

Space technologies are technologies of communications, transport, and violence. These technologies are maturing, not emerging, and are benefitting some whilst negatively impacting others. The militarisation of outer space over the past 70 years is not an abstract, non-human, or 'natural' force, but the result of cumulative choices made by certain people, organisations, and communities for their own reasons. Whilst technologies can impose some determining influences on people, those technologies themselves were built by some people to meet their own specific ends and are also products of social and political values and forces. The original sin of the Global Space Age was not inevitable; we must recognise it as the product of choice and not a victimless, abstract phenomenon that 'just happened'. The original sin of space technology is an omnipresent social reality in humanity's technopolitical presence in outer space today. Dealing with that reality, and perhaps changing it or mitigating its worst effects, requires an understanding that it is the product of human choices and their politics, not some blameless, non-human one-directional trend of 'technological progress'.

Space technologies and orbital satellite infrastructures provide power for those who wield them as they enable actions that otherwise could not be done, or not as easily. They multiply other forms of state power, such as military and economic power. This allows power over others as it provides control over the infrastructure others depend upon.[29] This is spacepower, 'a diverse collection of activities and technologies in space or to do with outer space... how any actor can use outer space and what it possesses' in space or to support space activities.[30] Satellites, rockets, control stations, and peripherals are manifestations or tangible, material expressions of spacepower. Spacepower can provide forms of 'structural power'[31] which limit or determine the options and actions of others so that the behaviour of other states and actors becomes more predictable along economically and politically efficient lines. In turn more users want to take advantage of that infrastructure and build their own technologies around it, such as the U.S. Global Positioning System (GPS), and come to depend on it. That kind of development can be influenced by those who control that underlying infrastructure, and those that do not want to suffer from that level of control by others must build their own systems, as Russia, the EU, and China have done with satellite navigation systems.

The economic system dominating Earth depends on several thousand active satellites in orbit, along with the elaborate terrestrial infrastructure that supports it all. Satellites provide a large range of communications and data-gathering services that make modern military forces more lethal and capable, economies more efficient, lucrative, and digital, and provide a number of critical infrastructure services such as weather, climate, and agricultural monitoring systems. Many newcomers to space industries are developing their own military or economic space systems or partnerships on the global marketplace, keen to take advantage of, and trying not to be taken advantage by, these ethereal tethers of modernity that reach up to 40,000km above Earth. Much of this has gone unnoticed not only by the public but by many scholars in the study of International Relations, as 'infrastructure is boring, even when it is civilian and open. One only notices its absence when it does not work.'[32]

Large technical systems such as satellite systems in space, and their supporting ground infrastructure, are firmly in the maturation and stabilisation phase today, as they literally took off decades ago and many core concepts were proven. Most space systems today are variations on long-proven space technological concepts and decades old. They are not simply material, physical systems either. Social structures and political values are imbued in their creation and are designed to take advantage of, or improve upon, other technological systems that help a central authority exert its control over resources and people, not unlike continental railroads and telegraphy systems.[33] Large technological systems in space give structural power on a literally global scale to those who control them. A small number of key space powers and corporations that reside within them provide these 'global utilities' for many others wishing to take part in it, benefit from it, or catch up themselves, replicating the political-economic divides on Earth in orbit. Only by viewing spacepower as part of a Global Space Age can we begin to grasp and investigate these political technological forces—or technopolitics—at work today.

Third, as military space infrastructure in Earth orbit has come to directly impact conventional military power on Earth, its potential as a zone for warfare resembles that of a cosmic coastline rather than the 'ultimate high ground', as often claimed in military writings and commentary. Earth orbit is just another environment that is used for warfare; there is nothing special or different about space in terms of military strategy and politics today. Space is not a special place that is a sanctuary from conflict. The USA has led the military exploitation of the advantages of space technologies in fighting numerous wars on Earth with space-supported armed forces. Russia has rebuilt and modernised its military spacepower, demonstrating some modernisation progress in Ukraine and Syria compared to the state of the Russian military of the 1990s. China has spent the same time running to catch up to where the United States was in the 1990s as it seeks to modernise the PLA to fight high-intensity, high-technology wars. Europe, Japan, India, South Korea, and many other military powers are consistently improving their own military power by tapping into their own and allied space infrastructures.

If a war was to happen between the major space powers tomorrow, it is not unreasonable to expect that the satellites they rely on will be the targets of disruptive and destructive attacks, ranging from jamming radio communications to destroying them with missile-lofted nuclear detonations in orbit. As space is so important for practical military and economic purposes on Earth, military forces are working to threaten hostile satellites in preparation for wars we all hope may never come. Most anti-satellite (ASAT) weapons are based on Earth and aim up to space, like coastal defences pointing into coastal waters. Taking out enemy satellites may help tip the scales in military operations on Earth because they allow military forces to network and coordinate over vast distances, more easily locate and target enemy forces and rearward bases and headquarters with precision missiles, and detect missile launches and other surprise moves. China, India, Russia, and the United States have tested kinetic-energy ASAT weapons (impact vehicles that ram into a target satellite after being launched from a missile on Earth) in recent years. China, Russia, and the United States have been conducting cat-and-mouse games of 'tailgating' or 'stalking' each other's satellites in the geostationary orbital belt. Many more states are able to jam the communications of satellites or engage in computer network operations, or 'cyber-attacks', on the computer systems controlling satellites.

Earth orbit is a cosmic coastline, an adjunct to the terrestrial environments of warfare that is quite constrained and congested in places. It is an environment defined by its proximity to somewhere else. What goes on in orbit will impact military campaigns and economic behaviour on Earth itself in a very intimate fashion. This is a superior analogy to the 'ultimate high ground' vision promoted by several military theories of space and now official U.S. military doctrine. The ultimate high ground of space is a banal metaphor that at best merely points to generic advantages of using it, and at worst invites predictable, mechanistic, and fatalistic thinking to the craft of strategy, military planning, and resisting hostile space powers. Commanding space is the crucial concept for strategic thought about space—similar to the concepts of commanding the seas and the air—where satellites fly and their communications streams travel. Manipulating and impacting the command of space, specifically Earth orbit, has direct impacts on high-technology military capabilities on the battlefield today, and could create environmental havoc in orbit and economic costs and humanitarian suffering on Earth. Humanity's Global Space Age has only opened up Earth orbit, up to altitudes of around 40,000km, to regular and routine use for war, development, and prestige. It is merely a coastal zone on the grand scale of the Solar System. Beyond that, there is little of any direct military or economic use in the near future. Until they become economically viable, the military and industrial uses of the Moon and beyond will remain the subject of fascinating conversations, but idle ones suited more to hubris and science fiction rather than serious political and military analysis of spacepower's impacts on world politics today.

* * *

A note on astrography—the 'geography' of Earth orbit—is needed as terms about places in orbit will be used throughout the book. Different kinds of satellites are built for different kinds of services, and end up in different kinds of places above Earth, as seen in Table 1. Many of the types of satellite services will be explained as they are encountered in the book. There are four classes of orbital altitudes above Earth where these satellites go, and these are outlined in Figure 1: LowEarth orbit (LEO), medium-Earth orbit (MEO), geostationary orbit (GEO), and highly elliptical orbits (HEO, or Molniya orbits). Generally speaking, the higher the orbit that is required, the more prograde velocity (acceleration in the direction of forward travel) and fuel that is required to reach that altitude. To achieve orbital flight, a velocity of 7.8km/s is needed—almost five miles per second, or 17,500mph, or 28,000kph. These are extremely high speeds requiring large amounts of thrust. The exact amount would vary depending on the mass of the satellite and the type of orbit required. Until the 2010s most satellites ranged from hundreds of kilograms to several metric tons in mass. Today there are many small and very small satellites measured in tens of kilograms joining the bigger satellites in orbit. Heavier satellites require more powerful rockets. Orbital inclination refers to how many degrees north and south beyond the equator (at 0°) the satellite reaches in its orbital paths. Changing orbital altitudes and inclinations (known as orbital burns) consumes fuel, meaning that satellites have a fixed lifespan that is partly determined by how long their fuel reserves will allow them to maintain their most desired orbital paths (station-keeping) or how many more manoeuvres it has left if the satellite is designed to be more mobile.

Figure 1: Overview of select Earth orbits

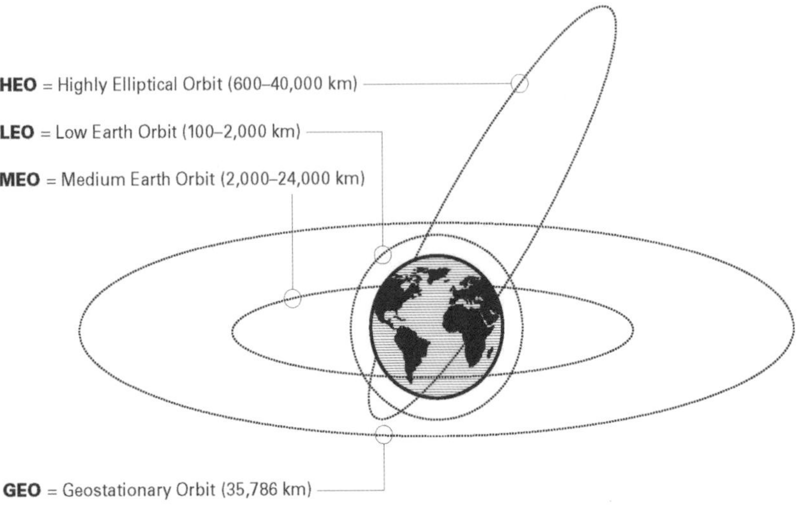

Source: NASA Global Change Master Directory.

Different orbits provide different fields of view, different loiter times (how long a satellite can see a certain area of Earth, and vice versa), and different tradeoffs for the type of service which they provide (e.g. relaying communications between terrestrial points, broadcasting a signal service, or gathering and transmitting information from satellite sensors). Imagery satellites (such as Keyhole) tend to be in LEO to ensure high-resolution images and sacrifice wide fields of view for close-up detail. Due to their low altitude and restricted fields of view or signals coverage, this increases a requirement for a higher number of satellites to ensure punctual 'revisit times' (when the same or another satellite will fly over the same area of Earth) or continuous, overlapping coverage from the constellation. Communications satellites in LEO (such as Iridium) can ensure greater signal strength because it has far less distance to travel due to their proximity to Earth's surface, but require sixty-six satellites in a constellation to achieve continuous global coverage so that one satellite can take over the communications relay when the first one passes beyond the terrestrial caller's horizon. Out in GEO, three satellites can see the entire globe, apart from the extreme polar latitudes.

Though there is no formal international agreement on where space begins, LEO tends to begin at around 100km altitude, above the Kármán Line, which is approximately the lowest possible altitude any kind of orbital motion can be sustained without additional thrust. However, most satellites fly above 300km altitude. LEO goes up to around 2,000km, but most LEO satellites orbit between 600–1,600km altitudes. Satellites in LEO will have an orbital period (the time it takes to fly around the globe) of around 90 minutes. At higher altitudes orbital periods will increase beyond 90 minutes. At MEO altitudes of around 20,000–24,000km, an orbital period extends to around 12 hours. A GEO orbital period is 24 hours. Counterintuitively, satellites need to increase their forward velocity to raise their altitudes, which makes them look like they are slowing down as they travel across the sky from the perspective of an observer on the ground.

Satellites usually orbit from the west towards the east due to the direction of Earth's spin. When launching towards the east, a rocket and its payload gets a boost from Earth's spin. Launching closer to the equator provides more of a boost and therefore there is a premium on launch sites close to the equator. The Kourou spaceport in French Guiana enjoys a 17% efficiency advantage over Cape Canaveral in Florida. Rockets launched from Kourou can launch 17% more mass with the same fuel amount or use 17% less fuel to launch the same mass as rockets launched from Florida. Satellites in a retrograde orbit are launched against the direction of Earth's spin, hurling satellites into east to west flight paths. This is the least efficient form of launch as it goes against the grain of Earth's spin, but is useful to provide alternative launch and orbital coverage options. An example of this is Israel's satellite launch capability which flies in a retrograde direction over the Mediterranean Sea. Any launch failures crash into the sea in the West and not above populated areas and other possibly hostile states to the east.

As well as altitude, inclinations are important in orbits. Those on a 0° inclination will travel west to east (prograde orbit) and fly directly above the equatorial line. Those on a 23° inclination will travel on an orbit that tracks no further than 23° north and then head to no more than 23° south as it travels west to east, moving from 23°N to 23°S alternately. High-inclination orbits allow satellites to

Table 1: Select Satellite Service Types and Examples by Orbital Regimes

	LEO *Low-Earth Orbit*	**MEO** *Medium-Earth Orbit*	**GEO** *Geostationary or Geosynchronous Orbit*	**HEO** *Highly Elliptical Orbit / Molniya Orbits*
Altitude (km)	100–2,000	2,000–24,000	35,786	600–40,000
Orbital period	90 minutes	2–12 hours	24 hours	12 hours
Satellite service types	• Communications • Imagery intelligence/Earth Observation • Signals Intelligence • Megaconstellations • Space surveillance • Crewed Space Stations	• Global Navigations Satellite Systems (GNSS) • Telecoms	• Communications • Telecoms • Terrestrial weather • Missile launch detection/infrared sensors • Space surveillance • Signals/Electronic intelligence • GNSS space-based augmentation systems (SBAS) • Regional Navigation Systems	• Communications • Missile launch detection/infrared sensors • Earth Observation • Signals/Electronic Intelligence

Satellite examples	• Iridium • Keyhole • LANDSAT • SPOT • Digital Globe • OneWeb • Gaofen • Lotos • Starlink • Helios • CSO • RADARSAT • IceEye • Sapphire • International Space Station • Chinese Space Station	• GPS • GLONASS • Galileo • Beidou • SES	• Wideband Global SATCOM • Skynet • Syracuse • Sicral • Inmarsat • Space-based Infrared System (SBIRS) • CLIO • TRUMPET • Geostationary Satellite Situational Awareness Program (GSSAP) • Gaofen • Quazi-Zenith Satellite System (QZSS) • Navigate with Indian Constellation (NAVIC)	• Molniya • Meridian • Sirius • SBIRS • Gaofen

travel mostly on a north-south axis with inclinations of between 60° and 85° with a slight west-east or east-west track on a mostly north-south flight path as Earth spins below. Due to their high inclination, they do not receive much of a boost from Earth's spin and therefore can be more efficiently launched from higher latitudes. This is why some satellite launches from northerly points such as Alaska, Scotland, Norway, and Russia are still feasible. A very useful form of polar orbit is a sun-synchronous orbit (SSO) which means that the satellite's orbit passes over the same point on Earth at the same time of every day. This kind of orbit is especially useful for spy and weather satellites which can easily allow changes over a specific location to be spotted. Large numbers of such satellites flying in sequence could allow several revisit opportunities in less than a 24-hour period.

MEO is not a highly populated orbit, unlike LEO and GEO. With a 12-hour orbital period and large fields of view, they are ideally placed for slower-moving (as they would appear to a terrestrial observer) satellites that broadcast continuous wide-area signals, such as navigation and television broadcast systems. MEO is also a relatively under-populated orbit category compared to LEO and GEO. Some constellations use several orbits and inclinations to provide more comprehensive capabilities, such as the SES satellite TV constellation located in MEO and GEO. BeiDou—China's GNSS—also has satellites in both MEO and GEO, with satellites in GEO providing additional enhancements for the BeiDou constellation (including a basic text-message relay system) and an increased service coverage and backups or redundancies over the Asia-Pacific region.

GEO is a relatively thin orbital space where satellites orbit Earth every 24 hours above the equator. This means that satellites in the GEO belt will orbit Earth at such a velocity and distance that it matches Earth's rotation. A geostationary satellite will appear stationary and motionless in the sky above the equator for an observer on Earth—the orbital speed matches the speed of Earth's spin. Satellites apparently loitering at a fixed point in the sky are extremely useful for tightbeam communications—when you know where exactly to point your radio antennae to send and receive signals. This is one of the most visible kind of satellite communications for most people: satellite television. Small dishes on private homes pointed towards the same point in the sky are receiving satellite TV broadcasts from above the equator, 35,786km away.

A slight variant on geostationary orbit is geosynchronous orbit (still using the GEO moniker), which involves satellites at a similar altitude at the equatorial plane orbiting with a slight inclination, meaning that their position relative to the surface changes a small amount during the day. Looking from Earth, a geosynchronous satellite will do a subtle 'figure-of-eight' meander in the sky, but roughly staying within the same small 'zone' in the sky and returning to the same point it started from every 24 hours. Orbits with slightly higher altitudes at GEO or greater inclinations (i.e. more eccentric geosynchronous orbits) tend to be used as 'graveyard' orbits, where satellites that are about to run out of fuel or otherwise have no further use are 'retired' there to free up room in the GEO belt for new satellites. The GEO belt is heavily congested as satellites in GEO cannot be placed in a GEO 'slot' too close to each other lest their broadcasts interfere with each other or satellites collide. GEO slots and specific radiofrequency bands are therefore highly prized 'orbital resources' in the international system and are allocated through the mechanisms of the International Telecommunications Union (ITU).

HEO (highly elliptical orbit) or Molniya orbital dynamics involve a very low perigee of around 600km (lowest orbital altitude above Earth's surface) and a very high apogee of up to 40,000km (highest orbital altitude above Earth's surface). This allows a satellite to appear to travel more 'slowly' over the surface of one hemisphere as it gains altitude and loses 'speed' relative to how much ground it tracks over on Earth, then almost loitering at its apogee before losing altitude and zooming around the other hemisphere relatively 'quickly' and returning to the first hemisphere. The Soviet Union, being a very northerly state, pioneered these orbits to maximise the time at which communications and missile launch detection (or early warning) satellites could spend above the northern hemisphere with a very wide field of view and spend the least time possible with a very narrow field of view above the southern hemisphere. This is particularly useful for monitoring the poles and extreme latitudes, as satellites orbiting at GEO do not have as good a view of the poles and the extreme latitudes of the globe.

There are more areas of potential interest beyond these Earth orbit types, such as lunar orbits and various 'Lagrange libration' points where the gravitational forces of two bodies cancel each other out and provide an energy-efficient loitering zone for spacecraft. These are returned to in Chapter 7 as some strategic thinking is (prematurely) turning towards dominating 'the ultimate high ground' of space beyond the four types of Earth orbit above. Until then, the book explores the origins and maturation of space technologies and the politics and strategies behind them. Military and economically significant technological infrastructures do not escape Earth orbit, and likewise fail to escape the power politics and military rationales that gave impetus to them in the first place.

* * *

I hope this book presents a useful, accessible, and broad perspective of the Global Space Age, how important military technologies and infrastructures came about, and how modern space activities struggle to escape their original sin. Part I, 'The Original Sin of Space Technology', shows how the Global Space Age began not as a story of two vain superpowers looking for less destructive methods of competition, but rather as part of the Thermonuclear Revolution—of the emergence of nuclear fusion bombs and long-range ballistic missiles. Waging nuclear war is a feature, not a bug, of space technology. The two superpowers opened up space for everyone else because the fundamental technologies of the Global Space Age were critical to the most effective forms of fighting nuclear wars. Beyond the superpowers, the linkages between the nuclear-missile military complexes and space continued, as France, Britain, China, Japan, and India sought their own advances in these new technological areas that were either directly linked to nuclear and missile programmes, or established a baseline of indigenous industrial capabilities that allowed a latent military potential. The competitive military and economic heart of the Global Space Age not only ensured the continuing military exploitation of space today, but took advantage of imperial practices of the space powers as the forces of decolonisation broke the European empires.

Part II, 'The Maturation of Spacepower', shows how military space technologies accelerated and stabilised in their types and forms into what we are living with today. Spacepower went from being an exotic luxury to a normal part of the military, political, and economic power of the major powers of Earth. Spy satellites, telecommunications systems, weather satellites, navigation systems, infrared missile warning sensors, and space tracking radars all assembled large technological infrastructures that provided proven templates for twenty-first century investments and modernisation. The 'stabilising' impact of spy satellites on the nuclear relationship between the Cold War superpowers is hailed as a major benefit of space technologies, yet the alleged stability gained did not come close to undoing the risks of nuclear Armageddon that space technologies helped create in the first place. These space systems became increasingly useful in the numerous 'proxy wars' and conventional war planning throughout the world in the 1970s and 1980s—hardly stabilising or non-violent in their implications and further entrenching the original sin of space technology.

Part III, 'Strategy in the Global Space Age', looks at spacepower's impact on the evolution of military power since the 1980s and presents a more accurate approach to thinking about space warfare and military strategy in space. Spacepower has gone from a warning and targeting system in nuclear war to an infrastructure that is relevant to any kind of non-nuclear or conventional warfare and counterinsurgency on Earth, for developed and developing states and their armed forces alike. Plugging space technologies into terrestrial military power ensures that there are sound strategic reasons for many states to develop anti-satellite weapons and other methods of disrupting and harassing space infrastructure. In many ways the 2010s picked up where the superpowers left off in the 1980s with an increased tempo of anti-satellite weapons development and testing. Despite this, space warfare is yet to be normalised in many professional quarters. Space warfare instead tends to be spoken of as a subset of nuclear war, deterrence, or ballistic missile defence which impose their own analytical blinkers on what spacepower is doing to military power. Space warfare should be approached in the same way we think about naval warfare, aerial combat, and how seapower and airpower contribute to modern strategy and military, economic, and political power. It should be approached as a geographic specialism of warfare in a place that resembles a coastal zone between places more than it does the 'ultimate high ground' or the open ocean. Such fighting is done to impact the control and denial of space infrastructure and its logistical impacts, which tend to impose the effects of dispersion on the modern battlefield and force postures. As space is now so useful in almost any high-technology activity, the spectre of space warfare will likely stalk the twenty-first century and beyond.

This book is a work of synthesis, drawing on research into a diverse range of contributions to astropolitics across the broad church of IR and its related fields. I hope this makes a significant, original, and rigorous contribution to IR and its intersections with technopolitics and material analyses of power politics, as well as linking a niche literature on spacepower and international and post-colonial space histories into a more comprehensive, global view of the space age we are living in. There is a wealth of literature out there on many military space technologies for academic researchers, and I hope this book introduces new audiences to the more specific and focused research that I reference throughout this book. I also hope it inspires others to pursue research on the many aspects of our Global Space Age, as there is so much yet to be written about it from any number of perspectives, approaches, and locations. This book is a materialist, technological, and strategic account and does not cover every way to investigate astropolitics—gender, race, diplomatic, normative, and postmodernist accounts are waiting to be tackled by those who are far more well-versed in those traditions than I am. I hope that this book shows how different approaches of IR should be seen as complementary, rather than in competition, as intersections between several scholarly traditions have come together across the following chapters to form a synthetic, broad argument and perspective. I hope the book serves as a route for others to pursue their own approaches to astropolitics, whatever those may be. As the historian David Edgerton writes, 'too often the agenda for discussing… technology is set by the promoters of new technologies'.[34] Space needs to be a discussed by everyone, not just those involved in pushing new space technologies into existence or those who stand to gain from it.

The book does not dwell on any single state or technology, but draws on a wealth of more focused research that does. I present a global 'big picture' that I hope outlines the broad contours and trajectory of global astropolitics from a relatively materialist perspective. It also aims to make a contribution to an emerging global space literature by promoting a global materialist view, looking at the connections among and within the space powers and their impacts on many places on Earth, whereas many other works are policy focused, country- case study-orientated, or examine identity and prestige aspects to spacepower.[35] This book intends to pay 'significant attention to the technological objects and large technical systems that comprise' world politics.[36] This is also a distinct effort compared to the existing literature on cultural interpretations of crewed spaceflight.[37] Operating on a planetary scale, large technological systems do not get much larger than space technologies and their impacts on life and politics on Earth. Space technologies continue to be neglected as part of our analyses of our globalised planetary political-economy, despite the fact that space technologies are literally global in scope and tether the globe with webs of communication channels.

Accepting the militarised and political nature of the Global Space Age matters because, as Daniel McCarthy writes, 'it is not that technology develops outside of human agency, but that it develops outside of *some* humans' agencies.'[38] Space technologies have always been developed by some people for specific reasons, and their impacts have not necessarily been to the benefit of all humankind. This is not an argument for pulling the plug on all space activities, despite the multitude of problems and harm that has arisen from or been worsened by the Global Space Age. This book is not a sledgehammer argument to 'defund space'. Space exploration, both human and robotic, and developing outer space in more equitable and sustainable ways are worthwhile endeavours, and our knowledge and uses of orbital space and beyond has and will provide benefits worth chasing for us all on Earth as we face ecological and climactic turmoil. But we should not be blind to the political, economic, and social consequences of ours and others' actions up there and down here on Earth. How, why, and with what consequences we do things in space are fundamental astropolitical questions everyone interested in space should be discussing. Changing how things may be politically done better in the future and moving on from space technology's original sin must begin by exploring how we got to where we are in the first place.

PART I

THE ORIGINAL SIN OF SPACE TECHNOLOGY

1

THE DAWN OF THE GLOBAL SPACE AGE

Heb ei fai, heb ei eni

They who have no blame are not born

At Rice University on 12 September 1962, U.S. President John F. Kennedy made his case to land men on the Moon, proclaiming that 'there is no strife, no prejudice, no national conflict in outer space as yet. Its hazards are hostile to us all. Its conquest deserves the best of all mankind, and its opportunity for peaceful cooperation may never come again.'[1] Yet space was already a realm of strife, prejudice, and conflict. Actions, not words, have set the tone for the Global Space Age. Women were deliberately excluded from the crewed spaceflight programme as Kennedy claimed there was no prejudice. Indeed, the project was exclusively to put *men* on the Moon.[2] The colder, self-interested aspects of humanity and power politics were not only there from the start but rather the impetus and driving force of opening up Earth orbit for practical use in the first place. The political philosopher Herbert Marcuse captured the original sin of space technology, arguing that 'Auschwitz continues to haunt, not the memory but the accomplishments of man—the space flights; the rockets and missiles. The wilful play with fantastic possibilities... testify the extent to which Imagination has become an instrument of progress. And it is one which... is methodically abused.'[3] The Global Space Age came about as a tool of the state—an instrument—in satisfying foreign policy, economic-industrial, and domestic political goals.

Many states pursued space programmes to meet military goals and to achieve some level of autonomy in key areas of new technologies so that no state had a monopoly on the fundamental space technologies which began to shape global economic development. This small number of states with a significant spread of core space capabilities, or the 'Space Club', 'is an expression of well-known international exclusionary politics on Earth', which is a far cry from the universal spirit of the Outer Space Treaty and Kennedy's speech at Rice University.[4] By the early 1960s, Earth orbit had become a realm of nuclear strategy, military power, and strategic intelligence gathering. The Global Space Age spun off the back of massive state investments in missile and military satellite technologies and government-led research and development (R&D) programmes, or 'Big Science'. It was an outgrowth of twentieth-century warfare 'which established state-sponsored and -directed R&D as a public duty and necessity' and helped bring about more technocracy for the Space Age, the 'management of society by technical experts'.[5] This is anathema to the space libertarians championing individualist homesteading on Mars.[6]

The Cold War drove tens of billions of dollars in spending on military space technologies, and that trend has not stopped since.[7] Space technology emerged from the government and military funding of science and engineering in the new post-war order, building new institutions that shaped international science development in ways that may not have otherwise happened.[8] Mutual nuclear annihilation and the genocidal war plans of the Cold War led both superpowers to rely on new advances in space technology to spy on each other, to verify that the other side was not cheating on arms control agreements, and to enhance their militaries' potential killing power. Rather than scientists being pawns in their governments' pursuit of the Cold War, they were active and sometimes enthusiastic partners.[9] The investments in those areas later led to numerous commercial and scientific satellite programmes. The Cold War not only brought about critical advances in intelligence gathering and command and control systems, but made the United States dependent on them. In turn, this became an important foreign policy tool as Washington selectively shared information and services from space with its allies.[10]

The superpowers enjoyed new capabilities but also suffered from new vulnerabilities and threats. Mitigating those vulnerabilities in part relied on international treaties and norms that enshrined the use of Earth orbit for military and intelligence satellites to support military forces and government agencies on Earth itself.[11] This international framework from the Cold War enabled the continuous military uses of outer space to enhance and modernise terrestrial killing power whilst proscribing overt competition in other areas, such as territorial aggrandisement on the Moon and beyond. Almost all states on Earth have signed onto this framework established in the Outer Space Treaty of 1967 that structures contemporary astropolitics and the development of an orbital political-economy. Technology rides the currents of political, economic, and military interactions, and space technology has spread well beyond the two superpowers, challenging reductionist accounts of the Cold War Space Age as a bipolar one, rather than a global one.[12] Understanding the ubiquitous, diverse, and continuing structural impact of space technologies on international politics today, and how to govern space today, requires a general understanding of why key space technologies were developed throughout the Cold War and how their modern versions continue to be developed and deployed.

The practical exploitation of Earth orbit for the purposes of war and development remains less well known even within elite policy institutions.[13] Easy as it is to puncture the wondrous narratives conjured by politicians on the pedestal, Kennedy's celestial rhetoric still resonates with widespread popular views of space as existing in an apolitical vacuum, free of terrestrial political sins.[14] Kennedy's speeches did much to shift popular perceptions of competition in space to that of reaching for the Moon in a 'civil' competition instead of the deployment of nuclear missiles on Earth and military-intelligence satellites in orbit.[15] Unlike the rocket and satellite 'race' of the late 1950s and early 1960s, the Moon race of the latter years of the 1960s had little direct relevance to the balance of military power on Earth. The most consequential advances in spacepower in both the Soviet Union and the United States in the 1960s and 1970s were in the development of satellite technology for a range of military, intelligence, and infrastructural needs. These technologies have had a bigger impact on human society and the power politics of international relations than crewed spaceflight. Though the products and finer details of space-based intelligence technologies were closely guarded secrets, the fact that both sides of the Cold War had deployed a range of spy satellite technologies into the 1960s was an open secret among military personnel and diplomats.[16] But this open secret rarely made it into the light of day in public discourse, and as such the reality of military space activities from the earliest years of the Cold War still remains unknown to general populations.

The original sin of space technology can be charted through the foundational developments in the United States and the Soviet Union to demonstrate the difficulty of distinguishing civilian and military uses of space and how the technologies proliferated and perpetuated injustices and inequalities on Earth. The foundations of space technology and infrastructure are rooted in imperialism. Both superpowers, despite their anti-colonial beliefs about themselves, conducted imperial practices or took advantage of historic territorial gains from settler-colonialism in setting up and supporting their space programmes. On a visceral and essential level, this chapter shows how ahistorical and incorrect it is to accuse certain states of 'militarising' space today. Space was always militarised. It is no coincidence that the major space powers also have significant nuclear and missile infrastructures. It is fitting that the study of a Global Space Age begins with the Thermonuclear Revolution—how nuclear weapons and missiles transformed geopolitics and major power relationships.

The Thermonuclear Revolution

Rockets and satellites were extremely desirable for military uses because they provided more accurate, reliable, and controllable methods of inflicting nuclear annihilation on their enemies. Yet the original sin of spaceflight was committed before the Cold War began. The origins of space-capable rocket technology and the opening up of space are in large part owed to the successes of Wernher von Braun and his team's efforts through the 1930s and the Second World War, specifically the V-2 ballistic missile programme. Funded by the Nazi state when seeking new military weapons that were unbound by the 1919 Treaty of Versailles, decades-old ideas and theories behind ballistic and orbital spaceflight could be realised because the funds had been committed to build powerful enough rockets to do so. The American Robert Goddard, whilst clearly capable and enthusiastic as a rocket scientist, never attracted significant funding in the inter-war period and struggled with a negative public image and a hostile press.[17] By 1932 Goddard had reached a technological dead-end with liquid rockets at the same time the Germans were making breakthroughs. Von Braun's early work for the military would soon enjoy significant 'scaling-up' as they demonstrated thrust capabilities far beyond what U.S. and Soviet rocket scientists had previously achieved.[18] The efforts of U.S. and Soviet rocket scientists in the 1920s and 1930s had provided little tangible military potential due to a mixture of pithy funding and political strife, with the latter being particularly acute during Stalin's Great Purge in the Soviet Union. Before returning to work on Soviet missile technology in the 1940s, Sergei Korolev—who would become the Chief Designer and fulcrum of the early Soviet space effort—spent years in a Siberian gulag after being denounced by colleagues at the height of Stalinism. Other Soviet engineers and scientists continued to face political instability and personal danger. During one of Lavrentiy Beria's pogroms, Dmitry Ustinov—the Minister for Armaments—had to step in to save the lives of some of his top engineers who were Jewish and targeted by Beria.[19]

At the end of the Second World War, von Braun defied orders to resist the Soviet onslaught at Peenemünde and fled towards the American front. He and his surviving team surrendered to the U.S. Army. As part of Operation Paperclip in 1946, von Braun and his 125-strong team's Nazi pasts were quietly absolved so that their knowledge and experience could be appropriated and put to work for the U.S. Army's nascent missile programme in New Mexico.[20] By the 1950s, von Braun and his team were just one part of a massive U.S. post-war aeronautical industry. Meanwhile, Korolev's teams had moved far beyond what the Germans had achieved in the 1940s as the Soviet military sought new ways of sending their new nuclear bombs to the United States.[21] These people and industries were set to work, in various ways and varying levels of intensity, on the early ballistic missiles and nuclear warhead designs through the 1950s, with their efforts gaining momentum from 1955 onwards on both sides, leading to the first satellite launch by the Soviet Union in October 1957 with Sputnik.

Long-range ballistic missiles that can reach orbital altitudes are far better at delivering nuclear bombs than aircraft, ships, and land forces due to their speed and the relative invulnerability of the warheads. As Intermediate Range Ballistic Missiles (IRBMs) and Intercontinental Ballistic Missiles (ICBMs) became more feasible as the lift capabilities of rocket engines improved, both superpowers seized on missiles as a better way to fight a nuclear war. Launching warheads into space at such speeds meant that the other superpower could be nuked in around 30 minutes, compared to the many hours aerial bombers needed to cover the distance over the Arctic. The technological families of ballistic missiles and Space Launch Vehicles (SLVs) are closely related. Delivering a satellite into a stable orbit requires some more velocity in the correct direction and a more refined launch trajectory than a nuclear missile which flies on a suborbital path through space for a short time before re-entering the atmosphere and dropping at many kilometres per second towards the target on Earth. The higher the altitude of the desired orbit, the more fuel and delta-V (velocity in the appropriate direction) is required.

Von Braun actively sold the potential of combining rocket technologies with atomic bombs as a transformative military capability.[22] Von Braun's commitment to spaceflight never wavered, but his apparent ease at taking funding from wherever it came and doing whatever it took to advance spaceflight technologies was criticised by the scientist and public figure Carl Sagan, who opposed the close relationships between space science, engineering, and exploration and the U.S. military-industrial complex.[23] Korolev too used the military-industrial requirements of the Soviet Union to achieve his own long-held goals of spaceflight, as seen below. The Redstone rocket which lofted Alan Shepard, America's first human in space, was an adapted V-2. The V-2 became the R-1 in the Soviet Union and was their first ballistic missile capability. R-1 designs and parts were transferred to China in the 1950s, useful in the very early days of China's own missile and rocket programme, and named the Dong Feng 1.

The missile is only one part of a nuclear weapon system. Together, thermonuclear bombs and space technologies threatened horrifying, abhorrent, and world-ending effects if used at scale in a 'third world war'. The thermonuclear revolution effectively makes major war a suicidal policy option to settle disputes between states, making total war a rather pointless tool of the state. Campbell Craig describes the thermonuclear revolution as 'the continued possession of thermonuclear weapons systems by several sovereign states, [in] an anarchical international order in which there is no authority capable of preventing these states from using them.'[24] The thermonuclear revolution locks the international system into a precarious status quo between the major powers:

> The combination of [interstate] anarchy and arsenals portends an eventual nuclear war, if one accepts the standard definition of [interstate] anarchy as precisely a condition in which major war is possible. It is a dilemma because the threat posed by this combination would logically seem to require transformative change if we are to avert a war that could exterminate humanity; yet, transformative change would, presumably, unravel an international order kept unusually stable at least in part because most of its major powers possess nuclear weapons systems. In other words, the kind of action necessary to rid ourselves of the danger of nuclear war threatens to increase the chances of it happening. We ride the back of a nuclear tiger.[25]

Bernard Brodie famously wrote in 1946 that 'thus far the chief purpose of our military establishment has been to win wars. From now on its chief purpose must be to avert them'.[26] Brodie may have been ahead of his time, writing in the second half of the 1940s where nuclear arsenals were measured in tens of kiloton-yield air-delivered fission bombs. The advent of thermonuclear—or fusion—bombs measured in megatons coupled with reliable intercontinental ballistic missiles (ICBMs) by the mid-1960s would make the annihilation of a good proportion of populated lands of the First and Second Worlds a nuclear hellscape within a couple of hours. Then, almost any war between the two leading powers of the international system would be 'too big, too all-consuming to permit the survival even of those final values... for which alone one could think of waging it'.[27] Yet many military forces continued to prepare to fight and win a nuclear war, bringing new space technologies online as they did so.

The Thermonuclear Revolution brought about the Space Age as we know it. Dozens, then hundreds of fusion bombs ready to be thrown across thousands of kilometres accelerated the end of 'free security' for the United States. For the first time in generations, the cities of the United States were no longer protected by vast oceans.[28] The detonation of a fusion bomb in 1952 by the United States at Eniwetok Atoll was 500 times more powerful than the fission bombs dropped on Japan in August 1945. Bombs were designed to be light enough to be carried by emerging missile technologies, and the relative invulnerability and speeds of IRBMs and ICBMs made them worth the expense compared to cheaper and more vulnerable crewed aircraft. There were still doubts in the 1950s about the potential of fusion bombs, yet tireless lobbying by hydrogen bomb and ICBM proponents within the United States helped push USAF (which had preferred bombers and cruise missiles) to eventually take ICBM development seriously.[29] With fusion warheads and a few hundred launchers, tens of millions could be killed in tens of minutes with reduced needs for accuracy given the blast yield.[30] Fixed military infrastructure would not need precise targeting when thermonuclear bombs could wreak destructive radii measured in tens of kilometres.

Korolev and his peers lobbied Joseph Stalin in the late 1940s to approve long-range missile development.[31] Stalin saw the potential of nuclear-tipped ICBMs to act as a 'strait jacket' for the 'noisy shopkeeper' Harry Truman, but before his death in 1953, he approved intercontinental cruise missile development as well as ballistic missiles.[32] The Soviet Union had made rapid and sizeable leaps in missile capability from the 300km-range R-1 (reproduced V-2s) in 1948 to producing weapons with ranges of 5,000km in 1953.[33] Shortly after the death of Stalin, Khrushchev eliminated Beria through arrest and summary execution by gunshot and sowed the seeds of his rise to power with military support. Korolev and many other senior engineers and managers scrambled to consolidate their bureaucratic positions as Malenkov, Khrushchev, and the rest of the Presidium came to grips with the massive armaments complex Stalin had personally overseen and excluded them from. Beria, whose star was in decline, was the only Presidium member who had managed to ingratiate himself into that system.[34] Khrushchev was the least familiar with the defence and missile bureaucracy Stalin built, which created more fluidity in decision-making within and among the design bureaus, and Korolev in particular, as the Presidium found its way in the post-Stalin Soviet Union.[35]

Until the early 1950s, the United States had focused more on cruise missiles and aerial bombers as a delivery method due to accuracy concerns with ballistics. In addition, each military service had *their own* ballistic missile programme. Soviet air defences were beginning to threaten the ability of the USAF's nuclear bomber wings to reach their Soviet targets intact, let alone survive the return journey. Similarly, U-2 spyplanes found it increasingly difficult to fly into Soviet airspace with impunity as time went on. Satellite reconnaissance was a potentially attractive alternative. By 1955 Eisenhower was receiving credible intelligence that the Soviet Union possessed a large arsenal of megaton-range bombs and would soon have a proven ICBM capability to deliver them.[36] This jump in Moscow's destructive capability with a fusion bomb was crucial in setting the Eisenhower administration and the U.S. Air Force down the path of significant ICBM research and development.[37] This was anticipated long before the so-called Sputnik 'shock' of October 1957.

The virtual invulnerability of nuclear warheads once launched, alongside their destructive power, led to arguments over the existence of the thermonuclear revolution where the expectation was that major thermonuclear-armed powers will not intentionally go to war with another nuclear power because of the 'omnicide' it might entail. The costs of fighting a nuclear war are so great that both sides are forced into an uneasy co-existence, with major interstate war between thermonuclear-armed powers becoming suicidal.[38] This by no means denies or downplays the significance of the many wars and millions killed and harmed since 1945 in non-nuclear wars. Nevertheless, it is the case that 'for the foreseeable future, and perhaps forever, the physical survival of vast numbers of human beings, and much of the nonhuman life on earth, rests upon the adequacy of the system to restrain the large-scale use of nuclear weapons.'[39] Everyone interested in international relations, security, and the future of the species must recognise this reality. Space technology was no panacea from this—in fact it was essential to it and remains so to this day. The original sin of space technology was committed.

U.S. and Soviet Spacepower Origins

Such missiles soon became a central, infrastructural technology that met a multitude of needs and interests not only within the military but also across government and the wider economy. Spy and communications satellites became more feasible with the increasing lifting capacity of ballistic missiles designed to lift nuclear warheads. A RAND report from 1946 had foreseen it.[40] Space technologies—from heavier lift rockets to guidance and satellite systems—played their parts in satisfying a 'network of interests' across government, the military, and industry and not merely solving engineering questions that characterised missile and nuclear weapons development, as Graham Spinardi argued over the Trident programme.[41] There was nothing 'inevitable' about these technologies and their development—their satisfaction of a multitude of interests enabled their large funding requirements as a product of a social network.[42] According to Donald MacKenzie, in the United States the military and political elites had embarked on a number of nuclear missile programmes without an agreed understanding of why it was doing so. These programmes were pushed from below, by engineers and technicians with 'career investment in and enthusiasm for particular lines of technological development'.[43]

By launching warheads on ICBMs, the road was also opening in the minds of other engineers, scientists, military strategists, and policymakers for SLVs to launch satellites into Earth orbit for a plethora of military and intelligence services. ICBMs and SLVs are not fully identical but they share the same military-industrial complexes and classes of technologies, skills, and knowledge. In other words, dual-use technologies. The Atlas was the first operational ICBM for the United States—but it also lofted America's first astronauts into stable orbits.[44] The U.S. Titan, Delta, and Atlas rockets continue as SLV workhorses today and are derived from ICBM designs. The Soviet R-7 which launched the first satellites and cosmonauts was primarily developed as the USSR's first operational ICBM—though it was limited in such a role and was quickly replaced in the 1960s.[45] The ICBM-SLV duality captures the inherently militarised, politicised, and morally ambiguous nature of humanity's uses of outer space. Christopher Gainor is right to argue that the Soviets and Americans only seriously pursued ICBMs due to the new possibilities opened up by thermonuclear—or fusion—bombs, meaning that without these ICBMs and their megaton-range warheads, space exploration and science would have developed at a far slower pace because their sheer expense just for science and exploration would have been a challenge to justify.[46] Therefore we would not have witnessed the Space Age as we know it had this original sin not been committed.

The 'race' for satellites and crewed flight demonstrations provided a convenient and acceptable cover for military-intelligence space and missile technologies on both sides of the Iron Curtain.[47] After 1958, the historian Michael Neufeld argues, space became part-civilianised in what was otherwise an entirely military realm. The militarisation of space was not something that intruded itself upon a scientific and 'civilian' activity—it was militarised from the very start.[48] The 'Space Race' of putting people and robots into space and onto the Moon is not the whole story of space technology in the Cold War. Historian Roger Launius writes that Moscow and Washington agreed on a policy which 'allowed free access to space for all, fostered unfettered rights of overflight by any nation, prohibited the placing of [nuclear] weapons in space... and barred nationalistic claims of sovereignty over celestial bodies.'[49] Importantly, they allowed the use of satellites to provide data and support services to military forces and intelligence agencies on Earth.

For Eisenhower, it was a useful public justification for investing in space with the U.S. Intelligence Community to use space-based technologies to prevent another Pearl Harbor kind of attack.[50] From the very start of the Space Age, the distinction between military and civilian research technological development was 'largely hypothetical as the need to maintain a nuclear lead trumped concerns about the negative effects of secrecy, technocracy, and federal priorities on scientific research.'[51] In January 1958 Explorer 1 became the first U.S. satellite to orbit Earth, taking measurements along its flight path and detecting the Van Allen radiation belts in low-Earth orbit, regions of Earth orbit where intense radiation is lethal to unshielded electronics. Both superpowers opened the funding taps and pursued the technologies of the early Space Race because it met many of their tangible military interests or objectives, not merely to pursue some idealistic universalist ambitions of space exploration or satisfying scientific curiosity.

Rather than viewing those early firsts in space as triumphs of a singular genius and the heralding of some new technological dawn and more benign competition between two superpowers, historian Walter McDougall declared the 1957 Sputnik launch 'a famous victory, an expression of much that is good in Russian culture', but also 'the distortion of technology, purchased at such a price, into a cold tool of the state.'[52] Another way to look at it is the demonstration of what was possible with the right investments in missile and space technologies for other states to follow—which many did. The Sputnik satellite launch was a sensation in the U.S. press and the Soviet Union had demonstrated it could hold American cities hostage with nuclear weapons. However, the shock, surprise, and desire to compete with the Soviet Union in space was far from universal in U.S. society. Alarmism, advocacy, and anecdote often substituted analysis and a conscious effort had to be made by space advocates to tie the importance of prestige in space engineering achievements with perceptions of U.S. power in the anti-communist struggle, both at home and abroad. Much of the Sputnik 'panic' in October 1957 was among the bipartisan elite, not necessarily the majority of the American public.[53] Excited more than shocked and perturbed perhaps, British scientists set about to track Soviet satellites and rocket bodies in orbit based on radio telemetry, ionospheric research, and optical observations. In short order the Kettering Group and other experts in the UK were providing accurate tracking data for the early satellites.[54]

The 'Sputnik shock' maelstrom among political elites kicked off popularised 'missile gap' fears in the U.S. media and placed a significant propaganda coup at Khrushchev's disposal (and Eisenhower's domestic political opponents), despite the then-secret fact that the actual military balance in nuclear missile capabilities lay squarely in favour of the United States. Nevertheless, Korolev's triumph with this new supposed utopian socialist technology prompted Khrushchev to brag that 'main-street Americans have begun to shake from fear [of nuclear war] for the first time in their lives.'[55]

The Soviet Union's missile and space programme was not the result of conscious choice to have a 'Space Race' to compete on the technonationalist prestige front with the United States. It was an opportunistic spinoff of Soviet nuclear missile developments. Korolev personally lobbied and persuaded the military and political leadership to allow a demonstration flight of the R-7 ICBM with a satellite on board rather than a dummy warhead, drawing on the value of beating the Americans who were working on similar lines rather than his own romantic visions of spaceflight to get his way.[56] According to Asif Siddiqi, the impressive 'achievements of the late 1950s and early 1960s were pushed by the chief designers, grudgingly approved by the Communist Party and government, and then used as propaganda vehicles by Soviet leaders for selling the virtues of the socialist system.'[57] The early spaceflight 'stunts' which intentionally tried to gain prestige from this newfound popular fascination in spaceflight technology, such as Valentina Tereshkova's first flight and developing a lunar programme to rival Apollo, 'had little to do with a rational program of space exploration—and even less to do with scientific research.'[58] The front-page successes of the crewed spaceflight programmes were piggybacking on the nuclear-missile-spacepower nexus in the Soviet military. The direct military applications of the race to the Moon (as opposed to the race for orbital spaceflight) were minimal, as the U.S. Secretary of Defense Robert McNamara made clear in rebuking comments by NASA Administrator James Webb that the 'moonshot' would provide some military value.[59] Beyond stimulating the aerospace industry and training a greater number of relevant PhDs and technical experts, the military value of Apollo was virtually non-existent.

Eisenhower's domestic political opponents escalated their criticism of his 'weak' policies towards the Soviet Union, and after Sputnik 'the trickle of criticism turned into a downpour'.[60] The Democrats played the supposed 'missile gap' for all it was worth. Yet the Eisenhower administration was far from complacent. They recognised the potential for Soviet satellites to spy on the U.S. from space to better direct their nuclear weapons, to deploy weapons in space (such as orbital nuclear bombardment weapons), and to transit warheads through space towards American cities on suborbital flights, bypassing air defences. This led to the initial work in developing the earliest space situational awareness (SSA) capabilities and concepts which led to what became the Spacetrack system.[61] For the rest of his administration, Eisenhower would not reveal the emergent military-intelligence space capabilities nor the U-2 aerial reconnaissance programme of the United States—despite encouragement to do so from John Foster Dulles.[62] By 1960 the federal budget accounted for 20% of university operating expenses across the United States, and the National Defense Education Act poured money into the sciences and arts, in response to the perception of 'falling behind' Soviet science.[63]

John F. Kennedy's decision to send American men to the Moon was driven by military-political rationales and implications. Following the Bay of Pigs debacle and the Gagarin triumph, Kennedy reluctantly conceded that the United States—or rather his credibility-starved administration—had to respond with 'something' in space. Historian John Logsdon argues that the Apollo project was a pragmatic choice from a President seeking a realistic way to beat the Soviet Union in a visible yet not outwardly militaristic project in space.[64] The Bay of Pigs and Gagarin proved providential for NASA's political and economic fortunes.[65] Rather than being a visionary of space exploration, Kennedy only reluctantly supported NASA's potential as a foreign policy tool to reclaim perceived losses in U.S. prestige, and once said that 'I am not that interested in space'.[66] Kennedy grew increasingly concerned with NASA's soaring costs and the diversion of funds from other, more deserving, spending needs. NASA's budget reached over 5% of the U.S. federal budget at its peak in 1965. In 2018 dollars, that would account for a NASA budget of $218bn, over 100 times NASA's budget of $20bn in 2018.[67] Through 1963 Kennedy continued to offer a joint Soviet-American crewed lunar mission to try to take the competitive sting out of the Space Race to the Moon. Khrushchev reportedly responded to the suggestion with the quip: 'Sure, I'll send a man to the Moon. You bring him back.' With the Moon race so explicitly tied to perceptions of national power, exorbitant spending that President Eisenhower described as 'nuts' was approved.[68]

Despite Korolev's long-held dreams of spaceflight and a lunar landing as seen in technical papers from the 1950s, the necessary resources could only come through demonstrating the military utility of the technologies such ambitions relied upon. The Soviet leadership were confronting more pressing military and security problems—how to counter a nuclear-armed United States and ensure the combat effectiveness of Soviet missile forces. Both civilian and military leaders simply did not take the stated U.S. goal of landing people on the Moon as a serious one, and did not fully fund Korolev's and Chelomey's heavy-lift crewed spaceflight vehicles. The R-7 ICBM/SLV, the R-5 and R-11 Short-Range Ballistic Missiles (SRBM) became technical successes, with the R-11 becoming the Scud missile, the variants of which would prove popular with many regimes in future. These missile and nuclear projects involved thousands of scientists, engineers, military officers, and bureaucrats, but Korolev was the 'heart' of the effort synthesising the abilities of thousands.[69] Without Korolev, there indeed may not have been a Gagarin given the former's ability to push through the satellite project in the face of political and bureaucratic opposition.[70] Korolev is better remembered not as an inventor but as a 'systems builder'—a 'distinctive type of modern personality with a knack for mobilizing vast human and technological resources into a completely new large-scale enterprise.'[71] Korolev proved himself a ruthless and cunning operator and manager of people, a dogged fighter within the competitive bureaucratic and political system to ensure a concerted effort—on his terms—on ICBM design and satellite progress between the various military design bureaus.[72] Like von Braun, he knew who to satisfy and how to pitch ideas to them, as well as how to organise and cajole large bureaucracies. Korolev's own managerial style was ruthless: firing and redeploying—or purging—workers periodically, assigning impossible deadlines, and using state surveillance resources to bug offices and monitor his empire. He also employed patronage and favours to develop loyalties where necessary.[73] Whilst he had many allies across the defence ministry and the missile programme, events outside his control such as Eisenhower's International Geophysical Year (IGY) announcement and the fall of Beria's group in the nuclear weapons industry helped seal the approval of the satellite effort.[74]

The R-7 ICBMs that underpinned the 'Sputnik shock' was still only a 'hypothetical' thermonuclear war capability against the United States in the late 1950s, with only four R-7s being deployed by the end of 1959.[75] These four missiles were prioritised as a retaliatory capability for one launch each, with the cities of New York, Washington, Chicago, and Los Angeles being lined up for annihilation in the event of an American attack on the Soviet Union.[76] The R-7's 'combat effectiveness' was deemed 'far from adequate'.[77] These were liquid-fuelled rockets and therefore vulnerable to air attack due to their slow preparation, fuelling, and launch time compared to solid-fuelled missiles. Liquid rockets are not usually left fuelled-up whilst in storage because they are far more likely to explode. However, this can be mitigated with storable liquid fuel systems which reduce this risk but impose other challenges on costs, logistics, and engine performance. Whilst solid fuelled rockets can be more safely stored, once lit, they cannot be turned off and the acceleration cannot be throttled up or down, like a firework.

Sputnik II, a bulkier and heavier satellite, proved that heavy Soviet bombs could be lofted into space and flung towards the continental United States once atmospheric re-entry shielding had been developed. This partly pushed the Eisenhower administration towards deploying IRBMs in Europe (including Turkey), which played no small part in causing the Cuban Missile Crisis in 1962 alongside a genuine Soviet desire to protect the communist regime in Cuba from U.S. invasion.[78] Chelomey's UR series of ICBMs provided higher-mass payload delivery into LEO in its SLV capacity, but also made improvements on liquid fuel storage over the R-7. Chelomey saw his ICBM booster projects as simply the means to place bigger, more capable military and exploratory equipment into space, including for Solar System exploration.[79]

The Soviet Union's highly popularised 'firsts' in space must be seen as tips of a military-industrial iceberg, and not just a symbolic gesture in a game of technonationalist prestige and countering images of Russia and communism as a technologically backwards people and ideology. Korolev's bureau was killing three birds with one stone: developing a rocket that could throw a nuclear warhead at New York, put a human in space, and deploy a spy satellite.[80] The resources needed for such achievements and subsequent advances in space scientific research were enjoying the sunk costs and dual-use technologies of the Soviet nuclear missile programme and the emerging military satellite exploitation of Earth orbit. Away from the first achievements that the world remembers today, more significant technological milestones were being realised in the militarisation of space for practical purposes. Based on the rapid improvements of the 1950s, some Soviet military space plans became ever more ambitious, as the historian Asif Siddiqi explains:

> In 1963, the breadth of the projects at Chelomey's OKB-52 [Design Bureau] was staggering. The projects included three new ICBMs (UR-200, UR-500, and UR-100), two orbital bombardment systems (GR-1 and GR-2), two space launch vehicles (UR-200 and UR-500), a nationwide strategic defense system (Taran), an Earth-orbital spaceplane (Raketoplan), a lunar and interplanetary spaceplane (Kosmoplan), plans for an automated anti-satellite project (IS), and an automated naval reconnaissance program (US). This was in addition to his old work on as many as ten different naval cruise missiles. All this was from an organization whose sole contribution to the defense industry by 1959 was a single short-range cruise missile.[81]

As the Soviet space and missile programme achieved many goals (though not all of those above), a greater amount of battles between the institutions and individuals involved in the missile and space sector would buffet Soviet space-technological development.

The military dominance of the logic behind such investments is telling when Korolev's N-1 rocket, which was meant to take cosmonauts to the Moon and back, was not fully funded by the Ministry of Defence (MoD). Its disinterest was more to do with the pursuit of nuclear missile parity with the United States, for which the N-1 was not needed. The N-1 programme did not clearly specify which military needs it could meet. Super-heavy ICBMs were ditched in favour of lighter ICBMs for nuclear delivery purposes in the Brezhnev era, terminating the UR-200 and keeping the UR-500 as a heavy-lift SLV only.[82] The UR-500 is also known as the Proton SLV whose variants are still reliable workhorses for space launches today, though originally conceived as a weapon of nuclear war. Miniaturised nuclear bombs and military-intelligence satellite designs provided upper reaches of how much lift could be needed. This then put an upper limit to the size of rockets the MoD would fund—the most ambitious and largest crewed and robotic rockets for Lunar and Martian exploration therefore would not find a sympathetic military constituency.

Korolev was not opposed to building weapons, and did not find the task abhorrent or merely a 'necessary evil' for realising his dreams of spaceflight. Korolev could be argued to be a patriot and believed in the Soviet Union's founding ideals, and had no inherent compunctions about building weapons for the Soviet state. It was clear for the Soviet Union that the direction and priority of space technology development would be military in nature, helped along by the rather bellicose language of prominent U.S. politicians on the domination of space and of Earth in the Cold War struggle. In the spring of 1960 Korolev's team outlined three major missions that required a large heavy-lift launcher that went far beyond what the R-7 had already achieved: first, military satellites in LEO; second, a global system of space-based communications and weather satellites, and finally the exploration of the Moon and planets. Korolev's 'big space plan', mostly approved intact by the Communist Party, was geared to military objectives and reflected 'the Soviet government's newfound interest in militarizing this new frontier.'[83]

The MoD dominated Soviet space policy and programmes. The Soviet Union's military space programmes did not suffer as much as its crewed 'moonshot' effort did. As seen in later chapters, the USSR developed a plethora of military and civilian space systems—including anti-satellite weapons—on a par with and in some areas exceeding the United States up until the collapse of 1991. For the duration of the Cold War, the overall direction of Soviet space activities, as well as their operation, would be largely subject to the whims of the military. The inter-service rivalries within the MoD plagued Soviet spacepower just as it did U.S. spacepower. It was not until 1981 that the Soviet Air Force and Navy, unhappy with the Strategic Missile Forces' monopoly over space, was able to create a more pan-Defence space organisation with the Chief Directorate of Space Assets that was not dominated by artillery personnel.[84]

In January 1966 the Soviet Union mourned Korolev's passing with a state funeral. His identity and role in the Soviet space effort was only revealed posthumously. The effort was centralised on Korolev until his death, and in his absence others tried to fill the vacuum and competed for the scarce resources Moscow would choose to lavish on the crewed spaceflight programme.[85] From the mid-1960s the Soviet lunar programme began to fall apart with two competing projects dividing scarce resources further.[86] A space programme is not merely a question of science and engineering: it is a battle of budgets, meeting political and important policy goals, managing people and egos, dealing with politicians and vested interests, and inspiring the right people to work in spite of such difficulties. The space technologies produced are the products of these social, political, military, and economic forces.

Civil and military space technologies and agencies remain difficult distinctions to make. It was not until 1965 that the Soviet Institute for Space Research under the Academy of Sciences was set up, but directed by an ICBM engineer. The porous boundaries between military and civilian were not unique to the Soviet Union. NASA was created following the Sputnik 'shock' when Congress found the overall direction of space science and rocket development somewhat lacking in stated purpose and ambition. It was decided that the National Advisory Committee for Aeronautics (NACA) would become a federal agency, renamed the National Aeronautics and Space Administration (NASA). This was an attractive option due to NACA's civilian character, its focus on technical excellence, and a 'quiet, research-focused image' which would not unduly antagonise the Soviet Union whilst keeping a close working relationship with the military services and their space projects.[87] It was decided that NASA had to have some high-profile space projects to be able to 'show off' U.S. technological achievements, and not merely become the agency that picked up projects the defence and intelligence communities were not interested in.[88] One of the pressing issues in choosing the first NASA administrator was not only finding a competent person, but one that was not a military officer so that it would not play into Soviet rhetoric about U.S. military space ambitions or 'frighten other countries'.[89] This is the engineered 'halo effect' observed by the International Relations scholar Deganit Paikowsky, where rockets for space exploration were a surrogate for military power.[90]

The U.S. Air Force was keen to take over most space projects, not only seeking to take Army and Navy space programmes, but also to prevent a large, separate civilian agency emerging to conduct 'civil' spaceflight and intelligence products for national security. The Air Force was led in this effort by no less than Edward Teller, the 'father' of the U.S. hydrogen bomb.[91] But NACA staff won the argument on Capitol Hill, and von Braun's U.S. Army team was transferred to NASA, with the USAF becoming the lead military agent for space, but losing space-based reconnaissance to the newly created National Reconnaissance Office (NRO) in 1961 which would take on 'national intelligence' activities with IMINT and SIGINT satellites. Crucially, deciding whether something was a military or civilian project, and whether it belonged in the military or intelligence agencies was a subjective political and bureaucratic choice by committee.[92]

By the mid-1960s NASA was the primary Federal Government agency for space activities other than space reconnaissance, ballistic missiles, and Department of Defense (DoD) space projects.[93] Founded in 1970, the National Oceanic and Atmospheric Administration (NOAA) is sometimes referred to today as the USA's second civilian space agency due to its extensive climactic and environmental monitoring duties, achieved via satellite data. Therefore, NASA has never been anywhere near the sum of U.S. space activities, though many tend to reduce U.S. space activity to NASA's programmes. NASA had been designed as a civilian and public agency for two important reasons. First, as an ostensibly non-military propaganda tool to counter Soviet space successes. Second, it served as 'an excellent smoke-screen' for American military and intelligence space activities and projects.[94] It deliberately obscured and downplayed the more important (for national security) cutting-edge military and intelligence space projects.[95]

Yet NASA's technological efforts and expertise were relevant to military space programmes, making an absolute civil-military distinction difficult to maintain.[96] Like any other space agency, NASA is not 'above' or beyond politics, nor is it purely civilian in its significance, influences, and contributions. It was not free from the original sin of military spacepower. In fact it 'dovetailed nicely into cold war rivalries and priorities in national defense',[97] and contributed to military technological development as much as the other way around. The politics of NASA and U.S. space policy is not simply whether something is partisan in Congress—the nature of its existence, the purpose and scope of its activities, its boundaries relative to other agencies, the locations of its offices, the companies to which it contracts work, whom is appointed to its leadership roles, and its international relations are all inherently *political* activities. There is nothing objective, 'natural', or apolitical about NASA—just like any other human organisation. It is the result of bargaining and power relationships between different socio-economic interests and pressures, what we can loosely call 'politics'. Tyson and Land pointedly argue against his scientific colleagues who do not see the politics of NASA's past and present activities, that it is not the space science community's 'personal science-funding agency. We are the wagging tail on a large geostrategic dog, which makes decisions without direct reference to the desires of astrophysicists.'[98] Space science, like all forms of science, does not exist separately from power politics in both domestic and international spheres, as well as other ideological, social, and cognitive factors in the development of scientific knowledge and the conduct of science diplomacy.[99]

More important for military power, superpower relations, and nuclear strategy was the creation of the top-secret NRO in 1961 and its technological systems and intelligence products. Eisenhower wanted to make 'damn sure' to leave the impressive space-based reconnaissance effort—the Corona project—with a nationally-orientated intelligence agency rather than the military due to the ubiquitous and strategic nature of such intelligence, fearing a rather narrowed vision that a military branch might impose on the technology, despite the best efforts of the USAF to take full control of Corona.[100] A major technological innovation and capability of the Space Age—IMINT from orbit—transformed the way top-level decision-makers received and requested intelligence. Such imagery became foundational to American National Intelligence Estimates (NIEs) and was according to former intelligence practitioner and historian Richard Immerman 'one of America's greatest intelligence successes.'[101] Corona was officially declassified and revealed only in 1995, though the general existence of space-based IMINT had widely been known since the 1970s. IMINT from orbit profoundly affected elite perceptions over the balance of power, nuclear strategy, and matters of war and peace in ways that deploying a dozen people on the lunar surface for short visits never could. Creating an overhead imagery infrastructure that was responsive to top decision-makers meant that whilst strategic intelligence benefitted, timely images or 'actionable intelligence' for tactical needs and battlefield operations suffered.[102] It would not be until the late Cold War that satellite imagery would start to become useful to battlefield manoeuvres and tactical war planning, as explored in later chapters.

Military and intelligence space technology bloomed and matured as the full potential of sensors and data relays in space were realised by the military complexes of both sides of the Cold War—but only after decades of investments in military and intelligence space technologies. Many of these technologies provided civilian and commercial spinoffs that define the modern space economy. These technologies are more familiar to us today as critical infrastructure from civil engineering, environmental monitoring, satellite television, and online dating. Yet none of these are free from the original sin of space technology; their heritage comes from the entrenched and intensified militarisation of space that began in earnest in the 1960s.

Hidden in Plain Sight: Militarised Space

The 'peaceful' purposes and uses of space have always been rather vague and contested principles. It is part of a Janus-face inherent to military space technology 'which makes it difficult to condemn or condone simplistically.'[103] Reconnaissance satellites eventually earned the right to stay in the military space programmes of both superpowers—despite the pacifistic statements by many governments regarding the peaceful uses of outer space. Moscow had initially tried to ban reconnaissance satellites at the UN because of their military utility for Washington. The ambiguities over peaceful versus 'military' or 'aggressive' purposes were resolved in practice in the language of the 1967 OST. It outlines broad pacifist and equitable principles, but in reality it has enabled and enshrined the military uses of Earth orbit due to the interpretation of 'peaceful purposes' to mean 'non-aggressive' military and intelligence reconnaissance from space, and how the self-defence principle in Article 51 of the UN Charter applies to outer space.[104] It does not prohibit the bulk of military space activities in space today, and only explicitly prohibits nuclear weapons in orbit, and any kind of weapon on a celestial body, such as other planets, moons, and asteroids. Non-nuclear weapons in orbital space are widely accepted to be permissible by the letter of the OST, though its spirit is another matter.

Early on, Soviet and U.S. officials floated the idea of prohibiting the use of space for *all* military purposes, referring to satellite services for terrestrial military purposes, and creating an international satellite programme. It was not only seen as a way to make spaceflight more acceptable to other states, but also as a way to hinder Soviet ambitions to deploy nuclear weapons in space.[105] Progress would be made on agreeing to prohibit the deployment of Weapons of Mass Destruction in orbit and was formally agreed to in the OST. Prohibiting ICBM flights through space and the use of space systems for military users on Earth was deemed too difficult to monitor and prevent, as well as denying what were useful military capabilities. Military communications satellites, reconnaissance, and missile/satellite detection and tracking were seen as more 'passive' and 'non-aggressive' uses of space for supporting military missions on Earth, and were deemed to therefore not violate the principles of the peaceful uses of outer space.[106] Nuclear explosion tests were carried out in space, three each by Moscow and Washington.[107] Fortunately, the Limited Test Ban Treaty of 1963 prohibited further nuclear detonations in outer space—they had witnessed the effects of enhanced radiation effects on satellites, a third of which had been knocked out by the Starfish Prime test in 1962. But they had also demonstrated the technical feasibilities of such high-altitude bombs which were then reliable plans for the first deployed anti-satellite (ASAT) missiles and anti-ballistic missile (ABM) systems.

The Eisenhower National Space Policy of 1958 deliberately set out to engage with the United Nations Committee on the Peaceful Uses of Outer Space (UNCOPUOS) and the wider international community in order to settle on a definition of 'peaceful purposes' that best served American interests in developing military space systems to support terrestrial warfare and nuclear weapons systems.[108] This interpretation of 'peaceful purposes' has stuck since the signing of the OST, and been widely adopted by all other major powers today given their eagerness to develop their own reliance on military and intelligence space systems. The convention of the

European Launcher Development Organisation (ELDO—a joint effort by Western European states to build a Space Launch Vehicle called Europa) states that its technology ought to be developed for 'peaceful applications', without much definition of what counts as peaceful.[109] As the European Space Agency (ESA) emerged from the European Space Research Organisation (ESRO—a joint effort by Western European states to collaborate on space research and science) and ELDO's merger in 1975, it interpreted 'peaceful' as 'non-weapon', meaning that military and intelligence satellites could be deployed.[110] This dual-use quality of space technology has allowed European 'military funding into civilian space programmes by maintaining space infrastructure for "peaceful purposes"'.[111] Making sure that reconnaissance satellites were not deemed to be 'aggressive', and therefore preventing a global ban on all military uses of outer space, meant establishing a looser definition of 'peaceful purposes' to include 'passive' and 'non-aggressive' military space operations such as spying and communication satellites for military forces.[112]

The concept of space as a 'sanctuary' from the threat or use of force was always a debatable and illusory one. The development of rudimentary terrestrially-based ASAT weapons signified that a significant body of opinion across the Cold War divide agreed that a major war on Earth would also involve orbital warfare. By the 1970s, the idea of space sanctuary was arguably a non-starter and one consigned to the history books or discussions of what else may have been had different geopolitical decisions been made in the 1950s and 1960s.[113] As early as 1958 the U.S. military was thinking of the control of outer space as a necessary capability for the military in order to ensure the peaceful use of outer space, much like the control of the sea and the air are necessary parts of military strategy.[114] As NASA was being formed, it had been accepted within the U.S. policy community that the military space programme's envisioned uses of space for missile early warning, space tracking, communications, and spy satellites were consistent with the 'peaceful uses' of outer space.[115] The Eisenhower administration's pushback against more aggressive visions of waging a nuclear war in space and developing crewed nuclear battle stations in orbit shows that the pursuit of military space technologies at that time was not a rush to develop every conceivable military space technology.[116] Many proposals were not pursued, including a plan to nuke the Moon as a demonstration of U.S. nuclear and missile capability, and to use the Moon as a base for the nuclear bombardment of Earth. Fortunately, the Lunar Based Earth Bombardment System which contained underground missile bunkers with personnel tours of 7–9 months on the Moon did not get very far.[117]

The military nature of the early technological developments pervaded events and initiatives that are remembered today as major milestones in space science and rocket engineering. The IGY was an international science project that helped stimulate atmospheric and space research over 1957 and 1958. Roger Launius interprets the IGY as both a cunning ploy by scientists to 'hoodwink' the political and military establishment to fund long-dreamed-of satellite science and engineering projects, as well as a tool used by politicians to recruit space scientists to the cause of military technological development and to contain the Soviet Union and its increasing nuclear and missile capability.[118] The USAF's own primer on military space called Eisenhower's stated desire to separate the military and civilian space efforts impossible in practice, as it was the military that had the scientists, funding, and hardware in the necessary areas. The U.S. Navy's Project Vanguard was the favoured effort to launch a satellite for the IGY until its failure and the U.S. Army's Project Juno—under the leadership of von Braun—succeeded with the Jupiter rockets to launch America's first satellite.[119] Smaller countries were not immune to this either. For example, Norway's space research and geophysics origins are due to their relevance in ionospheric research and radio communications in the High North and the Arctic, which played a direct role in U.S.-Norwegian efforts at containing the Soviet Union and monitoring its activities there. The *civilianisation* of Norway's space research occurred in the 1960s as satellite communications overcame many of the particular terrestrial radio communications issues that America and Norway needed to resolve in the 1950s.[120]

The United States attempted to secure Soviet 'agreement' in allowing satellites to have the right of overflight, as there was no agreement in international law as to where the altitude of sovereign airspace ends. Allowing satellites to fly over the Soviet Union—the

Open Skies principle—would open up the secretive continent-state to American sleuths' eyes from above on a scale that could not be matched in any way with aircraft.[121] Eisenhower was adamant that the IGY had to contribute to establishing the principle of satellite overflight, and not take resources away from the ICBM programme.[122] The development of an imagery satellite and the IGY effort received equal priority in resources from Eisenhower in National Security Council Action 1846.[123] The Soviet launch of Sputnik turned out to be a victory of sorts for Eisenhower. Sputnik's launch had established a practical demonstration of the right of satellite overflight, something that would later be established by both superpowers for reconnaissance satellites by the mid-1960s.[124] No longer did the administration have to face Soviet opposition to U.S. satellites flying overhead given the demonstration of the right of overflight *fait accompli* Sputnik provided.

The insidious and 'hidden in plain sight' nature of the original sin of space technology is arguably most dramatically seen with the 'civilian' Space Transportation System (STS) or the Space Shuttle. Developed and operated by the civil space agency NASA, it was designed in part to carry military satellites into orbit. Indeed, the U.S. DoD saved the STS programme as it stumbled headlong into budgetary problems.[125] The DoD's involvement was decisive in its development, surprising NASA with the large size of DoD satellites, their heaviness, and how few of them were projected to be launched. The DoD threatened that if the STS could not accommodate their needs, the DoD would develop their own new generation expendable SLV rocket system and not use the Shuttle at all. NASA was pushed by USAF to accommodate the DoD's needs for LEO satellite launches.[126] The Shuttle programme stalled U.S. SLV development in the 1970s, and it was only in the early 1980s that it was realised that relying on only one exotic and unproven launch system was not a good move for resilience and assured access to space.[127] The economics of the STS was 'a curious mix of fantasy and hard facts of life', and was a programmatic failure, but remains an icon of U.S. spacepower.[128] As a fixed fleet of four orbiters and one atmospheric glider, without investment in follow-on generations, it would not sustain a continuous production line like traditional expendable SLVs or rockets. The Shuttle led to the extinction of an important part of the USA's space industrial base in expendable, or disposable rocket-based, SLV design and production.[129]

This impacted U.S. military satellite launch capabilities well into the 1980s as the Reagan administration raced to develop a new generation of expendable SLVs and the surrounding production lines, almost from scratch. Fortunately for the DoD, they had kept some existing expendable SLVs in production for LEO launches until the Shuttle had proven itself, but they could not invest in any new rocket programme.[130] For GEO launches, traditional heavy expendable SLVs were still the only way to go. Whilst the Shuttle today is often remembered for its role in building the International Space Station and fixing the Hubble Space Telescope, though significant achievements they are, the programme was in part shaped by military requirements in its development, and saved from the axe during the Carter administration's budget cuts by the then-Secretary of Defense, Harold Brown. There was no escape from original sin.

Original Sin and Empire

The original sin of space technology not only refers to its militarised origins, but also its imperial heritage and postcolonial legacy. Space powers have benefitted from previous imperial and settler-colonial conquests, and built the material and cultural foundations of the Global Space Age upon them. This is manifest not only in metaphors of the 'final frontier' but in the actual practices and locations of terrestrial space infrastructures. As the 'Space Race' dominated the public imagination and perception of space with the 'halo effect', it continues to do so today to our collective detriment, but especially to those who are most often marginalised and dispossessed in the international system. The Space Race and Apollo nostalgia in particular, but many other spaceflight and exploration achievements too, bear the trappings of a religion. They are seen by some as a spiritual quest, a rite of purification for humanity, a road to absolution.[131] Similar to sincerely held religious beliefs, it makes criticisms of such activities difficult to hear, let alone come to terms with, make amends, seek reconciliation, and prevent repetitions. Popular U.S. discourse about space as a 'frontier', or something like the 'Wild West' of the nineteenth-century United States ripe for military, economic, and scientific conquest, betrays overly simplistic and highly debatable notions and beliefs about human nature, civilisation, progress, and consumption.[132]

We project ourselves onto the cosmos—warts and all. The language of imperialism is all too common in space with words such as 'colonisation' remaining in standard usage despite the horrors of the last 500 years of settler colonialism. The fact that there are no indigenous people beyond Earth, unlike the so-called 'New World' in the sixteenth century, does not sanitise that term. The words we use matter because imperial practices continued into the Global Space Age and in the direct service of spacepower development. Empire and colonialism are not only evident in how people, organisations, and states talk about space, but *how they act* as well. How we see Earth and its history shapes how we see space and the future. More often than not, popular narratives on the benevolent aspects of space disguise the imperial and colonial original sins of the Global Space Age.

The U.S. and Soviet turn away from territorialising the Moon and establishing sovereign colonies there was a signifier of the twilight of formal empire, particularly for the United States' imperial practices.[133] However, the continued and often uncritical use of the word 'colony' to describe extra-terrestrial habitats for humans, and some works which actively promote a U.S. domination of the Solar System and multi-planetary colonial system, are enabling overtly imperial and settler-colonist space ambitions. Some champion the European settler-colonist model of exploitative imperial capitalism for space. Colonies are described as a 'paramount' concept for the future development of spacepower as the 'European villages' were in the 'New World' for the production and exploitation of material and financial wealth.[134] Given the competitive, exploitative, racist, and genocidal nature of settler-colonial empires which resulted in the abject and intentional suffering of millions for centuries at the hands of distant, imperial metropoles, it is fair to ask whether such language should be so explicitly used to describe modern space infrastructure and models for future development if we want to do things better in space.

The very concept of the 'final frontier' betrays an imperial, settler-colonialist fascination with space. Many choose to view the frontier image in a positive light in terms of exploration, invention, discovery, human survival, and material progress: tropes that are all too common in space science discourse and popular media. In the 2021 State of NASA address, the Administrator Bill Nelson proclaimed that:

> America has always had a frontier. At the beginning of our country, that frontier was westward. Now that frontier is upward and it's out into the cosmos. It is a part of our DNA to be explorers, adventurers. That's in our DNA as Americans, and we will continue to push the boundaries of space exploration.[135]

Indeed, 'for nearly all of its history, American [sic] has been a frontier nation'.[136] Echoing the many frontier analogies from the Columbian Exchange and 'virgin lands' in the 'New World', the frontier approach to the American West and outer space is rife with historical myth and also glosses over the brutal reality of imperial violence, genocide, and slavery that underpinned those eras of exploration, discovery, expansion, metropolitan enrichment, and technological development. As a result, many advocates' imagery about space seems 'sternly paternalistic', not to mention ethnocentric, violent, and socially Darwinist.[137]

Whilst there are no indigenous peoples in outer space, it is because of the continued legacies of settler colonialism on Earth that the imagery and histories of colonial empires continue to be a poor choice of wording for space activities if their proponents sincerely intend to include all members of any society in that future in a consensual and equitable way, especially historically marginalised communities within postcolonial and formerly imperial states. As Natalie Trevino argues, the 'Final Frontier is the Cosmic Order of Coloniality because that is the order of exploitation of humanity and nature', and invoking the histories of settler-colonialism only sets up a cosmic future of exploitation and hegemony, rather than a future for all humankind.[138] The O'Neill space habitation cylinders popularised in the 1970s were a selective reproduction of white American suburbia—the nuclear family—in space. This was a particularly alien vision of 'the human future' in space for Asif Siddiqi, growing up in 1980s Bangladesh, recalling that O'Neill's 'futuristic gravity-free shapes floating in the cosmos accommodated living spaces within them, full of gentle suburbs, idyllic parks, and mall-like interiors populated by white, well-dressed people who seemed completely content. To me, this was as alien as the cosmos itself.'[139]

Whilst many may indeed be dreaming of appropriating resources and 'expanding America' by invoking the frontier metaphor, expansion into the American West of 1850–1920 is described by Richard Brown as the 'Western Civil War of Incorporation' which:

> pitted insurgent or resistant Indians against the political pressure and military force that concentrated them in reservations throughout the West… [it] also impinged economically and culturally on the traditional lifeways and livelihoods of the Hispanos of the Southwest, who fought back… In the mines, mills, and logging camps on the wageworkers' frontier of the West, employees resisted corporate industrialists with strikes that frequently ended in violence. An Alliance of capital and government fought back with paramilitary efforts to control the far-flung workplaces of the West.[140]

Racism, genocide, forced dispossessions, insurgency, capitalist exploitation, and banditry shaped much of the American West's history. Sadly, there seems to be plenty of room in space for such terrestrial baggage as national myths and perilous analogies are uncritically sent into outer space alongside humans, their machines, and political-economic institutions.

'Astrofuturism' remains a powerful force in shaping how we think we should act with regard to space development, and it is often rooted in hierarchical political views and perspectives designed to trigger a favourable response in targeted audiences, not some universalistic, egalitarian vision of humanity. The scholar De Witt Douglas Kilgore writes that:

> the idea of a space frontier serves contemporary America as the west served the nation in its past: it is the terrain onto which a manifest destiny is projected, a new frontier invalidating the 1893 closure of the western terrestrial frontier. But it is also the space of Utopian desire. Astrofuturist speculation on space-based exploration, exploitation, and colonization is capacious enough to contain imperialist, capitalist ambitions and Utopian, socialist hopes.[141]

Astrofuturism shows how the power and longevity of von Braun's deliberate pitch of outer space as a new frontier and property waiting to be seized was so appealing when the 'American dream' seemed to be under threat from communists in Asia and the civil rights movement at home, and influenced many of the major figures in U.S. space advocacy.[142] This is all the more troubling given the imperial foundations of many space powers and physical infrastructures.

The experiences of 'small', 'fringe', historically marginalised, or 'peripheral' actors, such as poorer states, specific classes, ethnicities, women, sexual minorities, and indigenous peoples on the material, physical receiving end of these technopolitical drives to exploit the heavens remain marginalised if colonial language continues to be used. Noting the colonial political structure of space technologies—or the imperial technopolitics of space—archaeologist Alice Gorman writes that:

> the distribution of space installations does not necessarily coincide with the location of the principal financiers, users and scientists of space exploration—the industrial nations of Europe and North America. Rather, launch facilities tend to be located in areas regarded as underdeveloped and remote from the metropole.[143]

Bernard Brodie wrote in the 1960s that the United States had 'long been a status quo power... uninterested in acquiring new territories or areas of influence'.[144] Yet this account does not hold true when reviewing the United States' securing of territory and resources from the 1940s onwards to support its increasingly complex chain of globe-spanning technological infrastructures, not least in the creation of its orbital satellite communications and surveillance networks. Through acquiring access via imperial powers during the chaos of decolonisation after 1945, or using claims to sovereignty from the era of Teddy Roosevelt, the Azores, Kwajalein, and Diego Garcia were witness to the displacement of local peoples, and their supplanting 'with layers of technological systems—nodes in a global network of power'.[145] The United States' pursuit of Space Age technologies necessitated acquiring or consolidating effective control of 'new' territories. The Satellite Triangulation Program in Diego Garcia was involved in a multitude of missions for American satellite systems, but also for NASA's crewed spaceflight missions. That island territory was unilaterally cleared of local inhabitants by the British to make way for American space, intelligence, communications, and other military infrastructure. The Cold War historian Gabrielle Hecht writes that 'the infrastructures and discourses of global Cold War technopolitics continue to shape the possibilities and limits of power, just as the infrastructures and discourses of empire do.'[146] Space technology was and still is part and parcel of this. Historian Ruth Oldenziel states the U.S. satellite triangulation program and its successors at Diego Garcia 'were part of American-based espionage, space exploration, and satellite systems thus anchored in an island empire that had come into being over the course of a century. Again and again, these large Cold War technical networks were grounded in colonized islands in an era of decolonization.'[147]

In Australia, the Pine Gap facility serviced the American Defense Support Program (DSP) missile early warning system, as well as SIGINT satellites such as Rhyolite and Aquacade.[148] It downloaded 'data dumps' from the DSP and other satellites and transferred them securely to centralised Pentagon systems. Space infrastructure critical to the military power and security of the United States was being built on indigenous lands and was hardly free of political opposition by local communities. According to one surveyor, were it not for the protection the Woomera site offered to traditional Soviet spying methods, the facility would have been located in a more 'civilised' place and in not such a 'dreary' locale at 'the end of the world'.[149] Indeed, such attitudes echoed the writer Ivan Southall's description of the region as 'one of the greatest stretches of uninhabited wasteland on Earth, created by God specifically for rockets'.[150] Viewing that place as uninhabited meant that:

> no-one asked what had caused the apparent absence of people. Since the 1800s, Aboriginal people had been alienated from their country by the usual array of colonial processes: massacres; removal into missions run by religious groups; theft of their land, and so on. Most Australians considered Aboriginal people to be a 'dying race'. In reality, the vast restricted area of the launch range overlapped with the Central Aborigines Reserve and the traditional country of several Traditional Owner groups and nations.

> A combination of drought, the expansion of missions and the dangers of the rocket range meant that many Aboriginal people were under pressure to leave their country.[151]

God or 'nature' did not create a territory 'ideal' for rockets and other space infrastructure. Generations of colonial practices by dominant powers did and still do today, such as in the case of the Thirty Meter Telescope at Mauna a Wākea in Hawai'i where indigenous communities are dispossessed of land which is then desecrated in the name of science and industry.[152] This reality and language jars with the rhetoric of outer space as a place free of prejudice and conflict.[153] In the pursuit of the lofty goals of space exploration, people had suffered at the hands of states through territorial acquisition, consolidation, expansion and technopolitical impositions. The colonial and anti-colonial perspectives demonstrate the original sin of satellite infrastructure that was built on the back of imperial, colonial power politics.

The Soviet Union faced contradictions in inheriting the colonial Russian Empire and committing to the anti-imperial and anti-colonial precepts of Marxist ideology, particularly where there was no 'indigenous proletariat' in the 'colonial, non-Russian periphery'.[154] Moscow's territorial expansion of the nineteenth century was the result of conquest and colonial settlement of the vast northern Eurasian landmass. Many of these areas housed valuable sites and industries within the Soviet space sector. Indeed, the Kazakhstan launch site at Baikonur—a sparsely populated region prior to 1955 and previously used to exile people from Russia[155]—became a focal point for the most spectacular demonstrations of the supposed superiority of the anti-colonial, counter-imperial, socialist technology and science. The Soviet Union's control and possession of such suitable launch sites, along with much of Central Asia, was the result of it being turned into an effective colony of the Russian empire.[156] In the 1930s, the Kazakhs' nomadic way of life was deemed 'backwards' and incompatible with Soviet development needs and was therefore to be eradicated. The ensuing 'settling' and collectivisation of the Kazakhs led to a massive, cataclysmic famine in which one quarter of the population perished. Those Kazakhs who did not starve or die of disease fled, abandoning ancestral pasturelands and their nomadic ways of life.[157] With space technology at the vanguard of socialist development propaganda, the centrality of the Baikonur spaceport in Kazakhstan should be seen as part of larger imperial Soviet development practices in Central Asia that were 'seemingly so promising even in the early 1960s, [then] appeared to be a dead-end two decades later'.[158] Baikonur, the crucible of Soviet space-power, is the scene of past calamities visited upon the region and its people from the Stalinist era, again stressing how 'empty lands' ideal for space and missile technology development are often anything but naturally so, but the product of imperial and disastrously cruel practices that benefitted the first space powers.

Alice Gorman pertinently asks: 'the development of space industry is embedded in colonial industry and economic relationships... Indigenous people paid a cost for the 'curiosity' of others. When are they going to get a return on investment?'[159] Indeed, one need not look to some dystopic, distant future of humanity's habitation of the Solar System or focus only on the spectre of missile-based nuclear war, as Daniel Deudney does,[160] to warn against the negative and harmful impacts of the Space Age on human safety, freedom, and dignity. People have suffered at the hands of technopolitical forms of imperialism and all manner of structural inequalities in the Global Space Age here, on Earth. This is not to unfairly demonise the United States of America or the Union of Soviet Socialist Republics and allow every other space power a 'free pass' for their own space programmes and how they treated indigenous, weaker, vulnerable, or disenfranchised people, communities, and classes, as seen in the next chapter.

The remarkable technical achievements in space were born from the original sin of the military genesis of the benefits of space technology. Many successes of spacepower rely on systems and infrastructures, based on relatively privileged power positions in the international system, that are often the legacy of imperial and colonial pasts and practices. Humanity's journey to outer space was never, and can never be, a road to absolution unless we confront the realities of the original sin of space technologies. As much of the world was decolonising during the Cold War, other parts of Earth were firmly within the grip of colonial practices in the name of this supposed 'next step in human evolution'.[161]

Conclusion

The Global Space Age brought about more technocracy, the institutionalisation of technological change, influence from technical experts, and driving state policies according to needs of technological development. Sputnik's launch from Baikonur in 1957 sparked a leap in the relationship between the state and technology, and Western capitalist states perpetuated more socialist models from the Second World War of continuous state-supported, perpetual technological revolution alongside the state-wide infrastructure and masses of funds required to do so.[162] The space technological revolution was afoot; its roots were in the military-nuclear dimension. From this, all other uses of space developed. Humanity's reasons for entering space were tied to the original sin of waging nuclear war and mobilising vast amounts of resources and people towards state-directed goals. The dark heritage of rocket advances prior to the Cold War cannot be forgotten, as the antifascist powers directly benefitted from the work of Nazi scientists, officers, engineers, and genocidal methods.

Walter McDougall portrays Dwight Eisenhower in a positive light, contrary to how he was characterised in the wake of the Sputnik 'shock':

> Ike was not out of touch. He understood the problems of the age perhaps better than his critics among the Best and Brightest. He feared the economic and moral consequences of a headlong technology and prestige race with the Soviets; he feared the political and social consequences of vastly increased federal powers in education, science, and the economy; he feared, as expressed in his Farewell Address, the assumption of inordinate power and influence by a 'military-industrial complex' and a 'scientific-technological elite'.[163]

This does not mean that 'Big Government' and 'Big Science' are inherently evil things to be resisted or that expertise is to be shunned, or that we are doomed to only make things worse through the development of space technology. Rather, it shows how space activities cannot be reduced to just technical questions of engineering and science. The blind idealism evident in so much astro-utopianism requires the understanding and acceptance of the negative and undesirable political and social consequences of certain modes of development and technological change in the Global Space Age.[164]

Often hidden in plain sight, the militarised nature of the Global Space Age continues today as it began: the pursuit of goals and interests within war, development, and prestige. The distinctions between 'civil' and 'military' space activities are often blurred and the boundaries between agencies porous. The language of 'peaceful purposes', the 'military uses of space', and 'non-aggressive' had been deliberately fudged in treaties, meaning audiences heard what they wanted to whilst the major powers continued to do what mattered most to them in the development of military space technologies. The 'Space Race' began as part of the missile and satellite 'race' of the Cold War superpowers—not as an alternative to it. Space was not some peaceful arena the superpowers chose to compete in—the exploitation of Earth orbit helped bring about the spectre of nuclear Armageddon and further develop the killing power of states, and furthered inequality and imperial exploitation on Earth. As a result, outer space, and Earth orbit especially, has been thoroughly enshrined as a militarised and politicised environment. Both superpowers engaged in a hidden competition to develop new military space technologies and services, and everyone had come to accept a 'de facto regime where space was militarized.'[165]

Historian John Krige argues that:

> The Cold War irreversibly politicized [nuclear and space technology]; both were intimately tied to national security, and interstate rivalry. At the same time both were also embedded in global networks through which knowledge in all its forms circulated. As Itty Abraham put it, 'No atomic program anywhere in the world has ever been purely indigenous.' Andrew Rotter called the atomic bomb 'The World's Bomb.' Asif Siddiqi recently emphasized that 'every nation engaged in [ballistic missiles and space] technology has been a proliferator and has benefited from proliferation.'[166]

Missile, nuclear, and space technologies churned in the Global Cold War. This did not begin with the superpowers, but it catalysed with them. Their technological advances and demonstrations only accelerated the interests of the 'middle powers' of the Cold War and they then strove to master fundamental nuclear, missile and space technologies that they could proliferate, transfer, or sell as they saw fit and within the political frameworks and institutions they had to operate within. A global perspective of international space history cannot overshadow the fact that the USA and USSR were by far the most advanced space powers of the Cold War, but to understand where spacepower is in the international system and global politics today, we must acknowledge the global nature of space history from the dawn of the Global Space Age and look at what the 'other' powers were doing at the same time.

2

BEYOND BIPOLARITY

A fo ben, bid bont

To be a leader, be a bridge

Neither superpower nor the Germans before them had a complete monopoly on practical and theoretical space technology. Europe, China, Japan, and India have their own journeys into orbit and beyond. This chapter explores the genesis of these space powers and entrenches a global space historical perspective that is so fundamental to understanding the major contours of power and technological spread or diffusion in the twenty-first century. As the superpowers led, Europe, Japan, China, and India developed, cooperated, and learned from them, as well as sought their own rocket and satellite programmes, eager not to usher in an era of complete dependence on Washington or Moscow. The 'Nazi aerospace exodus' led to German engineering expertise in rockets, guided missiles, and jet engines ending up in not only Britain and France, but also Argentina, Brazil, Egypt, and India.[1]

Many states saw the potential in the use of space for war, development, and prestige. 'Flag waving' exercises with astronauts or robotic probes are not restricted to the Cold War superpowers. Dimitrios Stroikos argues that China and India's space programmes were heavily influenced by a postcolonial technonationalism, eager to use space to demonstrate modernity and power within their new states and overcoming colonial legacies.[2] National, state, or even stateless nations' flags are prominently displayed in many space activities, particularly high-profile missions. Despite such nationalist overtones, such efforts were more often transnational, cross-border affairs as the flow of technology was manipulated by a global cast of humans, corporations, agencies, and institutions, sharing, obtaining, and transferring knowledge and technologies. John Krige and Asif Siddiqi set the scene for a Global Space Age, arguing that:

> the history of an 'indigenous' launcher is a prime candidate for a nationalistic narrative. A government's prestige, access to markets, and military potential acquired by having independent access to space inspires nation-centered stories that occlude the multiple borrowings and transnational interactions that are sustained by interpersonal, inter-institutional and inter-firm relations and that are eventually built into the hardware. As Siddiqi rightly points out, a global history of rocketry punctures these national and nationalistic narratives, and exposes the 'connections and transitions of technology transfer and knowledge production' that they render 'invisible...' A global space history must retain the national as a key analytic category—not as an autonomous but as an interdependent actor, whose scientific and technological practices are inspired by national interest and framed by foreign policy. It is that policy that determines, for a strategic sector like space, who a country will cooperate with, that defines the terms of the engagement, and that structures the channels through which knowledge will flow across borders—as well as which channels will be plugged to ensure that it doesn't.[3]

As the two superpowers demonstrated the feasibilities of space technologies, others realised that it was unwise to abandon this new geographic environment and technological avenue to Moscow and Washington and resign to total dependency—even among allies and partners.

From the Wreckage: European Decline and Integration

Similar to the superpowers, French and British space ambitions dovetailed with their nuclear and ballistic missile ambitions.[4] The decline of British spacepower followed its decline as an autonomous nuclear weapons power, whereas France became the European leader in nuclear, missile, and space technologies. The French and British empires were in tatters as the Nazi state and Japanese Empire were swept away. They relied on their imperial institutions and territories to assist in their efforts just as the United States and the Soviet Union relied on their own far-flung territories to develop and test their nuclear and rocket devices. Alongside well-known examples of nuclear imperialism that involved British-French uranium mining and nuclear bomb detonations across the colonial and Third Worlds,[5] 'astroimperialism' was made manifest as colonial sites became crucial to European missile, rocket, and space development.

NASA helped kick-start many satellite and space science programmes in the UK, France, West Germany, and Italy,[6] yet desires for some levels of autonomy in missile, satellite, and nuclear technologies were a necessary driver in establishing collective Western European space projects and organisations, today manifested in the European Space Agency (ESA) and the European Union (EU). Despite intimate cooperation on the Manhattan Project, the British Empire was shut out of the U.S. nuclear programme following the McMahon Act of 1946.[7] The UK Foreign Secretary Ernest Bevin proclaimed after an apparently terse conversation with the U.S. Secretary of State John Foster Dulles that London must have the bomb and 'put the bloody Union Jack on top of it'.[8] Britain's Tube Alloys nuclear weapons project was seen as an essential requirement of being an independent state in the post-war world, according to the official UK nuclear weapons historian Margaret Gowing.[9] In the mid-1950s it was the view in some corners of Whitehall that if the UK did not launch their own satellite it would be seen as an 'underdeveloped country'.[10] For British and French leaders, the nuclear possessor was the new coloniser, the non-nuclear states the colonised. Nuclear weapons and ballistic missiles were a technopolitical solution to an identity crisis, rather than a purely strategic move to respond to the Soviet military threat in Europe.[11] This pursuit of prestige alongside the more direct fear of nuclear annihilation was not some idle fancy, as the superpowers differentiated between 'haves' and 'have nots' in space.[12] Britain launched its first satellite, Ariel, atop a U.S. rocket, Scout, in April 1961 with its own 'indigenous' Blue Streak/Black Arrow ballistic missile and SLV programme in tow.

Britain's nuclear bomb delivery options were not particularly plentiful nor straightforward. Despite the first successful nuclear detonation off the northwest coast of Australia in 1952, the gravity bomb-carrying V-bomber aircraft would not enter service until 1955, and only then on an 'emergency operational capability' basis.[13] Britain would eventually abandon the V-bombers, and pursue the indigenous Blue Streak medium-range ballistic missile (MRBM) programme and cancel it as it completed, then pursue and subsequently abandon the U.S.-supplied Skybolt air-launched ballistic missile, then eventually settle on the Polaris submarine-launched ballistic missile (SLBM) purchase from the United States.[14] This reduced the need for an indigenous British ballistic missile programme. Opinion in Prime Minister Harold Macmillan's Cabinet shifted towards preferring a mobile nuclear launch system as part of a larger U.S. nuclear force, rather than having a UK nuclear force that was sufficient to deter the Soviet Union by itself. Skybolt then seemed like a politically acceptable and cheaper alternative compared to Blue Streak.[15]

Blue Streak was to some an overly-redundant capability once the British had access to their U.S. equivalents, the Thor MRBMs, though these were not fully independent capabilities. Thor missiles relied on targeting data from U.S. Strategic Air Command as opposed to UK-provided targeting data for the V-bombers and Blue Streak.[16] Fittingly, a Blue Streak and a Thor stand side by side in the Rocket Tower of the UK National Space Centre in Leicester today. In nearby Melton Mowbray, Leicestershire, three Thor MRBMs were based at an RAF facility there. Another Blue Streak found life not as a centrepiece of a spaceflight exhibition, but as a chicken coop in French Guiana.[17] Summarising one collection of views on the rationale for cancelling Blue Streak, historian Richard Moore explains that because it was liquid-fuelled (as opposed to solid-fuelled):

> Blue Streak fails because it is a 'fire-first' and not a retaliatory weapon, and because politicians see cheaper US alternatives, perhaps more likely to succeed technically and certainly more likely to reinforce Britain's most important international partnership. Opposition to Blue Streak, notably from the Admiralty and Treasury... at the end of 1959 enables the Chiefs, and then ministers, to damn Blue Streak as militarily useless.[18]

However, Moore disputes this view and instead pins the blame on the RAF for preferring an air-launched missile and new bomber aircraft. Blue Streak was a threat to the future of crewed aircraft in the MoD budget fights. He claims that 'the Blue Streak story is not one of mismanagement and huge cost and time overruns; rather, the MRBM was a successful product of Britain's "warfare state".' It was on time, working, and under budget. But it suffered when two of Blue Streak's main advocates retired from Whitehall in 1959.[19] The air-launched nuclear weapons alternative was not cheap nor invulnerable either, and according to some estimates even more expensive and vulnerable to Soviet defences, making the strategic utility of Blue Streak more promising than it initially appears in some accounts.[20] Moore explains that 'decision-makers made strategic points, but to diverse ends: politicians cared about cost, votes and international relationships; senior officers cared more about possessing ships and aeroplanes than developing concepts for their use'.[21] Technologies and lessons learned from Blue Streak would later find a use in advancing Britain's 'Moscow Criterion'—ensuring that British nuclear warheads could always get through potential Soviet anti-ballistic missile (ABM) defences with so-called 'penetration aids'.[22]

Macmillan noted in his diary regarding the UK's nuclear weapon systems that 'we were all right until the late sixties—after that we were not sure what to do'.[23] Regardless of whether it was strategic logic, the budget, the Admiralty, or the RAF that killed off Blue Streak, pursuing Skybolt and later Polaris meant that there was no military rationale for spending large sums of money on MRBM/SLV technologies. Blue Streak the MRBM enjoyed more success as Black Arrow the SLV. Britain's pursuit of nuclear weapons meant it had a foothold on becoming a spacepower, but the 'major watershed' of the Blue Streak cancellation meant the UK gave up on its own ability to develop its nuclear weapons delivery systems.[24]

Britain was torn between two camps: Atlanticists who favoured closer integration and dependencies with the United States, and Europeanists who favoured rapprochement with France to increase relative British autonomy and influence with its west European neighbours. Desires in the early 1960s to share Blue Streak to curry favour with the French fell foul of the Kennedy administration's non-proliferation concerns.[25] Blue Streak would eventually become part of Western Europe's quest for an SLV, showing yet another case of space technology's original sin. The British considered pursuing Blue Streak development for SLV and space research purposes which had direct diplomatic and face-saving benefits with the Australian Government. Woomera was home to the Blue Streak MRBM and the Black Arrow SLV test facilities after RAF Aberporth in Ceredigion, Wales had been ruled out.[26] At its peak Woomera town was home to 4,500 people in the middle of appropriated Aboriginal lands.[27] The Woomera site cut across territories used by the indigenous Aboriginal peoples who suddenly no longer had access to its resources or travel routes. The Australian Government itself was reluctant to be involved at all in the launcher development due to the financial costs.[28]

The Black Arrow successfully launched Britain's satellite Prospero in 1971—making it the sixth state to launch its own satellite on its own launcher. Britain remains the only state to successfully launch an orbital satellite and then cancel its SLV programme. Not even French missile successes could goad the British into keeping up. This demonstrates the power of the original sin of space technology in shaping spacepower development. As the military rationales vanished for the UK to have an indigenous nuclear missile capability thanks to U.S. missile system deals, so vanished Britain's potential as a space power. France today enjoys a far greater status as a European space power with more influence and power over western European space technological development owing directly to its more autonomous nuclear and missile choices in the Cold War.

France detonated its first fission bomb in February 1960 in the Sahara Desert in French Algeria. Its first fusion bomb was detonated in August 1968. The French nuclear programme encountered stiff resistance within the Defence Ministry, which wanted to protect the budget to continue prosecuting wars across the besieged remnants of the French Empire as well as meeting military requirements in Europe.[29] Like the British, the French nuclear weapons programme was not strategically coherent. Its stated strategic rationales were not consistent with a supposedly Gaullist, autarkic strategy given its reliance on external assistance and justifications based on contributions to NATO's nuclear deterrence posture. French nuclear weapons are sometimes interpreted as more of a 'trump card' for Paris in alliance politics at NATO, echoing Ernest Bevin's comments in 1946. Prior to the rollout of a nuclear-powered ballistic missile submarine

(SSBN) second-strike capability in 1974, severe doubts led to the nuclear weapons programme being called a 'military lemon of the first order' by Joseph Alsop and Raymond Aron.[30] Debate continues in academia as to what extent the French nuclear programme lacked a clear strategy. Regardless, France became a nuclear weapons state which required indigenous missiles. It set France on the path to becoming Europe's single-most capable space power and the centre of Europe's integrated space launch capability.

France and Britain pursued joint SLV development through ELDO, the joint western European effort to develop an SLV founded in 1964 by Belgium, France, the UK, Italy, the Netherlands, West Germany, and Austria. In an episode that may have some insights for Britain today, the British effort at ELDO with Western European states began in 1960 after they had failed to get any interest in a so-called 'Commonwealth' SLV from Canada, South Africa, and New Zealand.[31] The French were particularly eager to cooperate with the British on nuclear and missile technologies in ways that the United States had no control over. The French were interested in using Britain's SLV programme to assist their own development of its nuclear missiles, but this was a concern for the British in terms of the transatlantic relationship.[32] The British sought to widen European participation as a way to retain a technological capability but to get others to pay for it, whilst smaller European states were concerned that the bulk of the work—and therefore industrial contracts—on a European SLV had already been divided between London and Paris.[33]

France's SLV and missile effort accelerated on the return of Charles de Gaulle to power in 1958.[34] De Gaulle was determined to develop the Force de Frappe (strike force). The space activities of the French government also found a new focus by setting up *Centre national d'études spatiales* (CNES) in 1961 and ELDO by 1964.[35] Similar to the United States, 123 German engineers from Von Braun's programme in World War II ended up working on missiles for the French military, in particular on the Véronique, Diamant, and Ariane rockets.[36] France had plenty to contribute to ELDO and to gain through joint SLV development. The Agate, Topaze, Emeraude, Saphir, and Rubis sub-orbital rockets formed the various stages of the Diamant SLV which was capable of orbital launches and put France's first satellite, the Asterix, into orbit in 1965. This made France the third state to launch its own satellite atop its own SLV. These 'precious stones' rockets were part of De Gaulle's gambit to use nuclear missiles to regain French prestige and claims to sovereignty, at least in public.[37] The Diamant was a joint Defence Ministry and CNES project, illustrating the blurred distinctions between 'civil' and 'military' space in the French context. CNES would go on to dominate much of European space activities, with it becoming the largest contributor to ESA as well as the delegated, centralised management agency for ESA's launcher programme in later years.[38] The Saphir was developed as an MRBM and was fired successfully thirteen times at the Hamaguir launch site in Algeria. The linkages between European launcher capability and France's ballistic missiles remain today, with the Ariane Group simultaneously overseeing the Arianespace SLV company as well as the construction of the M-51 SLBM for the French SSBNs.

ELDO focused on developing the Europa SLV. It consisted of a British first stage rocket, a French second stage, a German third stage, and an Italian test satellite vehicle. As well as British and French missile programme rationales, they and other European states pursued SLV capabilities for the emerging satellite telecommunications sector, which appeared by the early 1960s to have some commercial potential on top of clear military and public service applications.[39] As time went on, London's more privileged relationship with Washington meant they could access military space services and intelligence that others could not—including 'raw' collected IMINT and SIGINT satellite data. The UK eventually withdrew from ELDO in 1966 due to significant technical and managerial issues, the fears of massively increased costs, and successive French vetoes on British membership of the European Economic Community (EEC).[40] In early 1967 the British allegedly tried a new approach to curry favour with France for EEC membership—supplying useful information on thermonuclear warhead design to assuage Gaullist fears that the British were too subservient to American interests.[41] Britain remained in ESRO as it focused on space science and satellite applications. Unlike France, Britain never set up an equivalent body to NASA or CNES, an agency to oversee space policy or be a major agent of it. It was not until 2010 that the

UK Space Agency was founded, expanding the remit of the British National Space Centre which was set up in 1985. Britain chose to focus on niche satellite components, aerospace industry, and space research at its universities through ESA rather than invest in more expensive launcher development programmes.[42]

Western European astropolitics and technological development were defined by intergovernmental cooperation and integration, but it was a very shaky start. ELDO and ESRO were 'poorly funded and not respected, [and] went down as a textbook example of how "not" to organize, fund, manage and operate a cooperative space effort that crossed states' borders'.[43] ESRO's science projects had to turn to the United States for launch services, and succeeded despite national space aspirations.[44] Yet this did not dampen the drive for western European integration in space. ESA was formed amidst the 'virtual chaos' of the European space landscape in the early 1970s with new, coherent intergovernmental structures and mechanisms to ensure members got what they wanted whilst also coordinating state activities towards larger, integrated goals and projects.[45] Scepticism on the value of a Western European SLV persisted even after initial successes with Ariane at ESA. The former Chancellor of the Exchequer Kenneth Clarke quipped that ESA was 'an exclusive club designed principally to put a Frenchman into space'.[46] Such comments belied an ignorance of the UK space sector's—in particular its universities'—successes in space science and engineering within the ESRO and then ESA.

As well as self-interested motivations, integration and economic interdependence were genuine desires in western Europe.[47] Spacepower became a tool of western European integration and techno-economic competition with the superpowers. As western Europe integrated piece by piece on Earth, in space the sheer expense of the required technologies meant integration was the only way to go if dependency on the United States was intolerable. In 1976 West Germany mostly ceased a 'national' space programme and poured almost all its efforts into the collective effort at ESA.[48] ESA was part of mounting a challenge to U.S. space industry's dominance, and autonomy in every aspect of space technology and industry drives EU space policy and investments today.[49] Some European scientists pushed ESA's activities as a strictly peaceful, scientific and commercial endeavour designed to foment European peace and others wished to ensure that Europe did not become geopolitically 'impotent'.[50] Whilst the U.S. was keen to promote such scientific cooperation in space to drive European modernisation and market access for U.S. goods and services, European states were keen to cooperate as a junior partner early on, but in the long term develop as a more equal partner, and avoid complete techno-economic dependency on the United States.[51]

Although ELDO, ESRO, and ESA each banned explicitly military activity, ESA and the wider European space sector developed space technologies with plenty of dual-use capabilities thanks to some loose wording and the interests of certain member states acting outside these institutions. ESRO was to be exclusively scientific and 'peaceful', with NATO and the Western European Union pushing for more military space integration among their members. To do this, ELDO, which used military-derived technologies, was kept as a separate entity from ESRO. The hostility towards ELDO by the more pacifist states in ESRO showed an inability to come to terms with the original sin of spaceflight.[52] NASA's cooperation with ELDO was severely limited in part due to the massive gap between U.S. and European rocketry, but cooperation was further hampered by the porous barrier between the military and civilian technologies and agencies across ELDO's bigger members, causing non-proliferation concerns in Washington DC.[53] Sheehan writes that:

> while there were those in the scientific community who sought a purely non-military space program for Europe, for the governments that created ESRO and ELDO this could never be the core consideration. As the latter part of the Cold War would show, European space technology was inevitably dual use and would be exploited for both civilian and military purposes.[54]

The 'European space system' would not prevent Europe's entry into military and 'security' space activities. 'European' spacepower is a patchwork of member states, diverse institutions, industries, and specialist communities. New technologies and industries were a potential tool for 'European' influence in the world.[55] Despite NASA's attempts to promote cooperation with ELDO whilst advancing U.S. missile technology non-proliferation goals, the success of the French national space programme by 1966 meant it could further its interests in solid rocket fuel technologies, precisely the kind of technology the U.S. did not want France to be developing. As such the United States tried to save ELDO by keeping the British in to dampen the attractiveness of a unilateral French space and missile programme, but to little avail. European booster technologies were then advancing rapidly without much outside help.[56] The Kennedy and Johnson administrations were generally supportive of unified western European efforts at space science and SLV development in order to help address the 'technological gap' between America and Western Europe, which was perceived to be a problem in ensuring technological capitalism could stand against communism. Yet, the French nuclear weapons programme as well as the Intelsat communications satellite company's preferential governance rules towards U.S. telecoms companies were stumbling blocks to significant technological transfers, cooperation, and sharing that went beyond specific space science projects and into military, intelligence, and commercially oriented space technologies.[57]

By 1970, Britain, Italy, and Germany abandoned ELDO, leaving only France and Belgium as the remaining participants. Yet two years later all but Britain would return to a new launcher project that would eventually become Ariane. France chose European cooperation for this launcher system but still paid the most for it, with the bulk of European space spending going on the Ariane project to 'provide independent access to space for Europe, and of course, France'.[58] A CIA assessment in 1983 viewed Ariane as a serious competitor to U.S. launch services, particularly in light of its marketing and the U.S. Government's decision to suspend production of expendable SLVs in favour of relying on the new Space Shuttle to deliver satellites into LEO.[59] More than a 'sovereign' capability, Ariane went on to become a commercial success—capturing 50% of the global commercial launch market for satellites by 1985. This was despite the awe and respect NASA enjoyed from the Apollo programme by 1969 and the genuine attempt of the Nixon administration to involve Canada, Japan, and western Europe in its post-Apollo exploration programmes. As it became apparent that involvement for these much smaller space powers would become dominated by the Shuttle and would leave few resources available for other space projects, enthusiasm for major international space cooperation abruptly evaporated. As European and Japanese rocketry and satellite capabilities advanced according to their own means and priorities, so did their industries' capabilities and ambitions. With each passing year the inability of the United States to provide launch services for western European programmes and interests became a major sticking point and held up any significant cooperation between Europe and the U.S. in space science in the post-Apollo era.[60]

Whilst Europe pursued its independent access to space and worried about 'European autonomy' relative to the superpowers, its influence prevailed across the imperial and increasingly postcolonial world. The Cold War witnessed a wave of decolonisation which happened alongside advances in spaceflight and increased industrial interests from European countries in some far-flung imperial territories. Following defeat in Algeria, France needed another launch site. France considered many imperially-acquired sites across the Global South. The final choice came down to Kourou in French Guiana, or Roussillon in mainland France. The former was chosen despite its extremely high humidity, in favour of its proximity to the equator resulting in a 17% increased launch efficiency over Cape Canaveral and the thousands of kilometres of oceans which lay to the east and the north.[61] In the 1960s, the French also maintained tracking stations in South Africa, the Congo, Burkina Faso, Algeria, and Lebanon.[62]

French Guiana and Woomera in Australia shared not only their status as colonies of European empires, but also shared a penal heritage. Both places became important test and launch sites—ground facilities—for space programmes:

> These empty places, former colonial frontiers, were stepping stones to new frontiers in the solar system. But the assumption of emptiness was mistaken; and so at Woomera and Kourou the space age begins, and continues, with challenges from those whom it displaces.[63]

The Woomera site disrupted Aboriginal ways of life that relied on the land it had fenced off. The end of ELDO also spelled the end for Woomera as a major site of SLV development as ESA and CNES turned to French Guiana as the home of France and Europe's spaceport.[64] French Guiana today is in part a product of slave labour from the days of the French Empire, and its use as a penal colony provided a sovereign French option for a geographically attractive launch site. After centuries of neglect Kourou suddenly had something of immense value to the metropole. CNES moved to expropriate land from the Creole peoples who did not hold legal title and were relocated to inferior lands and housed in unsuitable accommodation, practices which had disastrous impacts on the community and led to successive protests and demonstrations.[65] The 2,000 or so local inhabitants in the immediate vicinity of the Kourou launch site seemed like a 'minor affair' on the metropole's scale, and therefore local opinion on this new spaceport was less influential in its creation, though that opinion itself was varied at the time. The effects of the spaceport and its new surrounding infrastructure transformed French Guiana, with some criticisms that it had been 'implanted' without consultation with local officials, and the disconnect between the egalitarian visions and the local workforce revealed significant economic tensions and inequalities.[66]

Moving to east Africa, the San Marco Project was a joint Italian-U.S. effort to provide research and development on equatorial launch capabilities from mobile seaborne platforms which began in the 1960s. Whilst this project is celebrated today as a technical success, this was a 'Cold War science project whose history has had no place for Kenya despite being literally based there'.[67] The San Marco agreement was signed with the British Empire two months before Kenya gained its independence, meaning the Kenyan Government inherited the terms and conditions negotiated between London, Rome, and Washington. The agreement effectively sought to limit Kenyan participation. The San Marco project made the Italian Government and NASA participants in the ongoing political disputes between the Kenyan Government and the diverse communities in and around Malindi, and not for the better, including triggering a 'booming sex trade in underage girls'.[68] In later years the Kenyan Government and Parliament pressed the Italians for more integrated cooperation and benefits from the Italian presence. By the 2010s Italy-ESA-Kenya agreements were established on highly technical, highly skilled projects which include supporting the development of Kenyan space industry, research, and universities.[69] Power politics and local community politics still manifest in space science and technology development, and how 'cooperation' agreements are still a bargaining process and not inherently a shared, universal, altruistic quest to explore the cosmos and gain technical knowledge.

These technopolitical practices are not rare. As Alice Gorman writes:

> in common with other Western industrial concerns exploiting primary resources in the developing world, space installations involve the creation of technological enclaves, isolated from local life, but promising benefits from participation in the global economy[70]

The development of space technology is not spared from the inequalities in power relations and the legacies of Empire in the international system nor of the domestic politics inside 'cooperating' or 'hosting' states for space activities. Worse, space programmes can actively entrench and exacerbate such inequalities. The French in Guiana, the British-Australian Woomera detention centre, the forced relocation of the Chagos Islanders for the U.S. Government in Diego Garcia, and the imperial political-economy of the American-Italian San Marco Project in Kenya show the imbalances of political, economic, and technological power and a different tale to that of the European 'underdog' in space relative to the United States and the Soviet Union.

The idealism accompanying European integration through space is in direct contradiction with the colonial practices of European states in building the foundations of European spacepower. European states imposed their will on the global south and indigenous and Creole peoples in the pursuit of spacepower. Doing better, making amends, and avoiding repetitions in future space activities requires reckoning with and learning from such histories, looking beyond the narrow technical achievements of such programmes and onto their wider political-economic and social costs and legacies.

Wholesale 'cooperation' with the Americans came at too high a strategic and economic cost for Western Europe. Rejecting a U.S. monopoly is at home in the classic concepts of realpolitik and autarky in international relations, a far cry from the lofty rhetoric of

Europe's scientific community. Furthermore, many of the United States' European allies were fundamentally suspicious of a complete dependency in space services and techno-economic development. These were areas they saw military, commercial, and political benefits in doing things themselves. This outward European unity towards others is not representative of the internal bargaining and manoeuvring within the European space system where states and industries jostle for contracts and funds within the set rules of various institutions. Bargaining for contracts and resources between ESA's members takes place within agreed rules on proportionality and 'georeturn', and may be preferable to the domestic industrial political bargaining—or pork barrel politics—within or against the United States.

Today, European spacepower is shaped by the European Union, the European Space Agency, and the member states of both institutions on top of global cooperation and commercial agreements. By one account, the European political-military-industrial complex which directly involves the space sector is a hegemonic social force, making the goals of European techno-economic and technopolitical autonomy a 'common sense' approach to be adopted by other institutions, such as labour unions.[71] Between these, Europe as a whole is a source of military, intelligence, commercial, and civil space technological development and proliferation in the twenty-first century, not least in ESA's position as a major launch provider, scientific agency and technology innovator, and with the main contractor for the EU's flagship Galileo navigation system and Copernicus Earth observation constellation. As seen later in the book, collectively, Europe is today one of six major centres of spacepower on Earth.

Japan: Indigenous Rocketry and U.S. Influence

Japan launched its first satellite from its own launcher in February 1970 aboard an Osumi rocket, narrowly beating China into fourth place. Japan pursued the most essential technologies whilst enjoying a mixed record of interest, cooperation, and fear from the United States. Some elites in Japan shared European desires to become independent in some areas of space capability whilst carefully managing its dependencies and cooperative projects with the other space powers, mostly with the United States. Hideo Itokawa, or 'Dr Rocket' as he was sometimes referred to, is remembered as 'the father' of the Japanese space programme. Educated at Tokyo University, he was an aeronautical engineer for the Imperial Japanese Army during the Second World War, and after airplane and rocket development bans on Japan were lifted in 1953 he hoped that Japan could master V-2 rocket technologies as the Americans and Soviets had. Without much institutional support, Itokawa pursued solid-fuelled rocket engine research projects at the University of Tokyo, until the IGY announcement in 1954 catalysed the Japanese scientific community over atmospheric, space, and rocket research. After setting up the Akita rocket range in the north-west of Honshu Island, Itokawa enjoyed many successes with sounding rockets despite piecemeal funding from diverse sources. Early launches included the use of pack animals to carry equipment to the launch site—a remote beach. The Japanese Government became interested in funding Itokawa's rocketry, establishing the National Space Activities Council (NSAC) in 1958 and then the National Space Development Center. Meanwhile, the University of Tokyo set up the Institute of Space and Aeronautical Science (ISAS) in 1959.[72]

Also at Tokyo University and chair of the NSAC was Kankuro Kaneshige, who favoured a different approach to Itokawa's more nationalist, indigenous route. Kaneshige wished to pursue more commercially orientated developments but also to cooperate with the United States and western Europe. The principle of indigenous development meant that speed in technological development would have to be sacrificed, a trade-off Itokawa was comfortable with given the determination to overcome the numerous setbacks in his rocket programmes.[73] Yet Japanese leaders would be sympathetic to the prestige and status elements of a Japanese space programme, and this was not all too different to the potential Japanese nuclear programme that was discussed in the mid-1960s between U.S. President Johnson and Japanese Prime Minister Eisaku Satō. Like the nuclear question, an indigenous space capability was seen as a necessary path for Japan to avoid losing its perceived stature and rank as the U.S., Europe, and the Soviet Union were pulling ahead in these new technologies.

With the rocket programme gaining momentum, a new, safer launch site facing the Pacific Ocean was selected at Uchinoura, despite strong opposition from local fishing communities.[74] Following successes from the Kappa and Lambda sounding rockets, in 1965 NSAC gave the green light to develop a satellite launcher, whilst an orbit-capable Lambda SLV project was already underway. Despite several launch failures in the second half of the 1960s, Itokawa was adamant that Japan learn to develop indigenous SLV technologies, a position that displeased NASA Administrator James Webb who used NASA to achieve wider foreign policy interests. Itokawa's falling out with NASA triggered a campaign against Itokawa's leadership from the Japanese press. Itokawa was mistrustful of NASA and the United States, and held his domestic competitors for funding in contempt. These attitudes fed the negative publicity. Yet previous requests for cooperation from Tokyo on sounding rockets had gone unheeded in Washington.[75] Itokawa left the space sector and moved on to oil storage technologies, content that Japan had developed an indigenous SLV capability and his work there was 'done'.[76]

As well as satisfying domestic economic needs, having a visible space programme would mean Japan would be better able to participate in the discussions over the Outer Space Treaty, and with caution, consider participation in the Intelsat SATCOM consortium. In particular, the successful Chinese nuclear weapon test in 1964 strengthened a Japanese desire to secure prestige with space technology developments and eschew extensive cooperation with the United States, making plain that any interaction would be one of assistance rather than 'guidance' or 'domination'.[77] With successive Japanese governments already open to more space activities and responding to external political pressures, and industry also pushing a space agenda, the National Space Development Agency (NASDA) was set up in October 1969. Its remit focused on satellite development, SLV maintenance and development, tracking facilities, and creating the technologies required for each. Though Japan did not have an overt military aspect to its spacepower development for much of the Cold War, it remained intensely political as ISAS would compete with NASDA for resources and duties, but with the latter usually acquiring 80% of the available resources. NASDA set up another launch site at Takesaki, Tanegashima Island to rival and exceed ISAS' own by aiming for a heavier-lift capability to reach GEO. As a result of this institutional split and Itokawa's 'strident nationalism', through the 1960s NASA and Japan only collaborated sporadically with each other, and with difficulty.[78] By 1969, Japan had *two* civilian space agencies that were merged in the creation of the Japan Aerospace Exploration Agency (JAXA) in 2003.[79]

Though NASDA's original vision was to follow an indigenous route to satellite technological development and GEO communications satellite capabilities, the United States had already intervened in the 1960s by engaging with Japan (as well as Europe) on potential routes of 'cooperation' in order to attempt to control the development of Japanese rockets and regulating the supply of U.S. technology. ISAS' successful development of solid-fuelled rockets 'raised eyebrows' and caused consternation in the United States because of its latency as an IRBM capability. NASA only cooperated on very specific terms with Japan due to the National Security Action Memorandum 334 of July 1965. This memorandum prohibited technological assistance if it would enable other states to develop independent access to GEO for communications satellite deployment. NASA was in favour of such restrictions despite pressure from the State Department to the contrary, and reluctantly 'went along' with the transfer of Thor-Delta technologies and know-how to Japan but deliberately slowed down the pace and scale of the transfer. The U.S. offer of liquid fuel technology via Thor and Delta was meant to move Japan away from solid fuel technology—the propulsion technology that was more suitable for ballistic missiles. This was 'part and parcel' of a U.S. effort to contain potential Japanese nuclear ambitions, should they ever arise.[80] Again, the space-nuclear-missile nexus is unavoidable, even in the more civilian-orientated Japanese space sector. Japan's space development used advances in rocketry from U.S. nuclear and missile weapons designs.[81] Japan benefitted from the original sin of space technology through the licensing of Thor IRBM technologies from the United States, which then provided an acceleration in Japanese rocketry through manufacturing and using them—knowledge and experience the Japanese would deploy in developing the H-series SLVs.[82]

The U.S. effort to curb indigenous Japanese SLV development through sharing Thor-Delta not only failed, but backfired. It *acceler-*

ated advanced SLV development. The H-I SLV programme sought to establish an indigenous Japanese capability to deploy heavier payloads to GEO, and was jointly developed by ISAS and NASDA. Doing so required upper-stage cryogenic engine technology, the development go-ahead for which was given in 1982, and the first successful launch of the H-I occurred in August 1986.[83] Though many parts of the H-I were based on licensed Thor-Delta technologies from the N-I and N-II rockets, it paved the way for the indigenous H-II SLV which provided an even greater GEO launch capability by the 1990s. Approved in 1986, the first successful launch in 1994 was NASDA's 'declaration of launching independence' as the H-II used all-Japanese designs and components and could now offer launch services on the global market without breaching U.S. technology controls. The H-II proved so sophisticated that the flow of technology reversed. The United States began importing Japanese rocket technology! The U.S. Delta III SLV used Japanese fuel tanks, and the H-II could match what the Russians, Americans and Europeans could provide, though reliability became a problem. The H-II proved expensive at twice the launch cost of Europe's Ariane and the United States' Atlas SLVs, and had a troubled development history and a string of subsequent launch failures. The H-IIA, launched in 2001, was developed to address these issues, as well as the H-IIB to address further problems, but technologies were imported around the world in its development and construction in order to cut costs and offer a more commercially attractive launch option.[84]

Tokyo would not alter course on indigenous rocket development, so Washington sought to address the economic threat to the U.S. satellite communications sector instead. The United States threatened sanctions if they 'did not rein in their ambitions to be major competitors in the world market for comsats' but again Japan developed an 'immense in-house capability in the manufacture of geostationary satellites'. However, such threats did result in the next generation of advanced communications satellites being scrapped; and Japan's potential of becoming a world leader in the sales of satellite technology did not materialise.[85] The Japanese Government had a higher political priority than indigenous rocket and satellite development at this time—the return of Okinawa to Japanese control. It had secured this by 1972 in return for a Japanese licence to build and launch N-1 rockets based on U.S. Thor IRBM technology and giving up work on the completely indigenous Japanese Q and N series SLVs. However work on Itokawa's Lambda SLV series continued. American concerns about technological proliferation were not entirely unfounded. Japan had exported Itokawa's Kappa 10 sounding rockets to Yugoslavia and Indonesia under licence—meaning they could manufacture their own solid-fuelled rockets that could reach 700km altitude.[86] As the 1970s arrived, Washington became less sensitive to technological proliferation, especially in light of the improving relationship with Beijing and the extensive technological transfers that followed. As a result, Japan was invited to high-level cooperation in the post-Apollo civil space programme. Though such post-Apollo discussions came to nothing, it set a foundation for future discussion of the U.S. space station and the Strategic Defense Initiative (SDI) in the Reagan administration, which would evolve into the International Space Station (ISS) after 1991 to which Japan remains a major contributor to today—though Japanese participation in the ISS was not unanimously supported. Some feared the ISS' effect in diverting limited resources away from other domestically-orientated space investments.[87]

The 1998 'Taepodong Shock'—a North Korean test-flight of an IRBM across Japan—led to a shift in opinion on the desirability of developing more overtly military capabilities to address the modernising military threats from North Korea, and to a lesser extent China. After the launch, Japan institutionalised a national spy satellite programme to address its early-warning gaps so that future North Korean tests and launch preparations would be detected earlier, also without having to rely solely on allied information-sharing or extremely limited human intelligence capabilities in the DPRK.[88] This wider shift to developing more overt military and security space capabilities was encapsulated in the Basic Space Law of 2008 which:

> means that Japan's space policy has officially transitioned from one that exists only for peaceful purposes (a distinct definition that originally limited Japan from the development of any space technology that could be used for military purposes) to one with a strong—and,

> at long last, visible—emphasis on national security and the use of military space as a critical component of Japan's strategic defense.[89]

Industrial ambitions also furthered this appetite for militarising Japan's space sector and platforms because Japan's commercially viable investments were turning out to be false dawns, and as such increased external security threats ensured a heightened Government interest in funding the activities of the Japanese space sector. Japan's decades of investments in indigenous SLVs and satellite technologies provided the material base from which more military or 'national security' space capabilities could be developed. Japan's spacepower is increasingly tying into its conventional military power which has been wrongly interpreted as something of a minor military power in the quality and quantity of investments in the Japanese Self Defence Forces.[90]

What began as a more nationalist drive for Japanese spacepower in the 1960s led to American technology transfer in the 1970s to control Japan's technological development—but in fact only accelerated Japan's drive towards an independent SLV capability alongside consistent Japanese government and industrial interest in satellite systems and space technologies.[91] Japanese-U.S. cooperation in space, even in the civilian sphere, was subdued by Japanese interests in avoiding total dependency on the United States, whilst Washington feared Tokyo's rearmament and commercial competition through technological diffusion. Technopolitics and geopolitical, industrial competition therefore pervades even the less overtly militaristic nature of Japanese spacepower in the Cold War. Japan and the United States managed to secure a deal where Japan gained access to U.S. ballistic missiles for the sake of SLV development, showing how the original sin of space technology still weaves its way into more ostensibly civilian space programmes. Given the nationalist-autarkic drives of Itokawa, Japanese industrial and elite political interests, and U.S. commercial self-interest, it is remarkable that such cooperation on Thor and Delta rocket systems happened at all.[92] Whilst it was not until the late 1980s that Japan began deploying satellites for infrastructural services (as opposed to technology development and research satellites), the foundations for growth had been firmly laid in the Cold War era. Today, Japan is one of a handful of states in the international system with an independent launcher capability, an extensive satellite industry, and large, conventional military forces that are modernising with space-based support services from both the United States and its own 'national security' space sector.

China: Nuclear and Missile Primacy

China became the fifth state to build and launch its own satellite by its own means by April 1970, just behind Japan. China's rise as a comprehensive space power, second only to the United States, is based on foundations laid decades ago in its ballistic missile and nuclear industries. Its entry into space is therefore also marked by the original sin of space technology. Michael Sheehan argues that China has traditionally attached a high level of prestige to their space endeavours, and sought to overcome the 'brutal ravages of colonialism and imperialism' in the century or two prior to the Space Age and to 'rise beyond the imperialist legacy and to be recognised as a sophisticated and technologically advanced state.'[93] Yet prestige was additional to the real technological advances, development goals, and military applications that space innovations provided and represented. Its programme focused on essential nuclear missile technologies before expanding into an elaborate military and civilian space satellite infrastructure in the 1980s. Chinese elites invested in space technologies for a mix of military, prestige, and eventually economic reasons, meaning claims of a unique 'Chinese' route to orbit that deviated from those of other countries may be overblown.[94]

When examining Chinese space activities:

> many in the western media who ought to know better [respond] to Chinese space developments with a mixture of puzzlement, patronizing put-downs and dismissal. Chinese capabilities were often played down on the basis that their equipment was alternately primitive or imitative. If it worked, the presumption was that it must have been stolen or developed for sinister military purposes. There remains an extraordinary reluctance to concede to the Chinese the credit of having created, designed and built their own equipment.[95]

Harmless activities such as lunar exploration rovers are sometimes construed as a threat posed by China's general technological advancements. Like the other space powers of the twenty-first century, China invested in space technologies as part of satisfying a network of interests in the needs of war, development, and prestige. It cooperated where necessary and beneficial to do so, but perhaps more so than the other space powers China was isolated in the first decades of the Global Space Age. However, China and the U.S. cooperated in space during the Cold War, including in sensitive space-based intelligence capabilities and infrastructures. As reviewed by the researcher Cameron Hunter:

> Only a couple of years after the founding of the Chinese space program, American officials had already come to view it as a threat. This trend continued through the 1960s and into the 1970s, when dialogue after the Shanghai Communique led to US-China space cooperation becoming literally a 'normal' part of the relationship.[96]

This only raises the importance of integrating China's long space history with the other space powers as technologies ebbed and flowed across state boundaries in the Cold War. China has since become a significant source of spacepower technological development, proliferation, and commercially available space services in the post-Cold War world.

Sputnik's launch in 1957 made quite the impression on Mao Zedong, who lauded it as a major step in the 'progressive conquest of nature', and committed China to its own satellite and launcher programme.[97] Qian Xuesen, a U.S.-educated scientist and engineer from Hangzhou, Zhejiang, led the efforts in China's nuclear, missile, and satellite programme. Qian had been educated at the California Institute of Technology and the Massachusetts Institute of Technology, and worked on American rocket programmes and the Manhattan Project in the 1940s. Qian worked on solid rocket motors during the war years and drew up plans for a ballistic missile programme—work which was officially commended by the U.S. Army Air Force. He surveyed the V-2 rocket programme in Germany for the U.S. in 1945, and interrogated Wernher von Braun after his surrender to U.S. forces.[98]

In a twist of fate it would be von Braun, the Nazi SS Major who led the V-2 programme responsible for the lethal slave labour at the

Mittelwerk factory and designed the weapons used against the Allies, who would find a welcome home in the post-war United States. Qian Xuesen was detained and deported by U.S. authorities at the height of the anti-communist, anti-Chinese, and counterintelligence hysteria of McCarthyism. The Federal Bureau of Investigation (FBI) did not look kindly on Qian for sending freely available technical books to communist China. The Chinese United Front Work Department (UFWD), responsible for maintaining links with the Chinese diaspora and facilitating their return to China, did not have any plan to bring him to China, or risk any chance of Qian's imprisonment. Instead the U.S. expelled Qian themselves and sent him to China by 1955 in return for prisoners of war held in North Korea.[99] After arriving in China, Qian was quickly brought in to lead the missile and nuclear programmes. Therefore the man widely remembered as the 'father' of the Chinese atomic bomb and missile technologies was voluntarily sent to communist China by the United States—a remarkable own goal for McCarthyism.

That said, like all father figures of space programmes and 'great man' approaches to history, it is important not to overstate the role of Qian in the Chinese missile and space programme as he was not the principal designer for any of China's rockets. Like Korolev, his role was perhaps more influential in creating the political linkages between the dozens of rocket scientists, many of whom were also trained in the United States.[100] Whilst they kept their distance from Qian, the UFWD succeeded in returning 84 other U.S.-trained Chinese specialists.[101] A notable return was Wang Xiji, who had been educated at the Virginia Institute of Technology. Under the direction of Deng Xiaoping, Wang was set to work on developing China's first sounding rockets with the goal of eventually mastering ICBM and SLV capabilities.[102] Despite relative international scientific isolation in the Maoist periods, many Chinese scientists were in fact cosmopolitan individuals who had been educated in the West, Japan, and the Soviet Union—with almost 2,000 students and academics returning to China between 1946 and 1958.[103]

The fates of the rocket and satellite programmes—and most importantly the people working on them—were tied to Mao's own political fortunes. The space sector was a participant in the political struggles of the Great Leap Forward, the Cultural Revolution, and the internecine political infighting between the Chinese Communist Party and People's Liberation Army leadership. Like many of his counterparts around the world, Qian Xuesen and his staff had to be politically savvy to survive at the top of the space sector, not just relying on talent with machines and physics.[104]

Grand promises made by the scientists and engineers about the speed at which China could develop an SLV and satellite in Project 581 (in a matter of months!) helped inaugurate a serious missile and space effort in 1958. But it ran headlong into political forces as the Great Leap Forward's destructive practices demanded promises to be realised. The missile and space programmes were not immune to the Sino-Soviet split and the Maoist drive to 'reform' the managerial and bureaucratic system as it was allegedly re-establishing class and bourgeois hierarchies that had been witnessed in the Soviet Union. In the early years, Sino-Soviet cooperation on space and missile technology was not particularly comprehensive or enthusiastic, particularly as Mao detested Khrushchev. Both sides had sown the seeds of the Sino-Soviet split by the late 1950s.[105] Technical visits by Chinese scientists to the Soviet Union were not particularly cordial affairs and ended entirely after the Sino-Soviet split, where Moscow and Beijing ceased all cooperation and diplomatic exchanges based on a mix of an ideological split between Soviet and Maoist communisms, competing desires to be the leading communist power on Earth, and strategic disputes including border skirmishes in Manchuria.

Deng Xiaoping abandoned the satellite effort in Project 581 and refocused on the missile programme in the late 1950s as Mao's Great Leap Forward took its toll on people and the economy. At the Third National People's Congress in 1964 Zhao Jiuzhang, a chief designer in the rocket programme, suggested to Zhou Enlai, the Premier (or Prime Minister), that they 'combine the tests of our ballistic missile program with launching a satellite, and get the benefit of hitting two birds with one stone.' This was on the back of successful long-range missile tests by the PLA. Qian Xuesen advocated along the same lines, wanting to create a new defence organisation to pursue 'aerospace' projects.[106]

An all-of-government approach ensued following a 'legendary' meeting in October–November 1965 which:

> brought together representatives from all of the major institutions: the Committee on Science and Technology for National Defense; the Committee on Industry for National Defense (later merged with the former); the National Committee for Science and Technology; the general staff of the People's Liberation Army (PLA); the air force; the navy; the 2nd Artillery; the PLA Signal Corps; the 1st, 4th, and 7th Machinery Bureaus; the Ministry of Post and Communications; the Academy of Military Science; and thirteen other research institutes under the CAS [Chinese Academy of Sciences]. CAS Vice Director Pei Lishen chaired the meeting.[107]

The ubiquitous nature of space technology is on display here, and so is its original sin, putting Chinese space history firmly along a parallel road to all other major space powers. Reflecting the nuclear imperialism of the other nuclear weapon states, China too engaged in a nuclear imperialism of its own by testing and detonating its nuclear devices in Xinjiang Province, the home of the Uyghur people.[108] To this day, the rural poor in other parts of China have to put up with the risk of falling rocket debris as some Chinese launch sites are not remote enough from towns and villages.[109]

China's first satellite project, Project 651, was ambitious. It sought to develop an imagery, weather, and communication satellite. In a slight distinction to the U.S. and USSR, the Chinese rocket and satellite programme began as a civilian Chinese Academy of Sciences (CAS) project before becoming intimately involved with China's nuclear, missile, and military satellite programmes. Such technical progress collided with the Cultural Revolution, within which intellectuals were targeted as part of an entrenched bureaucracy and CAS became a 'major battlefield' where the young students chanting the aphorisms of Mao's 'Little Red Book' destroyed much of the satellite programme. The purges of the academics and technicians led CAS to ask Zhou Enlai to bring it under the auspices of the PLA where Zhou had succeeded, but not without difficulty, in shielding his people from the impact of the Cultural Revolution's purges.[110] The Cultural Revolution would change the objectives and emphasise speed and propaganda value from the first satellite launch, leading to politics trumping technological progression, hollowing out the scientific and learning value of the first satellite launch to the detriment of a more ambitious and sustainable satellite programme.[111] On Zhou's death, vast numbers of personnel from China's space sector turned out to mourn him despite Red Guard intimidation.[112] That vision of a range of satellite types would not be realised until the 1990s, after China had rebuilt from the loss of an entire generation of skills, knowledge, and talent that the Cultural Revolution had purged.[113]

Despite the retreat in its ambitious satellite programme, China's space technologies developed consistently through the PLA's ballistic missile and nuclear programmes. China's first satellite launch in April 1970 with the Long March 1 SLV was effectively a three-stage version of the Dong Feng 4 MRBM which could deliver nuclear warheads as far as the mid-Pacific.[114] The CIA and much of the policy elites in Washington showed a relaxed attitude to the Chinese satellite.[115] China also developed the DF-5A ICBM which was also used as an SLV.[116] China, like the other space powers, had much to gain from such dual-use technology as it 'effectively maximizes its returns, both politically and economically although the original purposes were clearly military.'[117] Given the limited economic and technical resources available, it was still worth investing the large sums in these crucial new technologies. China developed a range of DF missiles to begin developing strike options against U.S. bases in its near abroad, whilst the interior, remote sites of Chinese spaceports ensured they were secure from either U.S. or Soviet incursions. Given the massive industrial resources required and the clear security benefits of having a credible nuclear capability, it should not be entirely surprising that the PLA's nuclear and missile efforts became a safe haven for what remained of China's space projects. The pursuit of space technologies as military technologies means that China's entry into space is 'analogous to those developed earlier by the original space participants, the Soviet Union and the United States.'[118]

Though not as cutting edge to the same extent as Soviet and American space technologies in the Cold War, military technology need not be 'the best' to significantly enhance the power and prestige of a state. Demonstrating and mastering the essential technologies—particularly missile and nuclear technologies—provided a sound foundation for later developments and put others on notice that more advanced capabilities were always possible in future. It was also no coincidence that the nuclear, missile and space programmes not only addressed external threats, but also became a vehicle for re-establishing and maintaining Mao's political position after 1968, and later as domestic and international propaganda and prestige tools.[119] Mao and his cohort defined the scientific goals differently to how the Soviet Union did and sought to instil a confidence in Chinese science as an alternative model for Third World states.[120]

It was not until Deng Xiaoping had cemented his leadership after Mao's demise that China's space sector turned towards economic development and social needs. Qian Xuesen picked the losing side in the political battles of the 1970s and faded away from prominence after the first Chinese satellite launch. He sided with the doomed 'Gang of Four', and later spent many years trying to make amends with the victorious Deng Xiaoping. Qian was reconvicted by a U.S. congressional investigation in the 1990s for taking the Titan rocket design to China even though the design did not exist at the time of his deportation.[121] The Cox report led to Washington isolating itself from Beijing in space and rocket technologies in the wake of the Tiananmen Square protests and repression. By the late 1980s the Chinese space sector became crucial to 'the heart of China's efforts to develop a strong scientific and technological infrastructure as the basis of its future development.'[122] The seeds of China's growth as a comprehensive space power were sown, with advances across the military, economic, commercial, and scientific aspects of spacepower.

The original sin of space technology applies to China, and today 'the Chinese space program now has an extensive infrastructure, comprising four launch centers, with ground facilities for manufacturing and testing; a worldwide land, sea and space-based tracking system; a fleet of operational light, medium, and heavy-lift launchers; and a well–established institutional architecture.'[123] China's growth as a space power has taken a long time—but it is fully in swing as of the 2020s. As seen in the conclusion chapter, China now holds the second largest number of active satellites, at 467—higher than Russia's 164 but behind the United States' 2,939. There are limitations to what such launch numbers can tell us, as what is launched, why, and when also matters and is discussed later. Military space technologies are but one aspect of Chinese spacepower. China's rise as the world's biggest economy and a major industrial-technological powerhouse *includes* spacepower as part of this larger development and modernisation drive. Therefore China's space activities cannot be reduced to its military rationales and programmes. Those that do so risk seeing threats where there are none, such as in crewed or robotic exploration of the Moon and Mars, or the Chinese space station, and may fail to see the varied rationales and interests behind the wide range of space programmes conducted by China today.

There is a problematic tendency in U.S. commentary to portray all Chinese space activities as 'controlled', 'dominated', or 'run' by the PLA, implying that all activities are military or threatening in nature, even when there is no evidential basis for significant military programme management, such as with the Chang'e lunar science programme. Such claims 'had as much merit as concluding that the American Apollo program was "military-run" because the US Navy retrieved the astronauts from the ocean'.[124] While the PLA is involved in various aspects of China's space exploration programmes, its presence does not mean Chinese space exploration is inherently a threat or a singular concern above all others. Like other space powers, space policies are the products of many varied interests and institutions and its major space policy goals are decided by the civilian central government,[125] but that does not mean China has no military interests or poses no threat at all to others in space. Far from it, as discussed in later chapters. Rather, the governance and implementation of Chinese space policy is a dynamic mix of institutions and a multi-agency endeavour—government, military, and increasingly private sector—like any other major spacepower. China's space weapons programme is one part of a larger portfolio of programmes and interests.

For many, China's space capabilities and especially its advancing crewed space programme are seen as the zenith of techno-scientific power, a way to secure 'a place for one's mat', or a seat at the table in the international system after the collective traumas of the nineteenth and twentieth centuries. This is not necessarily a zero-sum attitude and reflects the process of joining an elite 'space club' with both domestic and international political payoffs and power within the accepted rules and norms of that elite club.[126] China had sought to join the ISS programme, only to be rebuffed by the United States at every turn. China's new space station is making good on their statements in the 1990s that they would pursue their own should they be excluded.[127] The Chinese space station will now become a new and alternative hub of orbital science. Beyond taking their cues from what they see as the most advanced technological countries and finding a place in the 'club', there is little other coherence or evidence of a singular 'grand strategy' or 'master plan' tying all of China's space activities together.[128] Rather, like any major economy and political entity, China is using space for a variety of reasons and purposes, some benign, some potentially threatening. China is not special or unique in that regard. It is pursuing spacepower for the needs of war development, and prestige, like many others and has not escaped the original sin of space technology.

India: Economic Development and Military Latency

Eager not to allow India to fall behind in essential technologies in the new postcolonial world, like Beijing, New Delhi invested as early as the 1960s in rocket and satellite technologies. India launched its own satellite into LEO aboard an indigenous rocket for the first time in July 1980 on the Rohini 1B launcher, becoming the seventh state to accomplish such a feat, and only a year later ESA placed an Indian satellite into GEO, demonstrating maturing satellite capabilities.[129] India's spacepower has been characterised as developmental in its approach, with a propensity to cooperate and partner in missions and technology transfers from a range of international sources much like its defence and armaments industry. Despite development being its primary goal India, like Japan, was also pursuing latent military potential and long-term techno-industrial autonomy by developing (as much as possible) indigenous space, missile, and nuclear sectors. Unlike Japan, India had a credible nuclear weapons infrastructure by the late 1970s, whilst Japan pursued nuclear energy only. Therefore the original sin of spaceflight and the military rationales of spacepower are still only at arm's length in the Indian case, despite the evidently civilian- and development-orientated approach of the Indian Space Research Organisation (ISRO), the lead agency for space in the Indian Government.

Before independence from the British Empire and during his first years as Prime Minister, Jawaharlal Nehru and other political elites established a science, technology, and industrial infrastructure in India.[130] Nehru was convinced of the transformative potential of science and technology for developing states and the early post-independence focus on science and technology programmes was not some great shift in direction for the new India, but rather the blooming of scientific institutions that had already been set up with foresight in the twilight years of the British Raj.[131] India's first dedicated efforts on space research originated within the Department of Atomic Energy's Indian National Committee on Space Research (INCOSPAR).[132] The British-educated scientist Vikram Sarabhai, remembered today as 'the father' of the Indian space programme, encapsulates the original vision of Indian space investments:

> There are some who question the relevance of space activities in a developing nation. To us, there is no ambiguity of purpose. We do not have the fantasy of competing with economically advanced nations in the exploration of the Moon or the planets or manned space flight. But we are convinced that if we are to play a meaningful role nationally, and in the community of nations, we must be second to none in the application of advanced technologies to the real problems of man and society.[133]

This echoes the words of Ralph Abernathy, the Poor People's Campaign leader in the United States, who wished for 'NASA scientists to tackle problems we face in society', drawing attention to the critiques of Apollo and the largesse in U.S. Government spending on the supposed stunts of crewed spaceflight.[134] India deliberately distanced itself from the Soviet stunts and the apparent vanity of the U.S. moonshot and stressed the practical benefits advanced technologies could bring from space development.

Although Sarabhai's expertise was not in rocketry, his skills and contributions lay in his von Braun-, Korolev-, and Qian-like managerial and political qualities. The writer Gurbir Singh describes Sarabhai as 'a great communicator, skilful manager, and highly energetic person'.[135] Born into a privileged life as the descendant of a wealthy merchant family, he eschewed political activism and grew his networks with the international scientific community. This was in stark contrast to his sister Mridula Sarabhai and aunt Anasuya Sarabhai who were politically active on suffragist, anti-colonial, and trades union movements, and also close friends of Mahatma Gandhi. Mridula and Anasuya were often at odds with the direct interests of the industrialist Ambalal Sarabhai—Vikram and Mridula's father.[136] Vikram sadly died at the young age of 52 in December 1971, possibly due to overwork and stress. But the political and institutional foundations of Indian spacepower were laid as ISRO was set up in 1969. Satish Dhawan led ISRO from 1972 to 1984, overseeing India's acquisition of several critical, foundational space technologies on an increasingly indigenous basis. Like the other space powers, India's space development was political and not devoid of statist self-interest and long-term strategic military interests, not fully escaping space technology's original sin.

India had to cooperate internationally on rocket and satellite development if it was to develop quickly. It engaged with the USSR, USA, Britain, France, and Japan, and hosted various rocket launches and tests from the Thumba launch site in Kerala. This site, now called the Vikram Sarabhai Space Centre, is 8.5° north of the equator, resulting in a significant equatorial launch efficiency advantage. Its location on the southern tip of India means that any debris or accidents are likely to crash into the Indian Ocean and not foreign state territory. India also set up a launch site at Sriharikota, today called the Satish Dhawan Space Centre, on the east coast of India at 13° north which was active from the early 1970s onwards and became a bigger facility designed to launch heavier payloads with more complex launch vehicles, learning from the experiences of Thumba. It also enjoyed the boom in elite and public support for space technology development following the successes of the U.S., USSR, France, Japan, and China with space technologies in promising development and prosperity.[137] Despite its more northerly position, it remains Asia's most equatorially proximate spaceport.

The international sounding rockets and atmospheric science projects in Thumba in the 1960s allowed India to not only master these new 'essential' rocket technologies, but also to develop applied science in understanding the monsoon weather patterns which directly affected the entire subcontinent.[138] The Nike-Apache sounding rocket launch from Thumba, with NASA's cooperation, was celebrated across India. Indian leaders and scientists 'saw space research as a harbinger of modernity in the newly decolonized state and as a symbol of prestige and development'.[139] Contrary to later rocket interests from India, the Nike-Apache sounding rockets were designated by the United States as low-end declassified scientific instruments. India rapidly learned from Nike-Apache and Scout rockets and built their own in the 1960s—the Rohini series—and moved towards bigger, heavier SLVs.[140] India's eagerness to develop as a space power may have been helped by Pakistan's activities. Pakistan's Space and Upper Atmosphere Research Commission (SUPARCO) was set up in 1961, and in 1962 a NASA-built sounding rocket was launched from Pakistan with the help of NASA engineers. Islamabad was therefore in possession of a new sounding rocket programme. As the years went on Indian spacepower went from strength to strength whilst Pakistan's languished as it simply could not match the resources required to keep up.[141]

India drew in cooperation from many quarters. Hideo Itokawa was hired as an adviser to ISRO at the invitation of PM Indira Gandhi.[142] Soviet M-100 sounding rockets were also launched from Thumba, and collectively these sounding rockets furthered India's grasp of solid fuel rocket technologies, beginning its indigenous sounding rocket programme in 1965. The first indigenous sounding rocket launched in 1967—Rohini-75—which featured a co-manufactured Centaur second-stage rocket motor with France. India's early rocket development programme was therefore extremely international, in stark contrast to the superpowers, not only in terms of visiting teams and hardware from the outside, but also in how many Indian technicians and scientists were educated in or visited American, European, and Japanese universities to complement the largely Indian-educated scientist base. Sarabhai himself maintained a high degree of international scientific and policy networks and Indian officials and scientists could tap into those global space technological connections to an extent China simply could not, at least not until U.S.-Chinese space cooperation took off in the 1970s and 1980s.[143]

In addition to the rocket work, India proceeded along satellite development and service design and purchasing launch services from the global market long before India had its own satellite launch capability. From the inception of INCOSPAR India's emergence as a space power proceeded on three major fronts—communications and remote sensing satellites, SLVs, and space application programmes—all at the same time rather than one after the other. INCOSPAR was set up by Homi Bhabha who also led the Department of Atomic Energy (DAE), and had ambitions to develop an Indian nuclear weapon capability, in opposition to Vikram Sarabhai's anti-nuclear views. In addition to satellites and launchers, ground infrastructure needed development too. As ELDO disintegrated and pulled out of Woomera in the mid to late 1960s, Sarabhai jumped at the opportunity and sent engineers to Adelaide to purchase a complete satellite telemetry ground station at a bargain price. This equipment would have been destroyed, but instead found new life in the Sriharikota launch facility.[144]

A major effort plugged space technology into India's ambitious plans for economic and social development, particularly in continent-wide telecommunications infrastructure, meteorology, and resource management.[145] This led to what Michael Sheehan calls 'in some ways the most cost-effective and successful space programme in the world... [by] using space as a crucial mechanism for lifting India's people out of poverty through education and social and economic programmes.'[146] Though China too was in a postcolonial, underdeveloped state, India took a more developmental and leapfrogging approach to space technology from the very start, such as satellite TV which is explored in the next chapter. Like Japan, India had less overtly militarised space programmes at the start. The civilian, infrastructural emphasis of Indian spacepower in the Cold War does stand apart from those of the USA, USSR, Western Europe, and China which all had either military origins or direct military interests animating them.[147] However, the military utility of spacepower did become more significant for India as it and its neighbours became more capable in military, nuclear, and missile areas.

India's progress in SLV development raised alarm bells in the United States due to what Sheehan notes was the:

> obvious military rationale that lies behind the country's development of both launcher and satellite technology. As with parallel criticisms of India's nuclear programme, there was a distinct element of hypocrisy in critiques that were sanguine about the American record in this area, but deemed a comparable [nuclear weapons and ballistic mis-

sile] effort by the world's largest democracy to be a potentially destabilising factor in international relations.[148]

Despite the common roots of India's SLV and ballistic missile technologies, they were distinct programmes. That said, they were parallel components of a comprehensive security and development strategy. New Delhi sought the state-wide development of an overwhelmingly rural country and the pursuit of security, military power, and reducing technological dependence on other countries. In addition, some Indian ministers tried to argue that its SLV programme would not go towards its development of MRBMs, and if they were joined up it would have slowed down the former and blocked much of the international cooperation India had enjoyed in its space-related ventures.[149] Despite some studies in the 1960s on using the SLV and the Rohini sounding rockets as IRBMs and SRBMs, the Indian government did not pursue a large-scale effort in longer-range ballistic missiles until the late Cold War.[150]

India needed technological advances in areas such as inertial guidance systems for SLVs, but these areas were also directly relevant for ballistic missiles. They provide a technological latency merely through being there, ready to be used and adapted if needed once the political direction is given to do so. The Japanese and Chinese indigenous satellite launches in 1970 were formative in triggering a greater SLV and latent nuclear missile effort in India. The United States did not wish to repeat the experience of the Thor-Delta deal with Japan which accelerated dual-use rocket technologies. Whilst the United States may have slowed Indian technological development by not cooperating after the 1960s beyond specific satellite missions, they could not halt a determined, unified, and patient series of Indian governments keeping up with China and Japan. The Rohini launch attracted global attention and the U.S. State Department 'expressed grave concern.'[151] The 1974 nuclear explosion test Pokhran-I openly declared India's nuclear weapons programme and the successful launch of the Rohini demonstrated India's maturing space, nuclear, and missile programmes.

The historian Odd Arne Westad argues that India was the Cold War 'wildcard' due to its mixed economic approach and non-aligned identity compared to the more ideologically rigid China.[152] In the late 1960s and 1970s Prime Minister Indira Gandhi carefully balanced closer cooperation with the Soviet Union whilst also condemning increased Soviet influence in Pakistan, whilst also receiving civilian U.S. aid in the shadow of frictions caused by U.S. support to Pakistan. Tensions boiled over in the 1971 Bangladesh War. India feared encirclement by two 'friends' of the United States—Pakistan and China—following Nixon and Kissinger's successful diplomatic openings in Beijing. Kissinger was convinced the 1971 Bangladesh War was a 'Soviet-Indian power play to humiliate the Chinese and also somewhat us... the friends of China and the United States have been clobbered by India and the Soviet Union.'[153] In this geopolitical context, Washington's suspicion over New Delhi's space launcher development should not be surprising.

India entered the higher ranks of technological powers with its first generation of solid-fuelled launchers, the generically-named Space Launch Vehicle (SLV) which conducted a series of flights in the early 1980s. ISRO went on to develop the Augmented Space Launch Vehicle (ASLV) by the late 1980s. These two systems established an Indian mastery of solid- and liquid-fuel technologies—with the former being more suited for ballistic missile uses. Many early launch failures led to derogatory names for the SLV and ASLV—the Sea-Loving Vehicle and the Also Sea-Loving Vehicle.[154] Nevertheless the SLV and ASLV allowed ISRO to master essential technologies for heavy satellite deployments and a technical ability to diagnose and identify launcher problems, instilling confidence in the organisation's ability to develop more ambitious rocket technologies.[155] The Polar Space Launch Vehicle (PSLV) first flew in the early 1990s and established variable-inclination and heavier, reliable payload capability for satellite deployments into LEO. For the PSLV, ISRO had to develop reliable liquid-fuelled engines to achieve the accuracies needed for satellite deployment that solid-fuelled engines could not provide. ISRO secured a technology transfer agreement with a French company for the Viking engine and other associated technologies developed for the Ariane 4 rocket.[156] As of early 2022, PSLV has successfully launched 51 out of 53 times—an impressive track record. The PSLV-XL variant (with solid booster strap-ons) can reach GEO, the Moon, and Mars with small payloads.[157]

Indian autonomy as a space power made another giant, eventual, leap in the 2010s. The Geosynchronous Space Launch Vehicle

(GSLV), first flown with Russian cryogenic upper-stage engines in 2010, provided India with a greater GEO-capable space launch vehicle, albeit with a small payload and reliance on Russia for those cryogenic engines. There were more geopolitical barriers to the development of the GSLV than its other rockets due to the sensitivities of upper-stage cryogenic engine technology which is required for far heavier lift capabilities into higher orbits such as GEO. Cryogenic engines need far less fuel mass to generate more thrust, resulting in far greater lift capacity. Such fuel needs to be kept at extremely cold temperatures though, and survive the harsh conditions of Earth orbit which can vary from extreme cold in the shadow to extreme heat in sunlight. ISRO had declined an offer of technology transfer on cryogenics from France on the back of their successful Viking engine collaboration, a decision ISRO chairman Satish Dhawan would later regret.[158] Subsequent U.S. and French offers on sales of cryogenic engines had strings attached and would not be approved by the U.S. over concerns about India's nuclear programme, whilst Japan did not respond to Indian requests.[159] The USSR had ideal cryogenic engines—the KDV-1—that it had developed for its lunar programme but never used. India signed the agreement on their sale in January 1991. Less than a year later the USSR no longer existed.[160]

The new Russian Federation reneged on the deal in the early 1990s. Part of another deal forged by the United States to ensure Russian participation in the International Space Station (ISS) and the Missile Technology Control Regime (MTCR) was to stop the transfer of its upper-stage cryogenic engine technology to India over ballistic missile proliferation fears. Washington offered Moscow access to the global satellite launch market, and funding for Russian work on the ISS and the Shuttle-Mir project. These would bring in far more hard currency than the sales of rocket technology to New Delhi.[161] This effectively saved much of the crewed Russian spaceflight programme and provided Moscow with a much-needed foreign revenue stream from the global satellite launch market. It saved the careers of many Russian space professionals and what was left of the Soviet space industry in Russia, Ukraine, and Kazakhstan. No amount of sales to India could do that. The MTCR was designed by its members to prevent the spread of missile and missile-related technologies they themselves often possessed—cryogenic engines was one of those. The deal agreed between Presidents Yeltsin and Clinton permitted the sale of six cryogenic engines to India, but not the wholesale transfer of the technology which is what India really wanted.[162] Russia invoked a force majeure clause in October 1993 and did not abide by the technology transfer elements of the contract with India, but some sales of engines and Indian work in some Russian facilities continued.[163]

Despite this and U.S. sanctions over its pursuit of cryogenic engines, India persisted with its own indigenous cryogenic engine development. By 2014 the GSLV Mk III, using an indigenous cryogenic engine, had demonstrated India's ability to loft much heavier payloads into GEO. Its dependence on Europe for heavy lift to GEO was eliminated, and the GSLV became a revenue stream as ISRO began selling heavier launch services for other states and foreign companies, complete with the commercial attractiveness of the equatorial boost in Kerala.[164] This also ensured an indigenous Indian ability to conduct more ambitious deep-space exploration missions. More importantly, GSLV Mk III lifted the bottleneck that was cryogenic technology on India's domestic demand for space launch services which held back the pace of ISRO's developments elsewhere.[165] ISRO's continued development of cryogenic technology created 'an atmosphere of political distrust, suspicion of industrial espionage and allegations of secret deals in the pursuit of commercial advantage and national defence.'[166] Again, space technology is hardly immune from terrestrial politics and not a matter of purely technocratic, scientific decision-making—indeed it is the continuation of terrestrial politics by other means.

Though U.S.-Indian relations on both nuclear and space fronts are relatively cordial and cooperative today with formal deals on cooperation and information exchanges in both areas, in the 1990s India's missile and space developments and U.S. perceptions of them were a major source of friction.[167] It was not until the 1990s that military objectives were explicitly added to India's aims in space technology development goals. Whilst work on the Agni SRBMs halted due to international pressure in 1995, the Kargil War of 1999 demonstrated ISRO's lack of space-based ISR and communications systems, and only accelerated military-orientated space technological developments.[168] India's strategic focus has been on Pakistan and China; the former could be addressed with MRBMs as they were neighbours and not separated by vast oceans. There was a less obvious drive to develop a nuclear-carrying ICBM for nuclear deterrence and warfighting purposes compared to the USA, USSR, and China. Pakistan did not become a serious nuclear and missile threat until the late 1990s, a long time after India's so-called 'Peaceful Nuclear Explosion' (PNE) of 1974. ICBMs, IRBMs, and SLBMs are needed if India wishes to hold *all* major Chinese cities and military sites hostage to its nuclear weapons, rather than MRBMs and SRBMs. IRBMs could reach a fair number of interior Chinese cities and military targets, but not eastern or northern China.

There is no obvious evidence of ISRO-DRDO cooperation on ballistic missile technologies.[169] However, as with other states, there is always a porous boundary between civil and military institutions as knowledge, data, and people can move around even if hardware does not. India's missile capabilities were already on U.S. radars by the time of the nuclear test of 1974, and indeed around that time Abdul Kalam, the SLV programme lead, went to work for DRDO to develop the SLV's IRBM 'cousins'.[170] The 1974 PNE showed that 'the technical infrastructure for a nuclear weapons program remained', but 'it was hidden away'.[171] India, according to nuclear expert Vipin Narang, 'would play fast and loose over whether it technically possessed a nuclear weapon', and adopted a posture of assured retaliation from 1974 onwards.[172] As Indian intelligence was becoming more aware of, and the military concerned by, Pakistan's maturing nuclear weapons programme in the late 1980s, more Indian military research resources were devoted to missile technologies as opinion was shifting towards becoming an overt nuclear weapon state.[173]

Today, India's nuclear missiles include the Agni IV and V ICBMs, with Prithvi II and Agni I, II, and III SRBMs. All are solid-fuelled and serve as India's fission bomb delivery systems. India's nuclear arsenal is kept under strict civilian control and custody, not military.[174] Also at an earlier stage of development is the K-15 and K-4 SLBMs which will be placed on the *Arihant*-class SSBNs.[175] Though Indian officials continually declare their peaceful intentions in the use of space, their military space programme is gaining momentum and is beginning to modernise Indian armed forces with more military-orientated communications, ISR, and satellite navigation technologies, partly in response to Chinese advances in these areas. ISRO has developed military communications satellites for the Indian Army and Indian Navy, though formal cooperation between ISRO and DRDO remains minimal and ad-hoc. This may be because ISRO is something of a model agency, and DRDO is anything but an agency to imitate.[176] India's advances in ballistic missile technologies over the last 20 years have also been used as the basis for its kinetic anti-satellite capability, which was tested with an interception mission in 2019. Therefore in contrast to the other space powers, India's formal pursuit of missiles followed its successes with SLVs. Its first 'official' missile programme—the Integrated Guided Missile Development Programme—did not come about until 1985.[177] But work on missiles of many kinds long preceded this programme.

Vikram Sarabhai, whilst leading the space effort and also heading the DAE, initiated a programme to develop missiles and rocket motors for the military at INCOSPAR in 1968, in what may have been a 'blow to his sense of integrity' given his very public stance on India's space programme maintaining a strictly 'peaceful' orientation. Sarabhai could not fully halt India's trajectory towards nuclear weapons development, but Singh argues his anti-nuclear beliefs were sincere. Some battles may not be worth fighting for the sake of others, and PM Indira Gandhi by November 1971 had made her intentions known to remove Sarabhai as head of DAE and place him in the new Department of Space.[178] Though Sarabhai died the next month, ISRO's move into the Department of Space formally delinked ISRO from the nuclear programme. The head of ISRO would also be the head of the Department of Space, and report to the highest political office and with direct access to the Prime Minister.[179] India's SLV and missile capabilities were already in breach of the MTCR when it was set up in 1987. The MTCR restrictions on India were formally lifted in 2010 but ISRO had been removed from prohibited organisations for cooperation with the United States by 2005 under the Next Steps in Strategic Partnership (NSSP) agreement.[180]

As India militarises more of its presence in space, the linkages between ISRO and Defence are still developing. There is only one formal institutionalised means of coordination between the Ministry of Defence and the Department of Space—the Integrated Space Cell through which the military acquires space capabilities for itself from ISRO as a capability developer and contractor. There is no integrated command structure linking Indian military space operations and activities with those of ISRO.[181] Nevertheless, India stands as a major space power with its own well-established nuclear and missile programmes, though its spacepower origins were markedly more civilian than the others with the exception of Japan. Indian officials have 'convinced' the United States to accept it as a de facto nuclear weapon state and therefore India's increasing military space potential can be seen by some as a useful additional method to counterbalance the growing Chinese economic and military projection.[182]

Whilst not at the same level as the United States, Russia, and China, India has a broader spread of space competencies than any single European state, and is comparable to Japan. Sarabhai's advocacy of 'indigenisation and self-reliance' still characterises ISRO culture today, and is manifest in India's increasing pursuit of 'essential' twenty-first century space infrastructure such as heavy lift capabilities, communications and imagery satellites, SSA sites, and PNT systems.[183] The postcolonial political and strategic significance of India's mastery of essential and increasingly sophisticated, complex space technologies cannot be understated. Seventy-five years after independence, India, the former colonial subject, can now access an environment of high military and economic importance by its own means that others can only prevent by declaring war and risking nuclear destruction. Meanwhile Britain, the former colonial master, cannot. British companies today must pay foreign providers—whether public or private—to put their satellites into orbit, including India. This is the equivalent of British ship builders having to pay Indian or other foreign powers and companies to put their ships into the sea for them. India has not only left Britain behind in spacepower terms, but persisted with a steady rocket development programme while western Europe's effort and ELDO stumbled and restarted within ESA. In the 1970s, India was on a surer, more stable footing in space development than western Europe, but France was ahead in ballistic missile development compared to India. This shows how the reality of technological spread and development in the Cold War is not simply a story of flows from the superpowers outwards to allies with other states being lower on the technological pecking order.

The crewed spaceflight ambitions and robotic exploration missions of India today are signalling perhaps a spiritual break from Sarabhai's founding vision for India in space. In 2008, there was a heated debate at the Department of Space over the Chandrayaan-1 lunar probe because it was hard to reconcile such a lunar science mission with national development goals.[184] Similarly, India's crewed spaceflight ambitions will not fulfil many development goals compared to its burgeoning communications, imagery, and navigation satellite infrastructure. Concerns about the direct returns from space exploration are not uniquely Indian—but the strength of socioeconomic development in the founding spirit of India's space sector is rather unique for a major space power.

Conclusion

The original sin of spaceflight does not just taint the dreams of Americans and Russians in space, but every major space power in the twenty-first century. Just as fears over falling behind in nuclear technologies in the post-war years led to the equivocation of the nuclear weapon states becoming the new 'coloniser', and the non-nuclear weapon states becoming the newly 'colonised', spacepower sustains a similar perception.[185] Whilst such language may be hyperbole, dismissing it entirely disguises the power politics and the imperial, exploitative technological practices that have been used and arguably continue to be used in the service of the space powers today.

Spacepower became a measure of practical and symbolic power in the international system in the Cold War, with 'haves' and 'have nots' in space shaping and being shaped by the fundamental rocket and satellite technologies held in the hands of a few states. Membership of the higher rungs of a global 'space club' is anything but declining in significance. The club is a socialising mechanism and an arena to rationally interact and negotiate over the distribution of power, conventions, status, and influence.[186] This happened in Europe, Japan, China, and India who could capitalise on U.S. and Soviet proofs of concepts and better target their more limited resources, and then bargain with the superpowers as they became autonomous enough in key rocket and satellite technologies. Three major Asian space powers are firmly in the space club today and are participating in the continuing development of space systems and applications, but also in allowing or restricting the spread and use of space technologies, rules, and customs on terms that are beneficial to themselves.

No single state or people has a claim to the story of outer space—it is a Global Space Age. People all over Earth have benefitted and suffered from it. London, Paris, Brussels, Beijing, Tokyo, and New Delhi have overcome attempts by both Moscow and Washington to restrict their development, but cooperated or even integrated where necessary and desirable. Trusting the U.S. to provide for all allies' needs and act as a steward was not a feasible policy for western Europe in the 1960s. This a direct rebuke to the arguments in favour of establishing a benign U.S. hegemony or dominated order in outer space.[187] Yet Europe could only do this as an integrated collective, working across state borders and nations.

If Washington's closest allies will not accept its monopoly on spacepower and a techno-industrial control of outer space, what hope is there for the rest of the world to trust in the U.S.? U.S. cooperation and technology transfer often came with strings attached that chafed its allies and increased their determination to assure a certain level of independent strategic spacepower capabilities and infrastructure. In the words of historian Walter McDougall, U.S. space technology transfers, cooperation, and competition behaviours amounted to 'giving away the store' at worst and at least a 'benign hypocrisy' in U.S. space diplomacy. The U.S. pursued:

> agreements in areas where the United States was safely dominant and advocacy of laissez-faire in areas where there was still a race... a hard-headed approach not quite up to American rhetoric. Yet U.S. statesmen would have been derelict in their duties had they acted differently.[188]

In each major space power, there have been military rationales or the latent military potential from dual-use technologies used in favour of investing in space technologies and satisfying a wide range of political interests to secure the high levels of funding needed for space technologies. A new space-technological economic base provided a new long-term potential in the military sphere that could be developed and deployed as desired by centres of power beyond the direct control of the superpowers. The dual-use and military considerations and motivations in the pursuit of these technologies were more latent in India and Japan, but not absent. In that sense, it is fair to describe India and Japan's journeys today as *militarising* space powers. Compared to the more civilian character of their space sectors in the Cold War, today they are pursuing more obvious military space capabilities and services. Militarising certain space programmes or actors needs to be distinguished from the militarisation of the environment of outer space which occurred at the very dawn of the Global Space Age. However, rocketry is only one aspect of this story. As important is what kind of machines they deployed in orbit and how satellite systems have evolved and matured into technopolitical infrastructures of centralised state power that few people on Earth can escape in the twenty-first century.

PART II

THE MATURATION OF SPACEPOWER

3

APPLIED WITCHCRAFT AND TECHNICAL WIZARDRY

Gwerth dy wybodaeth i brynu synnwyr

Value your knowledge to buy sense

The many twists and turns in the tales of global SLV developments are not even the half of the story of the Global Space Age. The exploitation of Earth orbit for military and intelligence purposes happened whilst global publics were captivated by the end of Apollo, the Soviet Union's space stations, and United States' Space Shuttle. Robotic missions across the Solar System or crewed orbital science outposts may indeed captivate interest and inspire many, but it was the spread of secret satellite infrastructure that was to have lasting consequences for politics, economics, society, and international relations on Earth. As space became accessible to humanity's machines, the world was fed a media diet of engineering firsts or scientific discoveries. Behind closed doors attention fixed on what remote machines could do in orbit to enhance military power, government communications, and technical intelligence gathering. 'Space applications' exploded from the 1960s onwards as the superpowers proved the first satellite concepts in imagery and electronic sensors. Together, satellite communications, navigation, and space tracking systems built a new technological infrastructure. By the end of the Cold War most forms of major types of satellite technologies and services had spread to and matured in many countries. They now wanted in on the Global Space Age and it put the U.S., Russia, Japan, China, India, and Europe in structurally advantaged positions in the development and governance of Earth orbit in the post-Cold War world. These ripple effects came from what dramatically began in the 1950s Soviet Union, now making waves in the twenty-first century.[1]

Spy satellite systems developed around the world during and after the Cold War have shown an impressive breadth and scale of military-industrial innovation, leaving their possessors in a state of technological maturity in space today. These technologies had emerged in several major powers before the Cold War ended. These satellites drove a dependency in the most technologically advanced states on the military exploitation of Earth orbit, entrenching the militarisation of space. The original sin of space technology does not stop with the SLV/ICBM duality and the thermonuclear revolution. Military and intelligence satellites were useful in the pursuit of nuclear war-fighting goals, then later to conventional non-nuclear military operations too as the twilight years of the Cold War approached. These military-technological infrastructures found other uses, some of which became a part of global civil and economic infrastructures. In some cases, such as satellite TV and mobile telecommunications, commercial as well as military interests co-existed and complemented each other to form powerful domestic and transnational political coalitions to fund and approve initial satellite constellations, though not without difficulties. Today's boom in commercial communications satellites is matched by a boom in commercial imagery platforms and products, again based on the successes of several Cold War military-industrial complexes. Whilst images of Earth from space can impress and provoke a spiritual response, as well as spark transformative capabilities for economic development, resource exploitation, climate monitoring, and disaster response on Earth itself, the military or state security-driven uses of these technologies begs the question of whether many states and companies are in fact building a planetary-scale panopticon.[2] This is not an idle question. Webs of satellites observing Earth in a multitude of ways are tethered to centralised state bureaucracies. Satellite technologies share the original sin of SLVs and continue to remain only at arm's length from the violent instruments of the twenty-first century state.

Space technology is not an emerging technology. It is maturing. It is in many cases several generations old and these cases long proven their worth. The first geostationary infrared missile launch detection satellites, such as the U.S. Defense Support Program (DSP), were launched 50 years ago, just before the progenitor of modern high-resolution digital camera spy satellites, the Kennen KH-11. GPS has been providing services for forty years and the Iridium mobile satellite telecoms is over 20 years old on its second-generation constellation. This should throw some caution to the common refrain that technology is changing or developing at an ever-increasing pace. The bulk of politics and strategy in the Global Space Age is about dealing with old, proven, and rather predictable technologies and concepts, as well as old ideas and failed technological projects that are revived from time to time. This is what David Edgerton refers to as 'the shock of the old' and the importance of looking at widespread but seemingly old technologies in use.[3] Innovation is less important than what finds in widespread and constant use over time. An obsession with fantastical, non-existent, or supposedly novel capabilities should be left behind on the launch pad.

Understanding the Global Space Age not only requires delving into missile and nuclear history, but also the world of Intelligence Studies and the study of covert methods, information collection, and analysis used by governments and military bureaucracies. Intelligence involves the art and science of collecting, analysing, and disseminating information about people, organisations, events, and trends of interest to leaders and decision-makers. Not only is the art of intelligence about gathering information, analysing it, and getting that analysis to the right people to inform their decisions, but also about preventing anyone else from doing the same or stealing your own secrets. Intelligence and counterintelligence are competitive and relative activities.[4] Intelligence, in ideal terms, 'will enhance security by bestowing on the wise collector, perceptive analyst and skilled customer a predictive power on the basis of which policy can be formulated.'[5] This is the grounded, professional world that many of the earliest space technologies inhabited not only because of their sensitivity and destructive power, but because of the nature of their tasks: to gather information and discover secrets about people, places, equipment, and organisations, foreign and domestic. These technical collection methods are in contrast to human collection methods because they rely on machines to collect information rather than people. The Global Space Age ushered in a new era of technical 'tradecraft' for terrestrial spooks.

Thousands of machines in Earth orbit survey our planet constantly; tethering political and economic leaders and institutions with the physical tools and material power necessary to enforce their will with ever-greater ease and speed over ever-greater distances. Spy agencies and government bureaucracies rely on these systems to collect and analyse their data to 'see' and 'hear' what is happening on Earth, practising techniques labelled in some parts of the U.S. Intelligence Community as 'applied witchcraft'. The historian John Lewis Gaddis spoke of the 'reconnaissance revolution' which made surprise attacks extremely unlikely to succeed and provided far more accurate information on either superpower's capabilities.[6] Whilst this is true, it is but one effect of these technological systems. Information is a double-edged sword. People can fear what they cannot detect or draw the wrong conclusions from an overload of information. Technical data can be ambiguous or only show one angle of an issue. Radio signals or infrared detectors never show the whole story by themselves, and they must be cross-referenced and corroborated with other sources of intelligence, both technical and human, at the meeting points of science and art.

These technologies have not created an all-seeing and all-knowing, omniscient capability, no matter what Hollywood thinks satellites can do or how much information satellites can actually provide. More information is not always a good thing, as it can lead to overload and indecision in people as to what to *do* about things—or analysis paralysis. People can create false information and find workarounds to fool prying eyes or deny information to gather. Technical intelligence techniques need to be coupled with human intelligence—HUMINT—so that technical data can make sense in the political and strategic world. Technology will rarely capture or predict somebody's intent, regardless of what physical behaviour can be observed. Even if good information is at hand, decisions can be fudged, bungled, or overtaken by events. The establishment of Earth orbit as the home of technically remarkable intelligence-gathering platforms produced greater degrees of transparency on Earth, but never complete transparency or omniscience.

It is important to set out definitions on various methods of collecting information or intelligence (INT). INTs are not set in stone and they differ in how they are used by various professional communities. What matters are the underlying technologies and techniques employed and whether the labels we use provide a good enough shorthand explanation of the kind of activities, information, and knowledge we are referring to at any given time. There are many INTs, as summarised by the CIA:

- Signals intelligence (SIGINT) refers to signals intercepts comprising, however transmitted, either individually or in combination, all communications intelligence (COMINT), electronic intelligence (ELINT), or foreign instrumentation signals intelligence (FISINT).
- Imagery intelligence (IMINT) includes representations of objects reproduced electronically or by optical means on film, electronic display devices, or other media. Imagery can be drawn from visual photography, radar sensors, infrared sensors, lasers, and electro-optics.
- Measurement and signature intelligence (MASINT) is technically-derived intelligence data other than imagery and SIGINT. The data results in intelligence that locates, identifies, or describes distinctive characteristics of targets, including telemetry. It employs a broad group of disciplines including nuclear, optical, radio frequency, acoustics, seismic, and materials sciences.
- Human intelligence (HUMINT) is gathered from human sources and officers. Collection includes clandestine acquisition of photography, documents, and other material; overt collection by personnel in diplomatic and consular posts; debriefing of foreign nationals and citizens who travel abroad; and official contacts with foreign governments.
- Geospatial intelligence (GEOINT) is the analysis and visual representation of security-related activities on Earth. It is produced through an integration of imagery, imagery intelligence, and geospatial information.
- Open-Source intelligence (OSINT) is publicly available information appearing in print, electronic form, and other media sources.[7]

IMINT, SIGINT, COMINT, ELINT, and GEOINT are the most important here. Perhaps with the maturation of spacepower and as the space environment itself becomes busier and more important, there should be a new INT—SPACEINT—to refer to the intelligence gathered from all sources about activities in the space environment and the characteristics of space infrastructures themselves, like aerial and maritime intelligence. This goes beyond 'Space

Situational Awareness' (SSA) systems which are mostly concerned with gathering space tracking data, or technical data about space activities. SPACEINT should include net assessments on spacepower and activities in space, anticipating intent and not just observing capabilities, drawing on political and economic analysis in space, and not only monitoring traffic. All intelligence gathered *from* space-based sources should not be called SPACEINT because the established INT types have traditionally been 'domain agnostic'. Whether a SIGINT platform is in space, the air, land, or sea is not important—it is just SIGINT. SPACEINT ought to be an all-source domain-specific intelligence about space and space-related activities. In short, SPACEINT is intelligence *about* space, not intelligence *from* space. Gathering and sharing knowledge, assessments, and understanding of what is going on in space itself, and Earth orbit in particular, should become a specialist branch of intelligence expertise due to the unique technological and geographic features in orbit, like aerial and naval intelligence specialisms. Such technical intelligence must be coupled with political, economic, and strategic analysts that can draw strategically significant insights from the technical information about space activities, particularly about aspects of behaviour that cannot be observed with technical sensors.

Seeing and Hearing from Orbit

Advances in camera technologies and flight stabilisation led to an ability to identify objects as small as 1.5 metre diameter on Earth by the mid-1960s for the superpowers, down from 6 metre diameter in the earliest satellites. A camera is useless if it is looking in the wrong direction or is spinning all over the place—its gaze needs to be fixed towards desired points on Earth as it orbits around it and take strips of images usually a few hundred kilometres wide. This wide field can be reduced for a high-resolution image but on a narrower field of view, or vice versa.[8] Many IMINT satellites carry more than one camera to provide a mix of wide and narrow views, often complementing each other in terms of geo-locating the detailed, narrow images. Cloud cover and the night can pose problems for visible light imagery satellites, but other techniques such as radar imagery can see through clouds and at night.

The bounty from the first successful space-based imagery missions bowled over the White House. Combined with other sources of intelligence and complementary net assessments, space-based imagery became the backbone of U.S. National Intelligence Estimates and often the centrepiece of Presidential Daily Briefings. Using the public cover of atmospheric science and equipment testing missions under the Discoverer scientific satellite programme (called Corona in secret), the CIA and USAF were able to develop and test their first photo-reconnaissance system whilst Eisenhower could claim the freedom of space and the use of space for 'peaceful purposes'.[9] Up until its first successful flight, there had been thirteen launch failures. Persistence can pay off. The first successful imaging flight of the Discoverer satellites in the Corona project yielded more imagery of Soviet territory and bases than the entire U-2 aerial reconnaissance programme had provided. Discoverer 14's imagery haul stunned Eisenhower. Seven orbits at 185km altitude in one mission had turned up more useful data than 4 years of U-2s flying at 20km altitude. Sixty-four airfields, 26 surface-to-air missile (SAM) sites, and a new rocket launch facility were identified across the Soviet Union and its satellite states. It was as if 'an enormous floodlight had been turned on in a darkened warehouse'.[10] Though the Corona project initially provided lower resolution images than the U-2, the volume, breadth, and impunity of its coverage was more than enough compensation to monitor nuclear facilities, missile sites, and troop movements.[11] A U-2 could not fly into and see the Soviet Union's interior without violating Soviet airspace. Satellites had no such restrictions once Sputnik had proven the right of satellite overflight over other countries.

It was not until later on in the Cold War that some IMINT satellites were able to see at night or see through cloud cover by using Synthetic Aperture Radar (SAR) imaging technologies, discussed below. Space-based IMINT and SIGINT allowed USAF's Strategic Air Command to better target their nuclear bombers for maximum effect in wiping out Soviet military installations, and also in anticipating the effects of U.S. nuclear strikes on Soviet urban centres through up-to-date imagery, especially as many critical military sites were close to cities. Therefore IMINT not only made U.S. leaders feel more secure as they had inflated the Soviet military threat through the 1950s, but it also made U.S. nuclear bombers more accurate and threatening to the Soviet Union.

With the creation of the NRO as a dedicated imagery reconnaissance agency in 1961, the 'civilian' and 'scientific' space programmes began to diverge from the military-intelligence dimensions of U.S. space policy between 1958 and '62. As NASA would take the lead in a 'civil' space programme, the USAF and NRO would reign supreme within the Pentagon on the military and intelligence space programmes. NASA was officially distanced from these 'black' or secret space programmes and focused on crewed spaceflight, space and planetary science, and engineering challenges in public view. Over the next decade, the CIA, USAF, and NASA would compete with each other for resources and the division of responsibilities as space technologies exploded in scale and diversity, and the budgetary largesse of the mid-1960s ended.[12] Intelligence historian Jeffrey T. Richelson referred to the contest between the CIA and the NRO as the 'space reconnaissance wars' with both agencies competing over the design, use, and products of IMINT and SIGINT satellites. This 'bureaucratic bloodshed' centred on issues such as the NRO's focus on the 'finished product' as a properly developed picture from space. What was actually in that picture was a secondary concern, which irked the CIA and its demand for relevant, processed, and analysed imagery, not just neatly produced photographs. The differing cultures not only caused clashes over the tasking of satellites, but also on satellite launch schedules. The conflicts were eventually resolved and a durable CIA-NRO partnership over satellite-based IMINT and SIGINT emerged by the late 1960s.[13]

The Soviet Union also had its own reconnaissance satellite programmes. Here, 'the intensity of the competition between the [superpowers]... was accelerating as surely as some of their rockets.'[14] By the 1960s and 1970s the Zenit and Yantar satellites became a reliable source of very high-resolution imagery for the Soviet leadership and intelligence services.[15] Demonstrating the pervasive military nature of space technology, the Zenit craft was designed in parallel with the Vostok craft that carried Gagarin and the other early cosmonauts into Earth orbit, and served as the foundational designs for proposed military space stations.[16]

The superpowers first developed capsule-return photo-reconnaissance satellites which required catching or picking up the film capsules after they re-entered the atmosphere. Images from a satellite could take days or months to make it to a terrestrial film development lab after being taken by the orbiting satellite, and longer again to make it to the desks of analysts and decision-makers. Once a satellite ran out of film it was useless. Demanding an early return of film capsules before they were finished shortened the lifespan of deployed satellites. Hexagon is a well-known U.S. example of this kind of satellite. Imagery satellites became extremely useful for arms control verification, and were referred to as National Technical Means of Verification in Strategic Arms Limitation Talks (SALT) I and the Anti-Ballistic Missile (ABM) Treaty in the early 1970s. The principle of non-interference with these systems as codified in Article XII of the ABM Treaty could be interpreted as only applying when spy satellites were used for treaty verification activities only, rather than a blanket prohibition on interfering with spy satellites of all kinds at all times.[17]

As well as using light to create images, detecting and intercepting radio signals were also established methods of 'applied witchcraft' that were simply lofted into orbit. In the early 1960s, the U.S. Naval Research Laboratory launched the Galactic Radiation and Background (GRAB) satellite. This was the first known spy satellite. GRAB was a SIGINT satellite designed to act like a 'bent pipe', redirecting terrestrial Soviet radio transmissions as they 'leaked' into outer space from the ground and bouncing them 'back' towards U.S. ground stations for analysis and processing. Much useful data on Soviet radar and anti-aircraft tracking systems was gathered in this way, especially before the Soviet Union encrypted internal radio transmissions.[18] SIGINT satellites are used by a variety of space powers today and tend to be placed in the GEO belt, although there can be applications for SIGINT in lower orbits. Whilst GEO SIGINT satellites can 'loiter' and pick up a wider area of signals, they also pick up more 'noise'. LEO SIGINT satellites can allow more targeted but fleeting eavesdropping efforts. Dwayne Day describes this as 'trying to listen in on a conversation between two people as you ride past on a bicycle'. LEO SIGINT satellites can provide detailed 'clues' as to the kinds and types of radio emitters over very specific locations to provide data for radio direction and radio traffic analysis from other sources.[19] Early large-scale SIGINT projects in the U.S. include the Rhyolite (CIA) and Canyon (USAF) projects, first launched in 1969 and 1968, respectively. The sleuths managed to keep finding ways of detecting and tracking radio emissions and signals, and then 'catching' them sometimes through direct signal receivers or even by catching the signals as they bounced off Earth's surface after being transmitted down by Soviet satellites. Such satellites detected telemetry data from Soviet missile test launches—information about an object or vehicle's flight performance broadcast to receivers a great distance away.[20] Washington did not reveal the existence of its SIGINT satellites until 1996.[21]

The Soviet Union developed electronic intelligence (ELINT) payloads to piggyback onto their earliest Zenit IMINT satellites. As a result the Tselina radar satellites were born, providing an early LEO-based SIGINT system for the Soviet Union in the mid-1960s. Throughout the Cold War, the USSR launched over 100 Tselina satellites into orbit in a constant replenishment routine as the satellites only lasted several months.[22] This contrasted with the U.S. approach where satellites were generally designed to last for many years in orbit and required fewer replacement launches. The Soviet Union named all their orbit-reaching satellites Cosmos and affixed a number, including 'military' and 'civilian' satellites and missions. In post-Soviet times, the Russian government ended the tradition of labelling civilian satellites as Cosmos, though civil-military distinctions between different space systems are still blurred at best.[23]

As well as tracking radio emissions, ELINT could be used to detect objects and vehicles on Earth's surface using electronic sensors and multispectral imagery to detect radiofrequency 'emissions' that devices produce, as opposed to radio communications interception or COMINT which looks into the contents of communications signals. By the 1970s the United States had deployed the White Cloud ocean surveillance system, made of a handful of satellites in LEO designed to track and identify Soviet surface ships and some submarines by triangulating radar imaging data as well as radio and infrared emissions.[24] However, detecting and forwarding ELINT data could take hours.[25] Another ocean surveillance SIGINT satellite constellation was named Parcae—after the trio of Greek mythological figures who in tandem sealed the fates of all—which worked as one with each satellite providing more precise data:

> the first satellite detected emissions from a wide swath of the ocean, a second further narrowed down the area from which the signal originated, and a third provided the precise coordinates needed for targeting information. It was a truly clever name for a trio of satellites intended to cut short the lives of Soviet warships.[26]

As Dwayne Day argues, the improvement and refinement of ocean surveillance capabilities by the NRO and Navy in the Cold War show a persistent interest in the NRO in applying 'strategic' space assets towards more 'tactical' battle requirements, well in advance of the 1991 Gulf War.

The political acceptance and technological maturation of satellite reconnaissance had a far greater and lasting impact on power in international relations, technological spread, and Cold War relationships than the technonationalist stunts of crewed spaceflight. U.S. President Lyndon Johnson quipped in March 1967 that:

> I wouldn't want to be quoted on this but we've spent 35 or 40 billion dollars on the space program. And if nothing else had come out of it except the knowledge we've gained from space photography, it would be worth ten times what the whole program cost. Because tonight we know how many missiles the enemy has and, it turned out, our guesses were way off. We were doing things we didn't need to do. We were building things we didn't need to build. We were harboring fears we didn't need to harbor.[27]

There was indeed a missile gap—in favour of the United States, not the Soviet Union. By 1961, U.S. intelligence estimated that the Soviet Union only had 10 operational ICBMs, in stark contrast to the 500 projected in the 1957 National Intelligence Estimate (NIE).[28] This reassurance no doubt helped alleviate some top-level American fears of the risks of a disarming first strike from the Soviet Union, increasing nuclear stability. Satellite reconnaissance may have saved the U.S. taxpayer billions of dollars as a requirement for 10,000 Minuteman ICBMs was downsized to 1,000 in light of the revelations of the Soviet Union's modest nuclear war fighting capabilities provided by the Corona project.[29] For the sake of every living thing on Earth, having up to 9,000 fewer nuclear warheads in existence was indeed a benefit.

An important leap in capabilities occurred with the emergence of digital photography. The first U.S. digital IMINT satellite was the Kennen KH-11 launched in 1976. Unlike film-based photography that sent capsules to re-enter the atmosphere, digital images could be transferred into binary data and beamed to ground stations for immediate processing and dissemination once the satellite was in range of a receiving station or communications relay satellites. The digital camera in modern mobile phones uses the same basic solid-state electronics, albeit far more miniaturised.[30] With increased downlink coverage after building more terrestrial stations, the NRO sped up the data-receiving process so that digital space-based IMINT became almost a real-time capability where images could be reliably gathered and transmitted on demand, subject to the satellite's orbital trajectories and planned coverage. Peter Hays explains that the previous KH satellites, using films and re-entry capsules, or more colloquially known as 'bucket droppers... were unable to provide timely information on fast-breaking events such as the 1968 Warsaw Pact invasion of Czechoslovakia or the Arab-Israeli wars in 1967 or 1973.'[31]

Four KH-11s were launched between 1976 and 1982 and came in under budget. They relied on the Satellite Data System relay satellites which flew in Molniya (or HEO) orbits and routed the KH-11 imagery data from wherever they were in orbit to ground stations in the United States instantly.[32] As it started making waves in the higher echelons in the U.S., the KH-11's details were 'compromised' to the Soviet Union. William Kampiles, a CIA employee, pilfered the KH-11 manual and turned it over to Soviet agents in Greece for a 'mere' $3,000 in 1978. The manual described most of the top-secret details of the KH-11 system, and disclosed how the United States could verify Soviet adherence to SALT and other arms control agreements. This helped Soviet planners learn how not to be seen by orbiting KH-11s.[33]

KH-11 made a significant impact on the Carter administration, the first to enjoy its benefits.[34] The day after his inauguration, Henry Knoche, Acting Director of the CIA, showed President Carter KH-11 imagery. They were not of some Soviet missile site, but of Carter's inauguration on the Mall the day before, much to his amusement.[35] The emergence of almost real-time photography from satellites played a strong role in ending the Manned Orbiting Laboratory (MOL) programme, which was a USAF programme to place a crewed IMINT space station in orbit that could take and process film images in space, then transmit data and analysis from the developed images.[36] Transmitting the raw imagery data to ground stations with no humans in space or the need to develop images from film was far more affordable and reliable. Modern U.S. electro-optical spy satellites are direct descendants of the KH-11.[37] After constructing the KH-11s, Lockheed Martin won the contract for the Hubble Space Telescope as both KH-11 and Hubble shared much of the same equipment. With some modifications, Hubble is merely a spy satellite pointing away from Earth and into the cosmos.[38] Not even the remarkable and humbling images of Hubble escape the original sin of space technology.

In the Cold War these satellites were useful for the superpowers not just to spy on each other—but on the rest of Earth, and to share that information when it suited them to do so. The Soviet Union and the United States relied in part on IMINT satellites—Cosmos 922 and Cosmos 932 for the former, Hexagon KH-9 and Kennen KH-11 for the latter—to monitor and track the progress of the South African nuclear weapons programme and clandestine Israeli cooperation with the apartheid regime in the 1970s. In 1979, two civilian U.S. Vela X-ray astronomy satellites orbiting at a distant 112,600km altitude detected what may have been an atmospheric nuclear weapons test detonation. It may have been conducted by South Africa based on the 'double flash' its X-ray sensors picked up between the southern tip of Africa, the Indian Ocean, and Antarctica. To this day it remains a subject of debate as to what exactly caused the 'double flash' explosions—usually characteristic of an atmospheric nuclear detonation—and who was responsible. South Africa and Israel are among the primary suspects, possibly having worked on a joint nuclear and missile effort.[39]

IMINT alongside other ELINT sensors in space and on Earth built an ever-more alert global monitoring system for the superpowers beyond the immediate design intentions of specific satellite systems. IMINT uses technologies that have applications beyond national intelligence's pursuit of military installations and missiles. Earth observation (EO), as space-based imagery of Earth is often called outside intelligence communities, can take many forms in the visible and invisible light spectrum. This 'multispectral' imagery has become a staple information and analysis requirement in civil engineering, infrastructure planning, and development, from monitoring vegetation health to determining air pollution patterns in urban centres. The Landsat programme, one of the best-known and oldest EO systems in the United States, is a civilian-oriented, land-imaging programme designed to meet civil infrastructure, mineral exploitation, and development needs. It came to satisfy self-interested military, foreign policy, economic, and infrastructural requirements despite being marketed as a purely civilian system for 'development' purposes in the Cold War. Landsat had military utility despite its low resolution compared to NRO satellites.[40] Landsat was a feature of U.S. technology transfer to China in the 1970s and 1980s with the building of two ground terminals in China, in something of a remarkable exception to the otherwise guarded or conditional nature of U.S. space technology-sharing.[41]

Elsewhere, Landsat was pushed by the U.S. Department of the Interior as part of their wider Earth resources and mapping agenda. Its development was intertwined with the United States' goals in the era of decolonisation to bring Third World mineral resources into 'global circulation', integrating more of Earth's economy into the U.S.-led capitalist system. According to the scholar Megan Black: 'despite Landsat boosters' claims that viewing the world from space would benefit humanity and the natural world, the satellite became a tool to further capitalist exploitation.'[42] Though many attached genuine hopes of global development and material improvements for all through the Landsat programme—and has no doubt proven useful in those goals—it assisted the U.S. oil and mining industry's exploitation of Third World resources on terms beneficial to the U.S. government and industry. The United States was not alone in this. In direct opposition to the universalist ideals of the Outer Space Treaty and the 'Overview Effect', as discussed in the conclusion chapter, space technologies have allowed states and companies to better target their extractive economic practices within the territories of other, often postcolonial states. Satellite promoters may try to win 'hearts and minds' but they need to 'win budget lines' from various government agencies too. In the internal funding debates Landsat's potential unilateral economic gain for the United States was particularly persuasive, not equitable global development goals.

Orbit sleuths and surveillance systems are not omniscient, despite all the data and imagery that they can provide. Soviet military units caught on to U.S. spy satellites and deployed decoys such as wooden aircraft and rubber submarines which taxed U.S. photo-interpreters and nuclear war planners.[43] Both of India's nuclear explosion tests took the U.S. Intelligence Community by surprise. Pokhran-I in 1974 (the 'Peaceful Nuclear Explosion' or PNE) worked despite U.S. efforts to monitor Indian nuclear activities. In the early 1980s, digging in preparation for underground tests was conducted under camouflage netting to prevent both Soviet and U.S. spy satellites from spotting any activity. However KH-11 satellites found evidence of the digging work and PM Indira Gandhi was pressured to cancel the planned test. Such imagery would prove useful in providing indications of possible test preparations in 1995 too, but for the 1998 tests Indian staff and workers meticulously planned activity so as not to arouse any suspicion. This included working at night to reduce IMINT detection, not providing any extra security around the bomb so as not to provide any unusual activity spikes, and only an extremely small number of Indian government leadership being let in on the secret plans at the last minute to maintain 'operational security', thereby minimising the chances of U.S. (or any other state's) HUMINT officers getting wind of it. The extraordinary efforts of Indian officials and technicians kept it under wraps until the bomb was set off.[44] The Pokhran-II tests in May 1998 caught Washington by surprise, despite the fact that technical imagery and monitoring capabilities had improved by leaps and bounds since 1974. In response to receiving word on an Indian nuclear test, the U.S. Director of Central Intelligence's non-proliferation chief reportedly asked 'is this some sort of joke?'[45] Such feats may be harder to repeat today, but even with almost constant commercial imagery surveillance now in orbit, facades and concealment are still possible. If one cannot completely hide, one can still decide what to show and how much ambiguity to present to imagery analysts trying to piece together actual activity from sometimes imperfect or circumstantial evidence.

Weather detection and tracking is also an important type of Earth observation and IMINT technologies. Even after sharing data from weather satellites with the international community was expected and became the norm, major powers sought their own guaranteed weather monitoring capabilities. Unlike most IMINT satellites, many weather satellites are lofted into GEO so that they can gaze at a fixed point and take in an entire view of Earth, which makes comparisons of large-scale change in one place over time easier. Weather satellites observe large cloud and weather formations that span continents and oceans. Alongside SATCOMs, weather satellites and their obvious military and civilian uses in forecasting were a major driver of heavy-lift SLVs that could reach GEO. However, LEO can be useful for weather monitoring systems. The early U.S. and Soviet weather satellites (Tiros and Meteor respectively) were LEO-based. The U.S. Navy was a major beneficiary of the earliest weather satellites developed by NASA, USAF, and NOAA.[46] The U.S. Tiros system paved the way for the GEO-based Defense Meteorological Support Program (DMSP), initially set up by the NRO, to provide critical weather forecasting improvements for the Pentagon from 1974. DMSP helped maximise cloud-free imaging opportunities as NRO satellites overflew their reconnaissance targets.[47] Later that decade, military field commanders could directly access DMSP data from their temporary HQ without having to request it through a fixed ground station in the continental United States.[48] Japan's Himawari and Europe's Meteosat reached GEO in 1977, and Russia's first GEO weather satellite, Elektro, was deployed in 1994.[49] China pursued a polar-orbiting weather satellite system in LEO—the Fengyun 1 series deployed in the 1980s and 1990s—in a sun-synchronous orbit (SSO) so that it would pass over the same point above Earth at the same time of day, making it easier to compare images day by day. It also built weather satellite data-receiving stations in the 1970s.[50] Beijing would later develop the Fengyun 2 series, which are located in GEO, as China mastered its own heavy-lift to GEO Long March rockets.

Satellites that could see and hear from orbit began to impact military operations other than nuclear war before the Cold War had ended. The Soviet Union's RORSAT and EORSAT systems—ocean radar sweeping and ELINT detection satellite systems, respectively—were designed to locate, track, and identify U.S. Navy vessels at sea to cue the Soviet Navy's anti-ship missiles. RORSAT was first launched in 1965, and EORSAT in 1974. Long-range missiles could strike ships that were beyond the sight of ground and aerial radars, therefore a space-based radar provided an 'over-the-horizon' targeting capability. This was part of the larger evolution of technical intelligence gathering such as IMINT and SIGINT from the air and into space-based platforms. This was a real concern to the U.S. Navy and renewed interest in anti-satellite capabilities. Powered by nuclear reactors, RORSAT and EORSAT were 'one of the most central components' of the Soviet Union's effort to find and target U.S. naval vessels with long-range cruise missiles. EORSATs were larger, in slightly higher and more stable orbits and long-lived missions, and carried ELINT sensors on board to detect and identify radio emissions and emitter types from ships, and then transfer the data in real time via relay satellites to ground stations where staff and computers could then process and analyse it to provide rapid targeting data. RORSAT and EORSAT proved their worth to the Soviet military in the 1982 Falklands War. Moscow tracked the progress of naval and missile engagements, and pinpointed the time and location of the embarkation of the Royal Marines onto the islands.[51]

The ability of the Soviets to better integrate their space-based detection and tracking of U.S. warships into their missile strike systems also foreshadowed the sensor-to-shooter cycles and 'net centric warfare' or the 'Revolution in Military Affairs' (RMA).[52] Indeed, it gave rise to what Soviet military analysts called the 'Military-Technical Revolution' (MTR). The Soviet Navy's space-enabled long-range strike capabilities led to an extensive SIGINT and IMINT effort by the NRO and U.S. Navy to locate, track, and identify Soviet cruisers, with the intent of striking them very early on in a conflict so that they would not pose a major threat to U.S. carrier battle groups.[53] A sensor-to-shooter cycle refers to getting the right targeting data to the right weapons system, firing the weapon, and then receiving new targeting data as quickly as possible. These arrays and collections of satellite systems were integrating with conventional weapons platforms on Earth. Spacepower had become relevant to tactical and operational capabilities, and not only for 'strategic' or nuclear wars. As Siddiqi notes:

> by the early 1980s [they] were part of larger Reconnaissance-Strike Complexes which comprised not only the satellites, but also

> naval reconnaissance aircraft, ship-board helicopters, attack submarines, surface combatants, auxiliary intelligence-collection ships, merchant ships, fishing trawlers, land-based signals intelligence sites, and covert agents.[54]

These satellites provided important methods of exerting Soviet influence on a command of the sea in the event of war. The NRO's Lacrosse satellites, deployed in the late 1980s and through to the mid-2000s, were an important source of SAR data and like the Corona imaging project became a staple of IMINT collection and GEOINT for Washington as well as increased 'tactical' uses in cueing long-range precision strike missiles.

Today, Russia is attempting to reconstitute its space-based SIGINT capabilities as many Soviet systems fell into disrepair in the 1990s. SIGINT systems such as Liana, which are made up of Lotos and Pion satellites, are designed to pick up signals from the seas and over land, and may even be useful in tracking submarines in specific situations. The deployment of some Lotos satellites has been ongoing for the past ten years whilst development and experimental work continues with Pion. Other possible projects such as Akvarel and Repei may be in development, yet according to Bart Hendrickx, 'Russia's space-based SIGINT effort is lagging far behind that of the United States and China', and 'leaves much to be desired.'[55] According to the Union of Concerned Scientists' definitions, the United States has deployed 31 ELINT satellites, China 44 (under the Yaogan label), and Russia is trailing with 7.[56] That said, Russia still enjoys a range of technological capabilities in SIGINT that many other states and some private companies are developing capabilities in. It is not necessarily the case that Russia has to keep up with the Chinese and Americans here in terms of quantity, as there may well be other priorities or needs reflecting Russian security priorities. The early phases of the 2022 Russian-Ukrainian war demonstrated fundamental problems in the Russian military's ability to fight a major war regardless of its space technology advances or shortcomings.

IMINT and SIGINT are better in some situations than others, depending on the location and target type. Terrain such as jungles and forests can inhibit radio communications 'leakage'; changing techniques and procedures can make direction finding and pattern recognition difficult for analysts. Going 'off-grid' is a way for guerrilla or geographically concentrated actors to get around advanced SIGINT collectors—using couriers and written notes instead of radio, though not without significant drawbacks of course. It has recently come to light that the Argentinian Air Force tried to track the Royal Navy taskforce as it headed to the Falklands in 1982. Despite the British fleet's radio silence, Argentina sought to monitor the changing Doppler shifts from Skynet and other SATCOMs which were beaming messages to the taskforce as it sailed south.[57] For every system there is a potential response to dampen or counter their effects and much depends on how they are used and integrated into larger infrastructures and suites of terrestrial capabilities that then meet feasible political goals. Efforts were underway to mitigate the power of eyes from above since their deployment in the early Cold War. Yet with each passing year more states and companies place more sensors of various kinds in orbit, making it harder to hide from every single kind of observation and detection capability—though not impossible.

Imaging Beyond the Superpowers

The late Cold War years became less bipolar as Europe, Japan, India, China, and even Israel were emerging as capable space powers.[58] West European states persisted with their first European military satellites designed for reconnaissance or IMINT, command and control, as well as commercial and civil infrastructural ventures. France and Germany worked together on the Symphonie communications satellites, whilst Sweden and Belgium joined France in the SPOT (*Satellite Pour l'Observation de la Terre*) imaging system. In 1986 France launched the first SPOT satellite which could take optical as well as thermal imagery, in a polar orbit. It was primarily for civil infrastructure uses as an Earth observation rather than a spy or reconnaissance satellite, but those functions can overlap. SPOT 1 was one of the first satellites to image the aftermath of the Chernobyl disaster and immediately provided insight into the scale of the calamity befalling the region despite the Soviet Union's efforts to conceal it. SPOT was also designed for commercial sales of the imagery within Europe and beyond, as its imagery would be useful in cartography, large-scale surveys, biome and vegetation monitoring, and infrastructure planning. By 1994, SPOT had ground stations for real-time imagery located in Canada, Spain, Brazil, Thailand, Japan, Pakistan, South Africa, Saudi Arabia, Australia, Israel, Taiwan, Indonesia, Italy, and Singapore.[59] This list only grew as the years went on and came to include China. Space infrastructure had already gone global in terms of end users, and with capabilities belonging neither to the United States nor the former Soviet Union.

By the late 1980s the SPOT programme was in full swing and paved the way for the Helios military optical reconnaissance satellites developed by France, Italy, and Spain and launched in the late 1990s. This is not to be confused with the joint ESA-NASA Helios solar observation satellites now in a heliocentric orbit. The first military Helios IMINT satellites took advantage of the work already done with the SPOT programme, and featured greater resolutions.[60] Helios's first generation used the SPOT 4 platform (or satellite 'bus'), provided 1-metre resolution, and went into LEO at around 680km, overseen and operated by CNES. The Helios programme is arguably an example of the opposite direction of travel to U.S. and Soviet space-based IMINT: from the civilian to the military. However, SPOT always had a clear military potential, which was not lost during its development and funding battles.

Cutting off end-user states from weather satellite data and imagery is particularly difficult given the widespread sharing of the data as global public goods to almost every state on Earth who could then 'leak' or deliberately share that information further to others. During the 1991 Gulf War, the weather was the worst it had been in over a decade. Almost half of coalition air sorties had to be diverted or cancelled because of it. Weather satellites provided freely available signals and data around the world which was shared via the UN's World Meteorological Organization (WMO). Weather data was also published by global media organisations. This data provided Iraq with opportunities to launch Scud missiles and conduct manoeuvres whilst the weather was at its worst. Even had an information blockade been implemented in the 1991 Gulf War, amateurs in the 1990s could build a basic weather satellite terminal with a personal computer, very high-frequency radio equipment, and plenty of PVC piping and wires.[61]

Uncertainty reigns over China's military reconnaissance satellite origins. A number of Ji Shu Shiyan Weixing IMINT satellites were launched in 1973–6. Their actual impact is debated and may have perhaps been more for propaganda purposes, or they may have been ELINT satellites rather than IMINT satellites.[62] China went without a large-scale EO or IMINT satellite system in the Cold War, but by the 1990s Beijing had amassed a 21-terabyte satellite IMINT database, relying heavily on SPOT and Landsat images. China did not begin to pursue its own EO or IMINT system until the mid-1980s under the name of Ziyuan and in cooperation with Brazil, known as the China-Brazil Earth Resources Satellite (CBERS). CBERS operated on a 70/30 China/Brazil share, and was in part an outgrowth of a particularly successful era in the Brazilian space effort, having built their first satellite in 1993. Where Brazil learned much in terms of satellite and launcher systems and gained a new source of crucial imaging data for its particular needs, China developed a foundational understanding of formal international, civil space cooperation methods and mechanisms. CBERS became operational in 1999, and the 2000s saw some accusations from the Republic of China in Taiwan, as well as the United States alleging that CBERS was gathering military-resolution imaging data.[63] However, such complaints are something of a moot point given almost two decades of space-based IMINT supplied to Beijing from Washington until 1989, and China's later rollout of its extensive Yaogan-class military-intelligence reconnaissance constellations which means China is doing what the other major military powers can with IMINT satellites regardless. This partnership continues today, with CBERS 4 and 4A orbiting Earth, with 5th and 6th generation CBERS satellites in the pipeline.

It was not until the first years of the twenty-first century that China had a significant military-oriented satellite reconnaissance series—Yaogan—which covers IMINT, SIGINT, and ELINT payloads. The first Yaogan satellite was launched in April 2006. Joining the Yaogan class is a small number of Shijian satellites which remain secretive in their specifications and mission types. They may be used for ELINT or SSA operations testing, especially as later Shijian (SJ) satellites reached GEO in the 2010s.[64] The Gaofen series of over 20 satellites is China's main high-definition imagery system which provides all manner of services from monitoring flooding and air pollution to tracking earthquake damage. Gaofen is also the first part of the China High-Resolution Earth Observation System (CHEOS), a programme designed to bring together all of China's imaging and sensing capabilities to provide synthesised information products for China and close the capability gap with the West. In a sense this would mirror the EU's Copernicus programme in bringing together European sensors and systems to create a single, more accessible database for users.[65]

India launched a string of Earth observation satellites in the Cold War: the Rohini RS-D1 (1981), the Bhaskara-II (1981) and the Rohini RS-D2 (1983). These had very little direct military utility, but were significant achievements in indigenous mapping capabilities, tailored and on-demand for Indian Government needs and growing capabilities in related industries, infrastructure, and higher education. Mapping and monitoring the subcontinent was more reliably and efficiently done with satellites, especially given the difficult terrain and lack of modern ground infrastructure in many regions. India since then has developed a range of observation and reconnaissance satellites, from the Indian Remote Sensing (IRS) Satellite in the late 1980s to the Cartosat, Oceansat, RISAT, and SARAL satellites in the twenty-first century. India's economic liberalisation in the 1990s also led to an interest in following France's lead with SPOT imagery sales. Antrix was set up by the Indian Government to develop commercial space applications, with its first activity being to sell remote-sensing data on the global market.[66]

These satellites have been instrumental in improving Indian resource management, as Gurbir Singh writes:

> India's domestic constellation of EO satellites guides Indian farmers on when to use fertilisers, where to dig wells, helps fishermen with information on when and where to go fishing, assists mining companies in locating minerals... and provides early warning to citizens who live in flood-prone areas... Industries, such as mining, oil, raw materials, farming, fishing and agriculture, have directly benefitted.[67]

Today, India maintains four RISAT and one Earth Observing Satellite (EOS) radar imaging satellites, as well as one optical imaging satellite, Cartosat 2A, in LEO for its military forces. In part, these investments were spurred on by the Kargil War of 1999 where Indian requests for U.S. satellite imagery were denied. The RISAT series were in part spurred on in response to the 2008 Mumbai terrorist attacks.[68]

IMINT is a rather ubiquitous technology and advances in one sector can feed into another. Like China, Japan, and France, India's increasing military IMINT capabilities are drawn from years of research and development in Earth observation and then 'militarised' for reconnaissance purposes once the shocks of certain events are responded to, or are the subsequent phases of a plan that sought to reduce costs by mastering commercial imagery first. In the case of China, it was a continually deteriorating relationship with the United States in the 1990s as well as successive, spectacular demonstrations of conventional U.S. military power; for Japan it was the 'Taepodong Shock' of North Korea's ballistic missile overflight of Japan. For all, the 1991 Gulf War and the so-called Revolution in Military Affairs (RMA) in linking high-technology weapons platforms with space systems for precision and networked warfare also pushed this drive for more tactically-responsive space systems among many military powers.[69] Japan's step-by-step mastery of Earth observation technologies has enabled it in the early twenty-first century to choose to wield those technologies for military purposes and reduce its total dependency on the United States for military reconnaissance from space.[70]

SAR and other IMINT technologies continue to spread. ESA developed the Earth Resources Satellite (ERS) which launched in 1991 and was designed to scan the surface of the seas and landmass, producing electronic 3D images through cloud cover and in darkness to measure wind speeds over an ocean, the height of water bodies, as well as tracking illegal dumping activities at sea.[71] Germany and Japan maintain SAR IMINT satellites, and Britain and South Korea are experimenting with their own, under Project Oberon and Project 425, respectively, with interests in developing future military SAR systems.[72] Germany's SAR-Lupe project operates in tandem with France's Helios IMINT satellites. Both states have a data-sharing quid-pro-quo agreement meaning Germany receives optical IMINT, and France receives SAR imagery.[73] A Finnish company called IceEye is now providing SAR imagery on the global market from its small satellites. China has deployed SAR instruments on a Huanjing imagery satellite, as well as rolling out the capability on Yaogan and Jianbing IMINT satellites. The private Chinese company Spacety's Hisea satellite has already begun selling SAR imagery on the market.[74] The U.S. company Planet operates a constellation of over 200 small satellites in LEO which provide an imaging capability that is more like surveillance than reconnaissance.

This is a boon to all open-source intelligence analysts and researchers as the work of the Middlebury Institute of International Studies and Bellingcat show, and as many have seen in the build-up to and during the new phase of the Russia-Ukraine War. However, the use and distribution of such images, even when from commercial companies, are bound by the laws of the state they are registered within. For example, the U.S. operates 'shutter control' legislation to ensure commercial imagery it does not want to be released into the public domain during a crisis or war remains secret until a later date. In the run-up to the invasion of Ukraine in February 2022, Western news and social media was dominated by satellite images of Russian forces, but not Ukrainian military manoeuvres and preparations.

Israel operates a number of optical and SAR IMINT satellites through its Ofeq series, a capability that takes advantage of the Israeli ballistic missile, nuclear, and space nexus. Israel uses its Shavit-2 SLV to throw relatively small payloads into LEO in a retrograde orbit at its discretion. The 1973 Yom Kippur War, or the October War, led to the near-defeat of Israel thanks to an initially successful surprise attack by Egyptian and Syrian military forces. No warning had been provided by the United States and Israel had found its own warning systems and intelligence agencies lacking.[75] As such, Israel subsequently resolved to develop its own satellite imagery capability, though this would take time. Ofeq-1 was launched in 1988 on the Shavit-2, taking advantage of Israeli knowledge and capability with the Jericho-2 ballistic missile, and ushering in an indigenous LEO IMINT capability for the Israeli Government and Defence Forces. Subsequent commercial spinoffs and applications emerged thereafter, mirroring the U.S. experience.[76]

The UK pursued but then abandoned a SIGINT satellite capability in the 1980s following the Falklands War—echoing the Blue Streak and Black Arrow debacle. Journalist Mark Urban claims that getting SIGINT support from the United States required 'special pleading' to get just a few hours' worth of coverage over the Falklands every day. The then-Director of GCHQ, Brian Tovey, claimed that 'we can ask the Americans to do things, but we cannot compel them... It brought going it alone back into fashion'.[77] The SIGINT satellite in question was probably Vortex.[78] It is not known what difference, if any, U.S. satellite IMINT and SIGINT made to the British campaign, with the weather being so bad traditional satellite imagery was useless. As strategist Alastair Finlan writes:

> Brigadier Julian Thompson, Commanding Officer, British Land Forces during Operation 'Sutton' (the amphibious landings) states categorically, 'contrary to rumour, there was never any friendly satellite coverage of the Falklands, or if there was, none of its intelligence, if any, was made available to those who could have used it.'[79]

Yet as mentioned above, the Soviets were able to track British marine embarkations with ocean surveillance satellites. It is not unreasonable to suspect that equivalent U.S. SAR satellites could have picked up useful Argentine behaviour and movements for the British task force. Whether such information or analysis was ever shared with the British is unknown.

Nevertheless, something convinced Thatcher and Tovey that Britain needed its own space-based SIGINT system. Zircon was proposed as a GEO-based SIGINT satellite to the cost of £500m, which would need replacing every five years, following initial studies approved by the Cabinet Office in 1986. This was a highly secretive study and proposal, and caused a stir when it was reported by a journalist, Duncan Campbell, in January 1987.[80] Police raided the offices of the BBC in Glasgow and the *New Statesman* in London in response. In the end, the project was deemed to be too expensive in the face of too little cross-Cabinet support, with only Thatcher herself driving it along. Since the days of Corona, the British had become accustomed to receiving raw, unaltered data from U.S. IMINT and SIGINT sources.[81] The UK Government abandoned Zircon and 'bought into' the U.S. SIGINT system to ensure a better stream of data from U.S. space assets to the tune of around £500m, the estimated price of one Zircon satellite (which would need a constellation of three to provide global coverage). Whilst the British paid for and received 'the best' equipment and a greater wealth of information, it reinforced a British dependency on the United States for SIGINT, and niggling questions over what would happen if the 'subscription' model was unilaterally terminated or altered by the U.S.[82] Intelligence historian Robert Ferris labelled the Zircon failure as GCHQ's 'most costly and embarrassing failure in acquiring high-technology equipment', on a par with the Blue Streak and Skybolt sagas.[83] In the words of Geoffrey Howe, the then-UK Foreign Secretary, Britain had decided to 'play the role of beggar rather than chooser in the world of high-technology intelligence-gathering'.[84]

As with SLVs, the French provide some contrast to the British. Around the same time, France began work on its own SIGINT space systems. Over the years since, several French SIGINT satellites have come and gone with the Cerise, Clementine, Essaim, and Elisa. Ceres is set to be the most recent generation of French SIGINT collection satellites. Whilst the number of on-orbit assets at any given time remains low compared to the United States, France has assured a degree of autonomy and independence from the United States in SIGINT satellite capabilities by retaining essential technological and operational footholds in those areas, especially when coupled with the role of Arianespace and ESA in guaranteeing an independent pan-European launcher capability. At the time of writing, the French military maintains two multispectral IMINT satellites and four SIGINT satellites.

Commercial and allied systems provide a reserve or in-depth imaging infrastructure for the state should dedicated government or military satellites go dark or be neutralised in a time of crisis or war. At such a time, it is only natural to expect that civilian space systems will be pressed into service to meet urgent military needs in future. This is indeed what has happened with 'civilian' U.S. and European satellites through the many wars and crises since the end of the Cold War. Despite their blurred distinctions in practice, it is important to distinguish between whether civilian or military agencies operate satellites, and whether certain systems are primarily designed to meet civil or military requirements.[85] Sharing the right data between large bureaucracies requires systems and contacts, and switching civilian systems for immediate military needs in a crisis is not a smooth, simple, or straightforward task. For the United States, these kinds of capabilities, especially SAR technologies for military and 'national security' requirements, remain within the purview of the National Geospatial Intelligence Agency (NGA), formed in 2003 to follow on from the U.S. National Imagery and Mapping Agency (NIMA), and going by the acronym of GEOINT. NIMA was formed from a smorgasbord of Pentagon and Intelligence Community departments and agencies.[86] This blurring of civil, military, and intelligence functions is not surprising or unique—in fact it is more the norm.

The Fires of War

Modern weapon systems, especially mobile long-range weapons that shoot over the horizon, need radio links to work and be responsive to commands. They risk being detected from orbit through electronic emissions. More sophisticated weapons systems will emit more radio signals and heat that can be picked up by many sensors. Autonomous and remote vehicles and weapons systems will be 'leaky' or produce electronic emissions that provide vulnerabilities if the right people with the right equipment are paying attention. Increasingly, military forces find it harder to conduct complex and large operations without giving away some footprint, emissions, or signature of their activities to the other side in the radiofrequency spectrum, through heat, or being big enough to be spotted visibly. This does not make concealment and deception impossible, but it further complicates the perpetual duels between stealth, detection, deception, and large-scale surprise moves.

Complicating any efforts at concealment are missile launch detection or 'early warning' satellites designed to detect the heat of rockets and missiles on launch. These systems use a specific form of ELINT—sensors which can detect infrared light or 'heat blooms' from things like missile and aircraft exhausts. These allowed Soviet and U.S. leaders to increase the window of time to determine whether a nuclear attack had been launched, and then to decide whether to retaliate with nuclear weapons or not before the enemy's apparent first nuclear strike hit. Early warning systems provide warnings and cues for terrestrial radar stations which then attempt to track the progress of the warheads in space as they approach military targets or cities. Traditional ICBM, IRBM, and SLBM launch warnings are provided with infrared early warning systems by detecting the heat generated by missile engines as they launch and boost their warheads into suborbital flights for a few minutes and then go 'dark' in infrared terms until the heat of warhead re-entry flares up on the sensors. On such a flight path, Soviet and American ICBMs could cover hemispheric distances in around 35 minutes.

Moscow began testing such systems in 1972, and what would become the Oko system would not be operational until 1978. This system was launched into a HEO or Molniya orbit with four satellites in the constellation, designed always to have at least one satellite within view of the United States at any given time. At its peak, the HEO orbits included nine satellites flying at perigees of 500–700km, and apogees of almost 40,000km above the northern hemisphere at intervals in such a way that allowed each satellite to fly along the same track, passing over the same points of Earth at regular intervals.[87] U.S. early warning systems, such as the Missile Defense Alarm System (MIDAS), DSP, and its contemporary successor, the Space-based Infrared System (SBIRS), are relatively well-documented systems. Jeffrey Richelson described the DSP and SBIRS systems as providing credible 'ground truths' for U.S. decision-makers.[88] First launched in 1970, the DSP satellites exceeded expectations and have proved their utility in a range of other roles, such as detecting shorter-range missiles like Scuds, short-range SAMs, fighter aircraft afterburners, SLBMs, nuclear tests, and forest fires. DSP was even picking up ammunition dump explosions and other fires in the Afghan-Soviet War, and tracked Soviet anti-satellite weapons flight testing.[89] Software tweaks were required to 'tune out' these blooms, including satellites burning up on re-entry, so as to focus the sensors on ballistic missile launches.[90]

From its earliest days DSP had been a priority mission, first for U.S. perceptions of nuclear stability by providing more warning time of an incoming Soviet nuclear attack. Later, DSP became tactically useful for military units on the ground by making them more responsive to conventional short-range missile attacks.[91] By the 1990s, DSP had become a crucial part of 'military force enhancement' from space. In the 1990–91 Gulf War, DSP became more active in a tactical support role for deployed forces in the Persian Gulf whereas 'strategic' nuclear missile warning was relegated to a secondary mission.[92] These technologies are not without risks and faults—Soviet early warning satellites on at least one occasion detected an apparent missile launch that turned out to be the setting sun and cloud reflections playing tricks on the sensors.[93] The people operating the systems learned to ignore such false alarms, such as in the 'Petrov Incident' in which the early warning system officer ignored multiple false alarms from the missile launch detection system. This incident and others like it illustrate the risks of accidental nuclear war based on faulty equipment and false positives.

At present, the U.S. SBIRS and Russian EKS are small early warning constellations—rarely going above six satellites per constellation due to their wide fields of view and accurate-enough resolution for heat emissions detection from GEO and HEO. Both have built corresponding ground stations and infrastructure to manage and use the data provided by these systems.[94] During the Cold War, the United States relied on Australian and West German ground stations to manage the DSP constellation which triggered no small amount of local opposition and protest.[95] Infrared sensors are a crucial part of making Earth more 'transparent' for military weapon systems and intelligence agencies. The most advanced technologies radiate large volumes of heat and are liable to be picked up by the SBIRS and EKS early warning satellites which use a combination of GEO and HEO/Molniya orbits. Hypersonic glide and hypersonic cruise missiles emit very large amounts of heat—possibly up to 1,600°C—throughout much of their in-atmosphere flights.[96] They are also slower than most ballistic missiles. They will therefore not be hidden from space-based infrared sensors and not convey much surprise other than in a narrow tactical sense of which precise location may be hit, though with nuclear warheads in use that may not matter much.

The U.S. Missile Defense Agency (MDA) is currently seeking to develop a LEO-based version of SBIRS under the moniker of Hypersonic and Ballistic Tracking and Space Sensor (HBTSS). The idea of a LEO-based early warning system was considered as part of the original SBIRS programme in the 1990s. SBIRS was meant to have 'High' and 'Low' variants, but in the end only the 'High' variant was built due to the high costs and relatively low capability gains from a LEO constellation. Whilst SBIRS-High was built with four GEO and two HEO satellites, SBIRS-Low and its constellation of 24 satellites and accompanying relay communications satellites never materialised and instead morphed into a testing programme. SBIRS-High development was not exactly smooth, but SBIRS-Low had extreme difficulties with ballooning costs and technical problems, as well as a less clearly defined mission compared to SBIRS-High.[97] In the 2000s, as SBIRS-High was developed and rolled out in the 2010s, SBIRS-Low was renamed the MDA's Space Tracking and Surveillance System (STSS), with three test satellites launched in 2009, a few years behind schedule. Despite being reported as successful in tracking ballistic missiles in flight, the MDA shelved the programme in favour of an 'alternative' LEO missile tracking system called the Precision Tracking Space System (PTSS) in 2009.[98] The PTSS was itself shelved in 2013.[99] Today, the MDA has resurrected its interest in these systems and have provided 'hypersonic' missiles as justification for yet another rebranded SBIRS-Low: the Hypersonic and Ballistic Tracking and Surveillance System (HBTSS). However, such missiles can be detected with the existing SBIRS system, therefore any advantages for HBTSS worthy of the investment would have to be made on grounds of redundancy, resolution and timeliness, tasking demands, and resilience. There is some uncertainty as to whether SBIRS can offer enough high-resolution and timely data to provide targeting solutions for a hypersonic interceptor, which HBTSS may be specifically designed to do.[100]

The SBIRS-Low and its subsequent permutations and rebranding exercises should remind us that the technological systems currently being discussed in many quarters have long and often troubled histories, and can receive new leases of life when political and military conditions are right. The planned successor to the currently deployed SBIRS (High) constellation is the 'Next Generation Overhead Persistent Infrared Block 0', or OPIR system. This is coming into criticism too due to delays and cost overruns,[101] but SBIRS and DSP were hardly without issues and were ultimately functioning, useful systems. Technological innovation and advancement is not necessarily something that is only accelerating or happening extremely quickly.[102] Slow, incremental improvements describe the generational upgrades in many space systems that are already at the cutting edge. Despite the supposed quality and usefulness of these systems, their data is still only of one kind and that data needs to get to the right systems and analysts. Ambiguity can still exist over what kinds of missiles are launched, such as regional or intercontinental missiles, ballistic or manoeuvrable, especially early on in the flight.

Whilst discussion in the United States may treat such capabilities as essential for their nuclear posture and the low number of SBIRS satellites may make them an attractive target,[103] it is not necessarily viewed with the same level of importance elsewhere. The Soviet Union did not rely as much on such early warning systems given the relative slowness with which it deployed its own early warning satellites. China did not deploy its first generation of an early warning satellite until recently, despite having spent most of its time as a nuclear weapons state. China is reported to have first become interested in developing early warning satellites in the 1990s and today fields the Outpost series of satellites which may include missile early warning capabilities. There are some suggestions that China's Outpost satellites, and specifically TJSW-2 and TJSW-5, may be as effective as the old U.S. DSP, but have not quite reached the level of quality as SBIRS.[104] As China continues to modernise its military power, and as space-based infrared systems help detect and track the flight paths of conventional missiles and aircraft, it is reasonable to project this as a growth and improvement area for future Chinese spacepower for both conventional and nuclear military operations as well as broader 'national intelligence' needs. This means a U.S. attack on a Chinese early warning satellite may not create the same political and military response as a Chinese attack on a U.S. early warning satellite because they have different values and uses towards each side. It also means 'like for like' responses may not be possible in a crisis or conflict.

Russian and U.S. nuclear missiles have to travel very long distances for both states to annihilate each other, so over-the-horizon sensors were useful. Pakistani and Indian nuclear weapons do not:

SRBMs and air-launched short-range nuclear cruise missiles are sufficient to hold the other side at risk from nuclear destruction. Space-based nuclear attack early warning systems are less useful here because a nuclear exchange could happen within a few minutes, rendering space-based infrared sensors too slow to provide any meaningful extra warning. An infrared sensor may have difficulty telling *which* aircraft may be getting ready to unleash nuclear hell as opposed to conventional bombs and missiles, especially in a busy air campaign. Ground-based radars are useful for warning systems in that context. A space-based sensor can only detect the heat plume of a missile once it breaks cloud cover. In the South Asian context additional warning time may only be a minute to 90 seconds at best out of a total flight time of 5 minutes for the longest missile ranges in a major Pakistan-India conflict. Those 90 seconds would not be much better than the warning provided by ground-based radar systems, with the difference measured in seconds.[105] This is markedly different to the U.S.-Russia situation, where space-based sensors increased the warning time provided to the U.S. from about 5 minutes before the warheads hit to around 30 minutes. That said, should India wish to develop a space-based early warning sensor system as part of its early warning networks, it can do so now that it has mastered GEO launch capability with the GSLV Mk III. The benign uses of infrared detection satellites—such as detecting and monitoring forest and industrial fires—could help create a coalition of interests in India around such a system. There are also a range of conventional military uses for such infrared satellites as they can track aircraft, large military activities, and other missiles, not only nuclear ones. Theatre missile defences are known to benefit from space-based early warning sensors, as seen in the Scud hunt and Patriot air-defence systems experience in the 1991 Gulf War.

Specific spy satellite systems must be developed for the specific needs of their users and funders. No single satellite system fits everyone's needs, and factors of technical feasibility, resources, the nature of the supposed threat, individual supporters of specific systems, and other internal or domestic political and bureaucratic factors shape what technological systems are built. Imitation is not enough—space systems are too expensive without identifying clear, specific needs in the cut-throat bureaucratic politics such capital-intensive projects entail. Despite the high entry costs, space technologies had irreversibly spread to multiple space powers by the middle of the Cold War, showing the importance of satellites to the emerging technopolitics of orbital infrastructures in the Global Space Age.

Conclusion

The Cold War was an era of technological terror as well as wonder. A new environment was opened up for exploitation and use by the state, and helped bring about the prospect of almost instant nuclear Armageddon. Whilst many forms of technological state power pressed ahead, other areas suffered. Collating his own research, Siddiqi argues that:

> Gagarin's flight did not, after all, happen in isolation from the political, economic, and social dimensions of the Cold War. And ironically… the same forces that allowed the Soviet Union to send the first human into space—the need to arm themselves with powerful new weapons—deprived the country of further national triumphs in the space race.[106]

He also argues that almost all civilian space systems were derived from military counterparts in the Soviet Union, and that purely civilian satellite designs were rare.[107] Whilst crewed spaceflight went through peaks and troughs of activity, one sector of space activity went from strength to strength: the use of space for the military forces, intelligence communities, and corporate or industrial interests. Deudney's question on whether this 'Earth net' of surveillance machines in orbit is building a planetary panopticon will remain an important one not only for military forces and intelligence agencies, but for citizens and civil societies. Such surveillance capabilities only enhance the power of centralised and hierarchical states to impose their will within and beyond their borders, tightening centralised control over continent-sized polities.[108]

The elaborate spying systems placed in orbit are the continuation of a long-running trend of state agencies attempting to reduce uncertainty and gather more information on potential or actual threats, both chronic and acute. But this does not provide omniscience because technical systems always have limits, human analysis and decision-making are still needed, and the targets of intelligence gathering can take measures against such efforts. The knowingly observed will change their behaviour in front of the observer. These space systems merely continued the art of intelligence gathering and counterintelligence into orbit. Reducing uncertainty was a major theme within the work of Chinese military philosopher Sun Tzu, who said that net assessments must be used to form strategic advantages, in 'making the most of favourable conditions and tilting the scales in our favor.'[109] More pessimistically, the nineteenth-century Prussian military theorist Carl von Clausewitz said that 'imperfect knowledge of the situation' could bring action to a standstill, and that 'chance, the accidental, and... good luck play a great part in war.'[110] No matter the information gathered, the final intention of the other side will never be certain. No amount of technical intelligence will provide clear predictions of what someone will do, though they can provide reasonable estimates of material capabilities which shape a range of possible actions.

There are some things that are inherently qualitative and cannot be assessed in an objective, technical manner. Sun Tzu warned of this by stating that a fundamental part of conducting net assessments was the quality and morale of the enemy forces and the skills of their leadership.[111] These qualities in particular seemed to have been lacking in the Russian military in early 2022, despite the masses of hardware on display and their advances in military space technologies as seen in Chapter 5. Paradoxically, greater knowledge may sometimes create more uncertainty, as 'knowledge of the circumstances has increased, but our uncertainty has not been diminished thereby, but intensified... Our decisions are unceasingly assailed by [knowledge and experience], and our mind... must always be "under arms".'[112] Whether in trying to see through the smoke of howitzers or the chatter of electronic emissions, leaders and commanders will still need to come to terms with incomplete, out-of-date, false, or just irrelevant information alongside uncertainty and the range of possible outcomes in shaping their own decisions.

No amount of applied witchcraft or technical wizardry in orbit will totally eliminate the 'fog of war' or the difficulties of command and decision-making at the top posed by the unknown and imperfect information and incomplete knowledge. Clausewitz's misgivings about information during combat are still worth heeding:

> A great part of information obtained in war is contradictory, a still greater part is false, and by far the greatest part is somewhat doubtful. What is required of an officer in this case is a certain power of discrimination… This difficulty of seeing this correctly, which is one of the greatest sources of friction in war, makes things appear quite different from what was expected. The impression of the senses is stronger than the force of the ideas resulting from deliberate calculation.[113]

Humans need to make decisions in the face of uncertainty, chaos, falsehoods, and fear. No doubt information from space systems have been useful and transformational for intelligence practices, but they are not a panacea that eliminates the unknown, nor are they foolproof technologies that can prevent disastrous decisions in politics, war, and strategy.

Space-based military and intelligence systems had some positive effects on international relations and in reducing the likelihood of war by making complete large-scale surprise attacks less likely and more difficult to achieve, as John Lewis Gaddis argued.[114] Norman Friedman explains that 'for the few navies which possess them, satellite sensors have drastically reduced the stealth which navies—and aircraft—used to enjoy virtually from the time they left port or flew beyond the radar horizon of the coast.'[115] The effects of this transparency can go either way, not necessarily towards stability or reducing the likelihood of (nuclear) war. Such transparency can reassure a potential adversary if potentially threatening ships are not being sailed around in a threatening manner. However, any provocative deployment of a surface fleet may be all too easy to spot and bring about a crisis or increase tensions in a way that naval deployments might have previously gone relatively unnoticed in real time in the pre-satellite age. It may have been better that Robert McNamara had no idea that the Soviets had already deployed 'tactical' nuclear warheads in Cuba in October 1962, and four times as many Soviet troops as he thought. Being constantly notified of the actions of others could only inflame a sense of paranoia, insecurity hypochondria, and disproportionate worst-case thinking; spamming one's own mental faculties with noise.

Academic James Clay Moltz makes a logical argument that the development of space-based sensors and keeping space an environmentally 'secure' place for those military and intelligence systems directly benefitted 'security' on Earth because space systems provided the means of verification both superpowers were happy to rely on to ensure they each adhered to arms control treaties.[116] Even though systems like DSP and SBIRS are 'nearly universally viewed as stabilizing factors in the strategic balance',[117] it does not change the fact that it is at best a mitigation against the fact that several individuals can choose to unleash nuclear Armageddon at a moment's notice at any time. Is that such a stable system, regardless of the mitigations brought by early warning and imagery satellites? Such is the condition of nuclear despotism where individuals have a concentrated and supreme authority over nuclear use, where satellites can only do so much against the requirements of 'the speed of nuclear use decisions; the concentration of the nuclear use decision into the hands of one individual; and the lack of accountability... represented at the moment of nuclear use'.[118]

When adding the ambiguous effects of space systems on perceptions of nuclear stability onto the existing inequalities exploited by various space projects around Earth, the political, social, and strategic effects of space technology appear paradoxical, diverse, and unpredictable rather than linear or a one-dimensional trajectory of technological development. Technological ambiguity is the lesson to be taken from this. Michael Neufeld argues that 'in short and on balance, the militarization of near-Earth space has been a positive force for global stability and the global economy'.[119] Today, those satellites themselves can be argued to have become a source of instability as they are tactically useful in combat missions that do not necessarily involve nuclear weapons. On the economic front, certain states benefitted more in the emerging space economy more than others, and some were simply the targets of extractive economic practices. The continued use of space for all military capabilities and economic functions complicates the contributions of spacepower to 'stability' and 'prosperity' far more than what many accounts—mostly from the United States—consider. Researcher Pat Norris recognised that the 'rockets that carried the politically calming Corona and Zenit satellites into orbit were the very same rockets that in the form of ICBMs were a major cause of destabilization to begin with'.[120] It is important that we take arguments, such as Daniel Deudney's, that space technologies have played a big part in presenting a catastrophic threat to humanity seriously.[121] The balance sheet of space technological development is not so straightforward. At best there is only meagre compensation, not absolution, from the original sin of the Global Space Age.

The same technopolitical effects and drivers are at work in the next chapter which looks at a collection of space services that have come to resemble infrastructural services: communications, navigation, and space tracking. These too were tied into the development of outer space as a militarised environment and met a multitude of state needs. By the dawn of the twenty-first century, many major states had built their own orbital infrastructures having seen the benefits of space technologies for the purposes of war, development, and prestige. It is not only spy satellites that have matured over the past many decades, but also orbital infrastructure which now feeds into almost every aspect of private and economic life, having first transformed conventional military power and enabled the state to engage in high intensity, remote, and precise mechanised killing capabilities.

4

TETHERS OF MODERNITY

Cyfaill blaidd, bugail diog

A wolf's friend is a lazy shepherd

Satellites and their ground infrastructures provide globe-spanning wireless communications and data-gathering services for a multitude of users, but often began with military interests and innovations. The military advantages and commercial-industrial gains these provide explain why there is no singular international satellite consortium, providing all users on Earth with a single service. A handful of the most capable states chose not to rely completely on the United States and the Soviet Union for space launch services. States, political interests, and economic actors beyond the Cold War superpowers wanted to secure a greater degree of technopolitical autonomy in essential military capabilities, a slice of the increasingly lucrative high-technology space and telecoms industry, as well as technonationalist desires to demonstrate an ability to keep up with the technological leaders of the international system. No power wants any other to completely dominate the technological tethers of modern political, economic, and military systems.

Spacepower provides what are now considered essential military and economic infrastructures, and they have matured and long moved on from their 'emergent' phases. With that comes familiarity, path dependencies, and higher costs of switching to alternatives as certain infrastructures become 'sunk costs' and shape successive behaviour and choices around their technical features and capabilities. Satellite communications, navigation constellations, space tracking, and the various ISR, EO, and rocket launch systems covered in the previous chapter constitute what I argue to be Herrera's 'internationally significant technologies'.[1] They impact and make up the material, physical structures of the international world and the domestic politics of states. Even though technological devices are ultimately human-made products and reflect social and political goals and values, they create their own material forces that impose themselves on human choices and agency once built.[2] Whether something is physically easy and cheap to do is in part the result of technological infrastructure, which is a material structure. Communications satellites are imbued with political goals and meanings, but the objective, technical fact they enable global wireless communications at the speed of light generates effects that others cannot ignore or they can exploit for their own interests. That objective technical fact creates further political, social, and economic impacts, for better and worse. Controlling and owning spacepower infrastructure therefore provides power and influence over the terms and ways in which people and groups interact with each other, and provides power in any global governance efforts on coordinating and regulating the use of outer space.

This chapter focuses on three general classes of space systems that make up major orbital infrastructures: communications satellites (SATCOMs), space situational awareness (SSA) and space tracking systems, and satellite navigation technologies. The technologies themselves are only part of the picture. These major infrastructures are forming around major political-economic blocs, which is no accident. They have become essential for modern military power, civil infrastructure and the digital networks that underpin the modern global economy, data-dependent systems, and computerised electricity and production systems. This again demonstrates the blurred distinctions between 'civil' and 'military' space technologies and platforms, and the pervasiveness of the original sin of spaceflight, but with perhaps an increasingly multipolar character as the twenty-first century progresses.

Planetary Webs

SATCOMs act as relays for terrestrial communications, providing important utilities for direct command and control (C2) of nuclear and conventional military forces, broadcast services, and satellite data relays. They were first developed and matured through the 1960s. A signal can be broadcast from one point on Earth, received and redirected by SATCOM, then sent to another point on the other side of the planet at the speed of light. Three satellites in GEO provided global coverage so controlling and receiving equipment (uplinks and downlinks) could be placed anywhere on Earth and receive the broadcast, with exception to extreme latitudes. This makes it far easier for military forces to wirelessly communicate with each other when on the move and over the horizon from each other, or on deployments where there is little terrestrial communications infrastructure. SATCOMs allowed military forces to spread out and disperse more, and increase mobility without losing the command-and-control tethers that kept them coherent, networked, and in line to carry out centralised orders and tasks.[3] As noted by space security scholar Joan Johnson-Freese in 2007, 'the increased incorporation of space-related technologies and capabilities into military operations until recently has occurred... as part of incremental modernisation efforts. Moving from landline-based communications to satellite-based mobile communications has evolved in the military as much as it has in the population at large.'[4] The reliance on mobile SATCOMs is more acute for the military due to the ability of much civilian infrastructure to rely on radio towers and ground cables, where mobile and expeditionary military forces have to rely on wireless technologies due to the lack of local infrastructure and their mobile nature on deployment.

Satellite television remains one of the longest running and most profitable commercial space sectors. The line between military and civilian/commercial worlds in SATCOMs has always been blurred: transnational organisations and companies such as Intelsat, Inmarsat, and Iridium are for-profit but are contracted by Government users and provide increased communications and broadcast bandwidth during military operations as 'secure' government and military channels. Telstar-1 was the first communications satellite, launched in 1962 and developed by NASA and AT&T. The Syncom television satellite was developed and launched by NASA in 1963, then the satellites were transferred to Pentagon control, again blurring civil-military distinctions, state and market forces, and agency responsibilities. The 1964 Tokyo Olympics was broadcast to the United States via the Syncom satellite.[5] The Early Bird system in GEO began service in 1965 and represented a massive leap in connectivity and bandwidth—it alone provided 240 voice channels whilst all existing transatlantic telephone cables only mustered 317.[6] Ground communications would not catch up in bandwidth until fibreoptic cables were laid and gave birth to the global internet.

Syncom made a big impression in India. Vikram Sarabhai pushed the concept of an Indian satellite TV infrastructure in a presentation called 'Television for Development' in 1969.[7] India, unlike many other developed and developing states, did not have a well-established terrestrial TV network, having only one transmitter with a range of 40km. Given the vast coverage provided by an orbiting satellite from GEO compared to the massive terrestrial network that would be needed to cover all of India, the Satellite Instructional Television Experiment (SITE) became a logical 'leapfrogging' move. The project was a joint initiative by NASA and the Indian Government, signed in 1969, and was supported by various UN agencies.[8] SITE allowed Indian Government use of NASA's ATS-6 satellite, 'parked' in GEO, for one year in 1975–76. It would develop tele-education services for basic reading, writing, and arithmetic educational programming, family planning, and emergency broadcasts.[9] Receivers could be assembled with minimal technical knowledge and with local materials, driving down the costs of receiving the broadcast for community tele-centres.

The SITE experiment drove demand in India for satellite TV broadcasts and led to New Delhi's purchase of INSAT satellites. These were manufactured and sold by the USA to Indian Government contracts. It was a positive public relations move for NASA in furthering the perception of U.S. spacepower in its Cold War propaganda efforts with the Non-Aligned Movement, specifically countering communist China's influence in India.[10] However, INSAT was approved by the Indian Government only four months after SITE began, suggesting that few were interested in the actual outcome of the year-long experimental programme and pressed ahead regardless.[11] The SITE project triggered other developments beyond the direct educational programmes—it led to electrification of many places and galvanised public opinion in support of India's nascent

ISRO, and helped push the development of mass broadcast media. SITE regularly reached over 2,300 villages, averaging 1,200 inhabitants in each. According to Ashok Maharaj, SITE was perhaps the most 'far-reaching effort to apply advanced Western technologies to the traditional problems of the developing world'.[12] SITE changed the debate about the uses of space technology in India, from whether India could afford a space programme to whether it could afford to go without a space programme.[13]

Despite the tendency of many in the space sector to invoke space technology's role in positive socioeconomic transformation and development in triumphalist accounts of its uses on Earth for the benefit of 'all humanity', the actual legacy of SITE remains mixed.[14] For Indian elites, there was no agreed position on why SITE was such a desirable project. Siddiqi argues that it was a technological dead-end that had little to do with existing national broadcast infrastructure plans. Dissenters needed convincing and therefore proponents of SITE developed their own rationales and visions for it, but common among all these visions of India on the desirability of SITE was the characterisation of India's poor and illiterate as a monolithic, passive population who did not have their own diverse needs, interests, and agency. As a result, SITE 'ultimately left... a deeply ambiguous and discontinuous legacy for precisely those who SITE was to benefit.'[15] SITE, and SATCOMs more generally, have parallels with technological and infrastructural imperialism, which are evident for Maharaj who writes that GEO satellites and their infrastructure in:

> postcolonial India can be seen as an extension of the terrestrial technologies that the British used to civilize/modernize a traditional society. In this case the United States replaced the erstwhile imperial power to bring order, control, and 'modernization' to the newly decolonized states through digital images using satellite technologies that were far removed from the territorial sovereignty of nation-states.[16]

For the United States, SITE was in part a tool to help steer India away from nuclear proliferation, epitomising the use of NASA as a tool of U.S. foreign policy, not just a 'science' agency or seeking to develop the world solely through charity.[17]

India was first a customer for Western providers as it moved towards its own indigenous systems through the later years of the Cold War and into the post-Cold War era. Successive Indian governments wanted to avoid such apparently useful infrastructure becoming a new pathway of dependency and external power and influence. India's first home-built experimental GEO communications satellite was launched for free as part of ESA's Ariane launcher testing programme. Having started the project in 1975, the first SATCOM mission lasted from 1981 to '83.[18] Before reliable and secure Indian SATCOMs were deployed, the Indian Navy had previously relied on the Inmarsat company's satellites, headquartered in London. Today, the Indian Government sustains 16 geostationary communications satellites in GEO, with the Indian Air Force and Indian Navy operating one each.[19] India's indigenous SATCOM capability would be impossible without the GSLV Mk III, which India strove to develop in the face of sanctions from the United States.

In the Soviet Union, SATCOM constellations helped knit together its far-flung military installations and its widely dispersed populations east of the Ural Mountains and in Central Asia. Strela and Potok were developed from the 1960s and sought to connect Soviet military commanders, the Soviet Navy, remote military facilities, and intelligence officers deployed around the world. Gonets was a civilian version of Strela which facilitated coded messaging systems, like mobile phone texting messages, for domestic Soviet uses.[20] Potok, Gorizont, Raduga, and Ekran satellites were placed in GEO. However, early Soviet telecommunications networks made use of the Molniya orbits, including the Orbita TV network which was declared operational in October 1967 on the 50th anniversary of the October Revolution. In 1971 the Soviet Union launched the Intersputnik International Organization of Space Communications (Intersputnik) organisation and SATCOM system after first being approved in the Ministry of Communications' Seven Year Plan in 1959. By the late 1970s, Intersputnik telecoms and TV downlinks were placed across the USSR, its 'satellite' states, and the Third World on a commercial basis in competition with Intelsat. Intersputnik relied on the Gorizont satellites in GEO to transmit its broadcasts.[21] Yet the international dimension was a backdrop to the rollout of satellite TV in the USSR itself. In 1961, only 35% of the population could receive TV signals, by 1966 it had risen to 55%, and then to 70% by 1971. TV satellites became an important part of Soviet economic development as Moscow sought to extend entertainment and luxury services across its territory, particularly to more remote and climactically harsh regions to improve conditions for its labour forces in Siberia.[22] Parallel to this however were concerns of Russification and politico-cultural hegemony of the workplaces and regions that enjoyed satellite TV and centralised programming coverage in the culturally, religiously, linguistically, and ethnically diverse places of the USSR.[23] Russian SATCOMs are as essential as ever today, with the Yamal and Ekspress satellites networking the vast Russian state.[24]

As Soviet spy satellites grew in number and sophistication in the Cold War, so did the requirement for communications between those platforms and their controllers and users on Earth. In a reinforcing trend, the more space-based systems that were developed to gather data and services, the more the appetite for greater quantities and quality of the data and services increased, and to handle the increased bandwidth more satellites were needed to transfer and relay the data between satellites, ground stations, and receivers. The increasing appetite for more timely information from SIGINT and IMINT assets further increased a reliance on SATCOMs to provide global connectivity and the ability to download satellite data as it came in and into the field for mobile units, rather than having to wait for the satellites to pass over fixed ground stations. The RORSAT and EORSAT spy satellites received support from the Tsiklon communications-navigation suite which was located at around 800–1,000km altitude in LEO. The navigation capability became the precursor to later navigation satellite systems in the Soviet Union.[25]

The United States fielded its first military and government SATCOMs in the 1960s, with the Defense Satellite Communications System (DSCS) launching in 1966 into GEO. By the mid-1970s both the Navy and the Air Force had their own SATCOMs—FLTSATCOM and AFSATCOM. In the case of AFSATCOM, it was not a series of dedicated satellites but rather allocated bandwidth from a collection of other satellites such as DSCS and FLTSATCOM to transmit secure messages between its central command headquarters and its nuclear forces. As U.S. military space expert Peter Hays explains, 'cumulatively, these new satellite communications systems revolutionized the U.S. military's command and control system by providing nearly worldwide and instantaneous communications with most types of major U.S. weapons systems.'[26] Today the United States government and military enjoys a smorgasbord of platforms for SATCOMs, including Wideband Global SATCOM (WGS), MILSTAR, the Advanced Extremely High Frequency (AEHF) satellites, and the Quasar relay COMSATs for IMINT and SIGINT data transmission. MILSTAR and AEHF act as critical relays in the U.S. nuclear command and control systems, designed to be survivable links in the nuclear command chain. Their bandwidth is limited but more secure and durable, with higher priorities and reliability for the most important transmissions in all situations including a nuclear war.[27] On 11 September 2001, Air Force One was struggling to receive updates as the U.S. President was kept in the air for safety. As they flew over ground towers, local TV networks would tune in and out. There was no satellite TV receiver on the aircraft. Advisors believed a major attack was ongoing as commercial satellite and terrestrial mobile phone systems had seized up and jammed due to overwhelming civilian demand. The only reliable system left to use on the aircraft was MILSTAR, the last link in the nuclear command and control chain which linked the President with the nuclear arsenals. One staffer asked, 'is this what a decapitation strike looks like?'[28]

WGS today provides more bandwidth for many kinds of communications requirements such as internet connectivity for deployed forces. WGS also involves the Canadian and Australian defence ministries, whilst AEHF was developed as an American-led multinational military SATCOM constellation, including Britain, the Netherlands, and Canada. As NATO military forces continue to become more data-dependent to manage their high-technology weapons systems and logistics infrastructure, the demand on secure communications bandwidth is only increasing, and U.S. allies have a ready way to contribute to such architectures if they are comfortable with Washington's 'leadership' in such projects. These communications systems are extremely expensive, even before applying the hardening technologies to make them jam-resistant and able to withstand the effects of enhanced radiation following a nuclear detonation or particularly bad solar storms. Costs and reducing duplication of effort is important for the U.S. and its allies, and clearly for many NATO members and other allies, U.S. leadership is seemingly a good option for cheaper access to high-technology military space infrastructure.

U.S. industrial dominance was not something Washington's allies were universally happy about, and even less so with deliberate attempts to cement that dominance. John F. Kennedy's signing of the Communications Satellite Act in 1962 led to the United States' dominant position in globalised telecommunications, wresting it away from the British who had controlled most undersea cables since the early twentieth century. The United States was using space technology to win the battle for 'freedom' against Soviet (or perhaps even British) 'tyranny'.[29] Intelsat, founded in 1964, was a satellite company majority-owned by the U.S. Government and tasked with providing telecoms for its member states, but also on terms favourable to the U.S. telecoms industry. Western Europe and Japan found this to be detrimental to their industrial interests due to the small print that they do nothing against Intelsat's financial and commercial interests. Italian-led proposals to ensure voting rights for European states in the governance of Comsat—Intelsat's precursor name—to reduce the domination of U.S. finance and industry were flatly rejected by the United States. Washington made it clear that Comsat would go ahead with or without European participation. At its inception in 1965 there were 46 states involved in Intelsat, none of them communist. European states had a share of 28% in votes in the governing system whilst the U.S. share would never go below 50.6% which ensured their preponderance over the governance, development, and contracting of the constellation. This was described by a senior official in the Johnson administration as an 'unusually attractive international vehicle for the US'.[30]

Whilst the United States was inherently supportive of western European integration during the Cold War, it wanted it to happen in a way that more directly satisfied Washington's economic interests. There was a limit to this strategy. Power relations is not merely about the powerful getting their way, but about how the weaker side can still get what they want or work to alter the balance in their favour over the long term. Though it would take time, Western Europe could develop its own systems as an integrated bloc. Some U.S. experts at the time warned that the Intelsat situation would only ensure the loss of Washington's control of the emerging European communications satellite systems and amended the organisation to be more open to industrial and economic competition from European companies. NASA could not cooperate with foreign space agencies if such assistance would lead to a capability to launch GEO SATCOMs, which would be a direct industrial challenge to the United States and Intelsat.[31] There was little appetite in Washington to accelerate the development of competing infrastructures, and cede the U.S. advantage in infrastructural, material power that Herrera highlighted above.

Despite the desire to stall allied development, these technologies emerged in western Europe. The Franco-German Symphonie communications satellites had to be launched atop a U.S. rocket on the basis that they provided only experimental services and would not benefit European commercial purposes. France and Germany learned a 'hard lesson' on the need for an independent European launcher. ELDO had failed to produce it at the time of Symphonie's development. If France and Germany wanted to develop commercially viable systems they could not do so through U.S. SLVs and the Intelsat organisation because there were too many strings attached. ESA's decision to pursue an independent European launcher proved to be the right gamble, as by the 1990s they had a reliable launch system as well as a commercially viable launch market for satellite deployment which could escape certain terms and conditions that U.S. launchers would have imposed on European space companies and agencies. In addition to this, by the end of the Cold War, several European states and their integrating military-industrial complex could muster the capabilities to develop their own high-grade telecommunications and imagery satellites. Italy began work on a domestic mobile communications satellite—Italsat—in the 1980s, but also developed a military SATCOM capability called Sicral, with a particular focus on providing military SATCOMs in mountainous regions.[32] Such specialisms demonstrate how space systems can be designed to meet specific terrestrial needs because terrain can alter how easy it is to receive and transmit various kinds of signals.

Requirements are different in urban areas, jungles, deserts, and extreme latitudes as equipment can only tolerate so much and the environment can alter how radio signals behave. Any serious military acquisition programme has to be specific about space investments so that in their most likely scenarios they have space infrastructure that can do the job.

The UK's Skynet military SATCOM was launched by U.S. launch services and Arianespace. One could not blame French policymakers if they were pleased that the British were now paying customers for the French-led European SLV after giving up on ELDO in the 1960s. Though the UK has a rather inglorious past with SLVs, Britain was the first state after the superpowers to possess a military SATCOM constellation. Construction work on the first satellites was contracted to foreign companies as no UK company had the capability at that time. The first generation of Skynet was a failure due to a variety of technical problems, but in 1964 Skynet 2B succeeded. Whilst Skynet 3 was cancelled due to costs in the 1970s, the growth in the MoD's and GCHQ's appetite for SATCOMs, new possibilities with small, mobile terminals and downlinks, the experience of the Falklands War in the use of Skynet terminals with Special Operations Forces, and the cancellation of the NATO IV satellite system drove Britain to develop Skynet 4 in the 1980s.[33] In the Falklands War the Royal Navy made extensive use of Inmarsat SATCOM terminals on its ships, which proved easier than linking them all up to dedicated military long-haul communications suites and provided a lot of (less secure) bandwidth in the process.[34] Today Skynet 4 and 5 satellites are in orbit, with the Skynet 6 generation currently in development tagged at over £5bn in cost, and remain something of an exception to Britain's general dependency on the United States for military and intelligence space services.

Japanese-U.S. relations on SATCOMs were more tempestuous compared to Britain. By the late 1980s, the trade balance swung heavily in favour of Japan and in 1988 the United States legislated their way to opening up previously protected Japanese domestic telecoms markets to U.S. telecoms companies. The Japanese Government was outraged that the United States 'was telling them to rein in their ambitions to be major competitors in the world market' for SATCOMs.[35] Major indigenous Japanese SATCOM programmes were subsequently cancelled and U.S. firms outbid Japanese ones for contracts in

Japanese industry. These serious trade and industrial spats did not extend so far as to undermine the core security alliance, as Japan continued to host U.S. military facilities including ocean surveillance SIGINT ground stations for the Naval Ocean Surveillance System (NOSS), and a cluster of radomes at Misawa that complement undersea and space-based ship-tracking systems.[36] Japan has also held a 'strategic' interest in military SATCOMs capability since at least 1985, and today its communications sector is being geared towards more military uses following decades of civilian and commercial development, like its imagery capabilities.[37] Allies can be friends, but they can and will fiercely argue over some areas of policy and industrial competition within certain bounds.

By 1987 Intelsat consisted of 114 member states, whilst the communist rival, Intersputnik, consisted of only 18. Military space analyst Bob Preston believes Intelsat 'gathered the developing nations into intimate communications with the industrial West, gave them access to essential means for economic development, and provided them with immediate hard currency returns on a very modest investment. It wrapped the Iron Curtain in an Information Net.'[38] This generous interpretation rides roughshod over the intense industrial and political wrangling between the capitalist countries over its construction and governance. Preston is correct, however, that the continuing fixation on crewed spaceflight in a contest for 'world leadership' is misplaced, given how satellite communications development in the Cold War had a 'far more pervasive and lasting impact than a flag and footsteps in the lunar dust'.[39] The geopolitics and competitive techno-industrial arguments, as well as U.S. industrial aggrandisement in the Intelsat experience shows how cut-throat and competitive inter-Allied relationships were during the Cold War in space. Space has never been free from the competitive and self-interested aspects of international relations and industry once machines could be placed there, because their infrastructural power on Earth was too great to leave fully in the hands of others.

Spying on Space

Like any environment, having a good sense of where things are and who is doing what is important. Space Situational Awareness (SSA) refers to detecting, tracking, and identifying objects that fly into, through, and from space. SSA capabilities, or Space Domain Awareness as the U.S. and UK militaries now refer to them, are useful information for anti-satellite weapon targeting as well as passively tracking orbiting objects so as to avoid collisions in orbit for mundane, benign 'everyday' use in space traffic management. Many of the earliest SSA capabilities emerged from nuclear early warning radar systems designed to detect nuclear warheads as they travelled through space to reach their terrestrial targets. Many states have developed such radars. The United States developed the Ballistic Missile Early Warning System (BMEWS) which had radar installations in many places in the United States, but also the Thule base near Qaanaq, northwest Greenland and Fylingdales in the North Yorkshire Moors in northern England. These are not dedicated SSA sensors, as they are prioritised for missile warning duties. France operates its own dedicated military SSA sensor—the GRAVES system. The Soviet Union developed radar bases along its northern approaches as well, given the fact that most nuclear weapons the superpowers would throw at each other would travel northwards over the pole, rather than east-west.

In addition to Greenland and Britain, the United States set up a missile launch and tracking system in Iran called Tacksman. This began as a missile telemetry intelligence (TELINT) station to gather information on Soviet rocket launches, and came to track objects in space as well.[40] This was lost in 1979 following the Islamic Revolution, and in light of the thaw in relations between Beijing and Washington a new facility was set up around the Tian Shan mountains in Xinjiang to help monitor Soviet missile activity in central Asia. The stations around these mountains gave the U.S. and China a dedicated missile launch detection and tracking capability between Baikonur to the west and Kamchatka to the east.[41] Such ground-based systems complemented satellites in verifying Soviet adherence to arms control agreements. The China-based stations, codenamed Chestnut, came about as part of the Sino-Soviet split with the U.S. offering SIGINT and IMINT data on Soviet military movements and assets on the Sino-Soviet border to the Chinese. The CIA set up a school in Beijing to train Chinese technicians to operate the ELINT equipment placed in the Chestnut sites in Xinjiang.[42] In

Ethiopia, the United States set up a deep-space TELINT station—called Stonehouse—in the 1960s, and tracked the Soviet Union's Lunar and Venusian missions from there. Stonehouse was hastily shut down in the Ethiopian Civil War in 1974.[43] A global technological infrastructure was built and sustained by the United States to gather as much technical information as it could about activities in outer space, and ensured that the Space Age's presence on the ground was truly global from its earliest years.

Radar systems are mostly used to track objects in LEO as opposed to optical sensors for GEO satellites. Radars are one method of detection, with optical sensors, radiofrequency sensors (like ELINT) being other methods of detecting and tracking who is doing what in orbit. Stuart Eves describes them collectively as a:

> wide area space fence sensor. The best known was… that operated by the United States—a chain of sensors across the continent that detected objects as they broke its beam. The United States also has large space surveillance radars with smaller fields of view that are used for collecting high-precision data on targets in orbit.[44]

These radar systems are also useful in detecting, identifying, and tracking many objects in Earth orbit, not only nuclear warheads. Satellites and pieces of debris down to as small as 10cm in diameter or perhaps even smaller can be detected. Some claim that S-band radars can detect 1cm objects out to 800km and 2cm objects to 1,500km altitude, transforming debris tracking capabilities.[45]

The vast majority of SSA sensors are terrestrial, but more SSA sensors are being placed in orbit. Canada deployed radar satellites in the 2010s called Sapphire to provide SSA data. The United States followed with their own shortly after. The United States has today deployed the Geosynchronous Space Situational Awareness Program (GSSAP) platforms where GEO-based satellites can fly relatively closely to other satellites in GEO and take highly detailed images and telemetry for inspection and intelligence gathering. The U.S. allegedly conducted a similar mission with a satellite named Prowler in 1990 launched from a Space Shuttle. GSSAP is part of the Pentagon's Space-Based Surveillance System (SBSS) which is rolling out more satellites explicitly designed to detect, identify, and track objects in orbit. Russia and China have also developed orbital inspection and loitering capabilities in GEO. China has conducted manoeuvres of various kinds involving the SJ-15 and SJ-17 satellites, conducting rendezvous and inspection tests that are not too different to U.S. operations and tests in the past.[46] Russia, building on significant Soviet heritage in these areas, has embarked on a co-orbital inspection and potentially weapons systems programme since 2010, with the Burevestnik system and the Olymp/Luch satellites' close flyby operations of other satellites in GEO as prominent examples. Some Russian satellites have deployed 'sub-satellites' at relatively high velocities in LEO which may indicate potential aggressive uses. These projects are supported by the Nivelir SSA programme.[47] Space-based SSA sensors and inspection systems are still in their relative infancy and are deployed on a small scale compared to terrestrial sensors, and therefore constitute a specific emerging capability that could be expanded in future, for both benign SSA reasons as well as for hostile anti-satellite weapons systems. Essential technological capabilities lend themselves to a large number of different uses, making arms control based on abstract technological categories a difficult task, as discussed in Chapter 7.

With big-enough space objects reflecting sunlight at just the right time and angle, optical telescopes can provide images of them. The Pentagon used optical telescopes on a mountaintop in Maui codenamed Teal Blue, and another in Florida called Teal Amber, to take high-resolution images of Soviet and Chinese satellites. These systems were used to inspect and monitor the Shuttle *Columbia* on its first mission in 1981. The orbiter's first flight also involved the use of the secret KH-11 imagery satellites to take further high-resolution images to inspect any damage to the craft to assess whether it was likely to survive re-entry or not.[48] The Soviet Union maintained similar systems. There was an extensive network of ground stations and facilities around the world designed to track orbital objects and communicate with Soviet satellites, as well as detect ballistic missiles in some cases. Soviet optical tracking sites could be found in Chad, Mali, Sudan, Egypt, and Somalia.[49]

Laser methods of SSA involve a beam directed at a 'cooperative' target which has the appropriate reflecting panel at the correct angle; the character of the reflection of the laser then provides extremely precise information of its location. This laser-ranging capability is widespread with laser-ranging sights all across Earth. The International Laser Ranging Service (ILRS) shares extremely accurate information on satellite locations by bouncing ground-based lasers off reflective panels placed on satellites explicitly for this purpose. The Apollo astronauts placed laser reflectors on the Moon so extremely precise measurements of the Moon's distance could be made. The drawback to this method is that whilst it is accurate, it requires 'cooperative' targets to bounce back the laser, and is a very specific method of acquiring the location of something that is already known to be there. Gaining accurate positions on non-cooperative targets with laser ranging and bounce back—or 'lidar'—as one does with radars is a tougher proposition. Passive wide-area systems like radars and more targeted passive systems like optical methods are able to 'scan' areas and detect things that may have been previously unknown.

A more novel and literally global approach to SSA is to use GNSS constellations (see the next section) to provide positioning data for LEO satellites which can then be transmitted to controllers and SSA networks. LEO satellites fly below GNSS constellations, and therefore LEO satellites may gain more accurate positioning data for themselves and their operators to within a few metres, further tethering space infrastructures to other kinds of space infrastructures.[50]

SSA capabilities have been 'globalised' in a territorial sense for a long time with the U.S., Russia, Europe, India, and China maintaining or sharing SSA stations and data well beyond their home territories. Ground systems of various kinds are installed in Turkey, Romania, Germany, and Poland. Systems such as Aegis can be installed on ships, providing a mobile warhead tracking system that can be deployed at sea. Echoing other nuclear, raw materials, and military infrastructures in the North-South divide, many southern hemisphere states host such stations because the major space powers' main territories and infrastructure are in the northern hemisphere. Terrestrial sensors can only see the volumes of orbit within their horizon, meaning many sites on different points of the globe are needed for complete coverage of all orbital regions around Earth. The same is true for deep-space exploration missions, such as NASA's control station in Australia, China's deep-space communications ships, and India's extensive ground facilities cooperation agreements that support its Mars Orbiter Mission.[51]

China has expanded its SSA efforts through the Asia-Pacific Space Cooperation Organisation, which it leads, in order to bring together SSA data from dispersed sites under a data operations centre in Beijing. The project is called the Asia-Pacific Ground-Based Space Object Observation System and links sensors from Peru, Pakistan, and Iran with China's own.[52] The Purple Mountain Observatory is China's main optical imaging method of SSA for both satellites and large debris objects, which was largely an outgrowth of more civilian projects, but the PLA operates at least five large-phased array radars for missile warning which can also provide SSA data. China has also signed ground tracking facility hosting agreements in Pakistan, Namibia, Kenya, Australia, Chile, Brazil, Argentina, and Sweden.[53] Japan and India have shown a particular interest in developing SSA radars.[54] In Europe, ESA's official SSA programme began in 2009, and shortly after in 2014 the EU chose to set up its own, the EU Space Surveillance and Tracking Programme. In both programmes, much work involves integrating existing member states' information and assets, as well as funding joint projects and facilities. As of 2018, the EU's SST programme involved 11 radars, four laser stations, and 19 optical scopes which then feed data into a Common European Database.[55] These expanding global capabilities, and more importantly their data, will be essential in any global space traffic management (STM) regime because having a system entirely dependent on U.S. data will not be politically or strategically desirable for many.

Most space operators today rely on the timely sharing of SSA data from the United States to avoid collisions between active satellites, defunct satellites, and bits of junk. Whether for missile defence, anti-satellite missions, hiding activities on the surface, building a space traffic management regime, avoiding or removing debris from orbit, SSA data is a fundamental requirement, especially as Earth orbit becomes more cluttered and congested with traffic and debris. This is reflected in the increasing scope and scale of SSA outside the United States. Russia is rebuilding from the losses its early warning and SSA networks suffered in the 1990s. As of 2019, Russia has twelve operational early warning radars and four more under construction or planned, though their role in SSA is unknown. It 'possesses a large network of ground-based early warning radars which has been spectacularly upgraded in the past 15 years or so to provide coverage of all potential attack zones, a capability achieved not even in the Soviet days.'[56]

The United States' main method of tracking satellites became the Space Surveillance Network and today remains the single most comprehensive high-quality source of data on orbital traffic. However, much of the data is often classified. Analysts piece together necessary bits of information to maintain a catalogue of objects—what they are, who they belong to, where they're going, and what they may be doing. According to expert analysts Brian Weeden and Victoria Samson, Russia's SSA capabilities are second only to the United States, with a smaller LEO catalogue but a more 'robust' catalogue of HEO and GEO objects.[57] More SSA sensors are being deployed by more states and non-state actors, both on Earth and in space itself. A private company, ExoAnalytic, has placed several sensors in the southern hemisphere by spreading 200 optical sensors across twenty-four sites as of 2017.[58] The proliferation of commercial SSA providers is challenging traditional notions of whether SSA as a whole should be a secretive activity or not, with the international private sector demanding more transparency and predictability in the space environment for their own safety—and perhaps lower insurance premiums.

States lacking the most advanced hardware can acquire some SSA capabilities. Even basic hardware or observation techniques can provide some basic SSA capability, as amateur observations with optical scopes and radio direction-finding techniques have demonstrated for decades, such as with the Kettering Group, an association of amateur satellite observers that tracked the earliest Soviet satellites.[59] In the twenty-first century it has never been easier to assemble open-source SSA data of some kind with databases such as the Union of Concerned Scientists' Satellite Database, SpaceTrack.org, and CelesTrak providing data accessible via the internet. SSA data can be acquired by cooperating with others, compiling and analysing open-source information, or pilfering such information through more traditional and clandestine intelligence methods.

As Earth orbit became militarily important on both sides of the Cold War, so emerged the anti-satellite weapons designed to interfere with or destroy enemy space systems, only increasing the importance of spying on the actions of others in space whilst also providing targeting data to possible ASAT systems. Space historian Aaron Bateman writes that:

> determining the location of Soviet satellites was essential for establishing whether or not the USSR was going to use its space systems to carry out hostile actions against American satellites. Thus, astronomers in the United States and abroad became important members of the national security community in their work to develop effective technologies for observing and targeting Soviet space systems.[60]

Even before Sputnik reached orbit, plans were afoot to track enemy satellites in orbit above the United States. In the late 1950s radar stations were set up as part of the Minitrack system across Latin America to detect satellites in orbit.[61] Extensive SSA capabilities are needed for any 'counterspace' or 'anti-satellite' technologies and missions. A weapon is no good if you do not know where to aim it. Many sensors, radars, and weapon systems have been developed under the guise of 'missile defence', especially 'midcourse' BMD intercepts or where a vehicle is sent to hit a nuclear warhead as it travels through space towards its target on Earth. The reality is that midcourse intercept weapons have been proven as anti-satellite weapons, less so as effective midcourse missile defence interceptors. There is a persistent doublespeak when it comes to the development of anti-satellite weapons as missile defences, and this is not unique to the United States; the spread of such technologies is examined in Chapter 6.[62]

Individual sensors can only provide specific forms of data, and therefore a diverse collection of technical readings needs to be collated. Satellite locations are often presented as an 'ellipsoid', a volume or 'bubble' of space of where the satellite is believed to be at any given time, rather than a point in orbit. SSA analysts continually cross-reference datasets and readings to try to reduce the size of the ellipsoid to provide a more accurate location probability.[63] If the ellipsoids of two objects overlap, there is a significant probability of collision and warnings may be issued to the satellite operators so that they can manoeuvre. Even with relatively accurate tracking, SSA data does not provide all the necessary information about the nature of the object (especially classified satellites), and more importantly, intent behind suspicious activity.

As technical as SSA is, understanding behaviour and activity in space remains ultimately an art form that practitioners and students of Intelligence will be familiar with. SSA is not the same as net assessments about activities and intentions in orbit, or SPACEINT. SSA is the more technical aspect of SPACEINT, whilst SPACEINT includes SSA data alongside qualitative and interpretative analysis by placing the data in wider strategic, economic, and political contexts and HUMINT to outline the less tangible aspects of the activities. In April 2021 some U.S. observers were alarmed by the Chinese space station launch—the Tianhe module—which was 'too close' to the ISS. An open-source orbital traffic tracker noted that:

> the Tianhe core module for China's new Tiangong-3 space station was launched from Wenchang on [April] 29. The 22,600 kg module was placed in orbit by a CZ-5B rocket, separating from the empty 21,200 kg core stage at 0331 [universal time]. About 5 minutes later, both the module and the core stage passed only 300 km from the International Space Station, which is rather close to come given that there was no coordination between CALT [China Academy of Launch Vehicle Technology] and NASA on the launch trajectory. This close pass appears deliberate given that really careful timing was needed to achieve it.[64]

Tracking actions with SSA data is one thing, but interpreting the motivations and intent behind actions remains a human endeavour and requires contextual political, bureaucratic, and strategic knowledge. To claim a deliberately provocative act based on technical data alone makes certain assumptions: (1) some groups or individuals in the relevant Chinese ministries wanted to send a political message; (2) that message would be received clearly by the intended recipient; and (3) the deliberate action or threat had any strategic significance or significant gain for China or the individuals behind it. Considering the strategic and political contexts, there is ample scope for doubt in these assumptions that underpin the *interpretation* of behaviour gleaned from technical SSA data as 'deliberate' given that there is no substantial evidence at the present time to strongly support such claims.

The political context matters in shaping intentions behind understanding technically observable behaviour. Without agreed norms of safe distances in space, some actions may be coincidental, a careless mistake, bad luck, or unrelated to the perceived threat or infringement. The 300km distance may be too close for some, but given the different altitudes of the Tiangong space station and the International Space Station it is not necessarily a shared understanding of 'too close'. Without agreed 'keep out zones' it is unfair to unilaterally declare that a flyby of 300km in very different orbital altitude is inherently deliberate without explicit evidence of such intent. An equally plausible explanation other than a deliberate move is that nobody at CALT had thought this would be interpreted as a hostile move, or may not have even considered it. Many objects pass within 300km of each other without incident.

If the flight path was a deliberately provocative act, it then raises the question of what the intended message behind it was. No clear message can be discerned from such an act without more political intelligence, an important part of SPACEINT. At present, there is nothing to help us identify any messaging. It is worth remembering that the ISS is not solely a U.S. project. It is the International Space Station, host to a multinational astronaut crew, built by an ensemble of states and carrying the experiments of many more. The ISS is a major plank of Russia's civil space programmes, China's major partner in its future lunar exploration projects. Was China deliberately threatening a multinational laboratory to send a message to—presumably—the United States, even though Japan, Europe, Russia, and Canada hold strong interests over its continuing operation? Would the U.S. Government correctly interpret a threat to the ISS as a message to it and no-one else? If so, what would be China's preferred response to that threat, if any? These larger political questions cast doubt on the strictly technical analysis above regarding deliberate action.

None of these questions have clear or objective answers because there is little evidence to inform us at this time over whether it was a deliberate act and what a message might be. This would be a particularly bad way to send any threatening message as China would face more international opprobrium and condemnation for threatening an international laboratory than it did following its 2007

ASAT test debris disaster. In addition, coordination is often difficult between China and the United States due to U.S. Congress's legal prohibition on NASA engaging with or talking to any Chinese officials and staff. The Wolf Amendment of 2011 requires a legal exemption to any NASA-China dialogue, meaning that low-level, routine, and mundane exchange of information—including space traffic data—does not happen on a regular basis. With that prohibition in place from the U.S. side, it is rather unfair to portray a lack of coordination on this event as circumstantial evidence of deliberately nefarious intent from China.

Any possible threat posed to the ISS seems negligible and ultimately pointless. The ISS is not a meaningful military or economic asset to any of its participants. It would be only a symbolic target and any debris would pollute a large swath of LEO. This was not the 2007 Chinese ASAT test which demonstrated a military capability (and created over 3,000 pieces of debris in LEO) but rather a space station meant to be an early capstone of China's crewed spaceflight programme. Given its importance to Xi Jinping's narratives and soft power claims furthering China's rise, Tiangong is a prized technopolitical achievement, less a cudgel to casually threaten another space station. Indeed, China has proved its ASAT capabilities and therefore could already threaten the ISS with weapons if it really wanted to. As such, no new ASAT demonstration was needed, least of all with China's own brand-new space station as the proposed killing vehicle. It raises the legitimate question of what purpose threatening the ISS serves other than to further tarnish the Chinese Communist Party's international image and severely undo its soft power efforts with its civilian and scientific space programmes. Leaving the rocket booster in such an orbit could well be described as negligent from an environmental point of view, but again there are no commonly agreed standards for this. Evidence and political analysis is needed to answer these questions before the act can be described as deliberate. Otherwise interpretations of this incident as deliberate action remain a worst-case assertion. We simply will not know without any further evidence one way or another.

SPACEINT, which incorporates SSA data with political, strategic, and human intelligence, might help us understand what is going on in Earth orbit, and more importantly, *why and to what end*. SSA data does not provide simple 'hard facts' when SSA data can sometimes be proximate, but especially when the parties involved in this event have no common understanding of what constitutes appropriate behaviour. Technically gathered data can be ambiguous when it comes to detailed information of hardware, and is almost always ambiguous when it comes to divining the intent behind technically observed behaviour. Astropolitical and strategic analysis is needed to understand what is going on in Earth orbit, not solely technical SSA data.

The blurred distinctions between the intelligence, military, and civilian worlds common in other sectors of space activity are evident in space surveillance as well. SSA analytical capabilities are varied across many U.S. Government agencies, for example. In Operation Burnt Frost, the United States shot down one of its old, decaying satellites. The mission was just over a year after China's ASAT test of 2007. An Aegis-equipped U.S. Navy Destroyer—USS *Lake Erie*—'reprogrammed' an SM-3 missile to hit the target satellite in LEO as opposed to attempting to intercept a nuclear warhead on a suborbital trajectory in its traditional missile defence role. Whilst much preparation was done across several DoD entities and the Intelligence Community, NASA was also involved through its Orbital Debris Programme office. NASA analysts were able to bridge specific gaps in knowledge about orbital dynamics and the behaviour of debris in space. However, the number of NASA staff able to work on the project was limited, and those who could work often had to do so with highly abstracted or compartmentalised information, due to the lack of security clearances held by some NASA staff. The satellite intercept was a success, and many interpreted this as a direct-ascent ASAT mission with a kinetic kill vehicle, though the U.S. Government maintains it was a shoot down to prevent the satellite from surviving re-entry and spilling toxic fuel.[65]

The promise of using the Shuttle for a large number of military satellites never materialised, but it was used on occasion for military and intelligence missions. The Shuttle deployed one satellite for the DoD. It also, however, may have deployed the Prowler in 1990, which flew relatively close to target Soviet satellites in GEO to identify hardware and behaviour, and may have carried out a laser test as part of the downscaled SDI programme. Though the U.S. Government continues to deny the existence of the Prowler satellite, amateur observers have documented that its behaviour does not match that of a rocket body—which is how official U.S. SSA data classifies the object—but instead has behaved as an inspection satellite would.[66] The NRO was involved in the design requirements setting of the Shuttle, using the Hexagon IMINT satellite to determine the maximum required length of the Shuttle's cargo bay, and used the Shuttle for six missions between 1985 and 1992. There were also some studies to consider using Hexagon cameras on the Shuttle and use it as a crewed space reconnaissance vehicle. In the 1970s the envisioned use of the Shuttle for military and intelligence purposes was much greater than what actually happened, as the limitations of the Shuttle and over-promises over its mission frequencies became apparent in the 1980s.[67] NASA's cooperation with the DoD and Intelligence Community in a satellite shoot-down is rather typical, and should temper any U.S. criticisms of other countries' military-civil space agency connections and networks. This again demonstrates the porous nature of 'civilian' and 'military' distinctions in space.

There is a large amount of information that can be shared from sensors, but sensitivity in the U.S. military on SSA data effectively prevents much sharing of data. Should a private company create SSA data about a U.S. military satellite entirely through its own means, the fact that it holds data about U.S. military space systems acts as an electrocuting 'third rail' in efforts to share greater amounts of timely SSA data with external partners, especially non-U.S. entities.[68] The demand on SSA will only grow as more countries wish to attack or interfere with satellites, avoid or remove debris and junk, or want to put up megaconstellations of thousands of satellites each creating far more 'conjunction risks'—or potential collisions—in orbit. Any transparent and international Space Traffic Management (STM) regime will need a shared SSA database. This is not an idle concern—in 2009 an Iridium satellite collided with a defunct Russian satellite in LEO, creating a vast field of debris in a highly populated band of LEO. With more objects in space, these risks only increase and create a greater demand for SSA data and collision risk analysis. More infrastructure is being built in this, and the United States in the Biden administration has continued the work on STM policy started in the Trump administration. STM will be reliant on agreed SSA data and transparency, rules of the road or 'traffic regulations', as orbital locations are often a matter of opinion, not hard fact. STM will struggle compared to air traffic control as orbital space is not territorialised like 'national' airspaces. Creating an impartial and trusted traffic controller will be a supreme political, not technical, problem in arranging a global STM regime as its rulings and procedures will directly affect sensitive satellites and military operations.

SSA capabilities have improved within states, and they have also spread to more states and non-state actors. With more data and transparency about satellites and junk in orbit, the technical foundations are present for safer everyday operations but also for acquiring targets and planning nefarious behaviour around spy satellites. The largest challenges will be political, rather than technical, in sharing SSA data that manages orbital infrastructures as everyone wants a relatively safe environment but without restricting their freedom to operate and deploy systems as they wish.

Guided by Navstars

The original sin of space technology has provided one of the most prominent, prevalent, and pervasive elements of global infrastructures today: Global Navigation Satellite Systems (GNSS). Perhaps no other space technology encapsulates the original sin better that has made its presence felt in everyday life for billions. GNSS not only guide your evening takeaway dinner to your home, help you find a love interest through location-based software and mobile devices, track wards and vulnerable people through smartphones, or allow a pest-scanning drone to find its way around a crop field by itself, but also allow military forces to bomb with increased accuracy—including for nuclear war-fighting—and synchronise with other units in real time across continents and oceans. GNSS makes military forces far more efficient and responsive at their work, a major feature of 'force enhancement' through military systems. Rather than bombing an entire city district to hit one target building, specific buildings can now be targeted by linking GNSS services with guided munitions after mapping the target with space-based IMINT and SIGINT. Desert manoeuvres become a routine affair with GNSS systems, ensuring military forces rarely get lost so long as the signal holds up.

The most well-known GNSS is the USA's Global Positioning System (GPS), which traces its origins back to parallel programmes by the U.S. Navy (Navstar) and USAF (Timation) in the 1970s, and Transit in the 1960s prior to those. The very first Transit satellite was launched in September 1959, less than a year after its proposal was filed and funding approved.[69] The first Navstar GPS satellite was launched in 1977, and the constellations did not reach full operational status until the mid-1990s.[70] The Soviet Union developed the GLONASS system, a GPS equivalent that became operational in the 1990s as well but fell into disrepair as the Russian economy collapsed. Today, China has deployed the BeiDou system and the EU has fielded the Galileo constellation which are only in the 2020s coming online with full operational status. With at least four GNSS satellites of the same constellation in view from an observer on Earth, a Position, Navigation, and Timing (PNT) signal can be received by the appropriate receiver device on Earth. GNSS technology uses atomic clocks and the laws of relativity to provide accurate positioning signals across Earth, which today allow precision bombs, drones, autonomous vehicles, and military forces to find their way around on Earth with high degrees of accuracy and timeliness. As everything in the universe is in motion, mastering the relative nature of spacetime, including the fact that time flows differently depending on one's own velocity, is crucial to pinpointing a 'relative' position on Earth below a web of GNSS satellites. These signals also provide a universal time reference which the financial system depends upon, as well as the base signals for the navigation software applications for both military and commercial uses that have exploded in the twenty-first century as mobile IT systems have continued to miniaturise.

Before satellite navigation systems, the only methods of getting an accurate position reading beyond 'dead reckoning' by humans and handheld aids like sextants and ballistic measures were inertial mechanisms or terrestrial radio towers that would broadcast their locations to those equipped with the right receivers. The former were very useful in situations where radio signals were unavailable or too risky in giving away emissions to enemy ELINT sensors, whilst the latter are most useful in fixed infrastructure for permanent military bases, civil airport landing systems, and port navigation services. Fixed terrestrial systems like Enhanced Long-Range Navigation (eLORAN) only work in friendly or safe territory with pre-existing local infrastructure, whereas inertial systems do not retain accuracy like satellite systems do as they drift and gradually lose their bearings over time. Navigation signals from satellites help 'reset' or even surpass inertial guidance systems without ground or local infrastructure for a mobile user (e.g. special forces in a desert, a submarine in the mid-Atlantic). Inertial guidance systems use accelerometers to estimate its location as it moves, but these drift and become less accurate over time unless they are given a new external reference point to 'reset' their bearings. As a short term stop-gap, they can be a useful level of redundancy if GNSS is briefly disrupted. GNSS can surpass other navigation alternatives because they can provide positioning information without the need for local ground infrastructure. Like any satellite constellation a central ground infrastructure 'at home' is needed to manage it, but users in the field can access the service without terrestrial infrastructure and can manage with handheld or integrated receivers on vehicles.

GPS, Galileo, GLONASS, and BeiDou are all types of GNSS constellations, whereas GNSS is a type of PNT signals mostly based in MEO but also GEO in the case of BeiDou. There are other kinds of PNT services that can be based on Earth or LEO, but they cannot fully replicate GNSS capabilities. Transit was originally built to provide navigation data to U.S. SSBNs which would approach near the surface to receive a location signal which they would then use to update the inertial guidance systems on the boat. Nuclear missile submarines need to know their launch points to accurately target their missiles. Transit used the Doppler shift in radio broadcasts to provide positioning data, which is a different method to GNSS, and was in use until the mid-1990s when GPS replaced it.[71] Satelles is another service which is a system that is piggy-backed on the Iridium SATCOM system in LEO. Satelles provides signals suited for less accurate needs than GNSS users, but with increased signal strength that can penetrate buildings and shallow depths underground. As an augmentation for localised radio location systems, systems such as eLORAN use fixed radio towers or buildings to broadcast their positions to local vehicles.

The original LORAN system was used to try to update the inertial guidance systems of passing SSBNs by the United States, but the use of foreign shores to provide this data to Polaris SSBNs caused open political spats. Space-based navigation systems such as Transit and later GPS avoided such political difficulties.[72]

Like all technological devices and systems, GPS is also the product of social and political processes, not merely the outcome of 'rational' or objective technological optimisation processes. GPS had to suit the needs of a multitude of users across the U.S. military, government, and industry to secure funding, not unlike the Space Shuttle. However, unlike the Shuttle, GPS went on to become a great programmatic success and became 'black-boxed' as a technology that simply 'happened' and 'worked' in the minds of most users and decision-makers, and was not up for further political and financial debate as it became so useful in the 1990s.[73] Yet GPS had many opponents in its development, which may be hard to imagine given its proven importance, its positive public image and role in critical military, economic, and state infrastructure today, and its infiltration into everyone's lives with location-based 'downstream' applications. GPS had a secondary mission after providing PNT services—to locate nuclear detonations anywhere on the globe with high accuracy designed to enable a 'look-shoot-look' approach to nuclear warfighting. This approach and use of GPS was resisted by some alongside a general opposition to nuclear warfighting strategies as opposed to those that emphasised the 'deterrent' effect of nuclear weapons.[74] These were political, economic, and strategic debates over nuclear strategy which then impacted the technical designs of GPS.

GPS is in part a manifestation of the social imperative of extreme accuracy favoured by certain military-technical communities and industrial interests in the United States. Donald MacKenzie's landmark work on accuracy in nuclear missiles dispels the often-held notion that greater accuracies in weapons is a 'natural' trajectory of technological development. In fact it is an outcome of socio-political processes and subjective human choices. Technological knowledge and development are outcomes and continuing acts of sociological and political interactions, interests, and relationships, not simply the discovery and dissemination of objective 'hard facts'.[75]

Technological development is about choice, persuasion, and compromise, not fate. When dealing with nuclear weapons and threatening cities with nuclear fire, missing a target by a kilometre is not a problem. The initial view of the U.S. Navy and USAF's nuclear war strategies were to annihilate cities, meaning accuracies measured in a few nautical miles were sufficient. Much of the initial drive for more accurate nuclear bombs with gyroscopic navigation were pushed by a single engineer 'who was obsessed with the idea' and convinced the military services to continue funding his research.[76] Eventually nuclear weapons gained metre-level accuracy. GPS enabled what some call an excessive amount of precision for nuclear weapons, relative to scale: 'the raw violence of a nuclear explosion is delivered with the precision… of a surgeon performing the most delicate of operations.'[77] A drive for greater accuracies in nuclear weapons emerged in the 1960s partly due to the USAF's response to the U.S. Navy's city-busting nuclear strategy which looked to take over the entire nuclear mission. USAF therefore focused on more accurate 'counterforce' aspects of nuclear warfare—destroying specific military installations rather than cities—to protect the nuclear weapons programmes and budgets in the USAF.

Reminiscent of the RAF's death blow against the Blue Streak SLV, this inter-service struggle over budgeting and control over nuclear weapons led to debates in U.S. nuclear communities over whether accurate nuclear munitions were necessary to engage in 'counterforce' strategies. GPS was a later participant in the quest for accurate 'counterforce' nuclear warfare in the inter-service competition in the Pentagon.[78] It was not developed intentionally for non-nuclear war or civil infrastructure purposes. Yet it is these latter areas where GPS and now the international GNSS industry has had a profound impact, again demonstrating technology's ability to generate forces and impacts beyond the original subjective intentions of its human designers and political-economic advocates, but then in turn seized upon and exploited by people to create new political and economic coalitions within bureaucracies and society as the infrastructure matured and entrenched.

By the end of the 1960s the effort to consolidate Navy and Air Force navigation programmes meant finding a compromise between the perceived needs and budgets. Finding that agreement 'on the design and deployment of a single shared system produced what Washington does best—multi-layered committees, subtle infighting, budget wrangling, secret meetings, and afterward, differing accounts of what was accomplished and who should get credit.'[79] Whilst the Air Force wanted a counterforce targeting capability to fend off the Navy's potential to become the sole nuclear weapons service, the Navy later became interested in more precise navigation for its ships and boats. After the Vietnam War, proponents of GPS used the effects of laser-guided precision weapons in Vietnam to secure funding and development.[80] Replicating such precision effects at a distance without needing close-up laser pointers would make conventional missile arsenals far more efficient. Such promises of precision navigation and targeting capabilities would be realised through tests in the 1980s and 'operational demonstration' in the 1991 Gulf War, showing how GPS could satisfy many different battlefield needs. According to Scott Pace et al., writing in 1995:

> although GPS can support U.S. and allied military activities, it can at the same time create a dependency. Furthermore, enemy uses of GPS can threaten U.S. forces and broader security interests. This dual aspect of GPS—its utility in American and allied hands, along with the risks of dependency and enemy use—highlights a fundamental dilemma for decisionmakers seeking to maximize the benefits of GPS technology while minimizing its risks.[81]

That dependency yielded benefits to conventional military capabilities and economic services that were judged to be worth it.

GNSS systems can provide many kinds of navigation signals for different users. The two most important classes are: (1) signals for military or approved users with specific receivers which enjoy a far more accurate, jam-resistant, and encrypted signal; and (2) freely available 'civil' signals that can be accessed by commercially developed and sold receivers with a weaker and less- or un-encrypted navigation signal. For GPS, the precise 'military' code was the P-Code, which is currently being replaced by the M-Code on the third generation of GPS satellites. The publicly and commercially available signal, Coarse/Acquisition, or CA, is the base signal that most people use every day and features a range of terrestrial signal augmenters around Earth or in orbit to make the CA signal far more accurate than the base signal for civilian users.

The surprising accuracy of GPS signals during testing in the 1980s meant that early on the CA signal was intentionally degraded out of fears of what hostile actors could do with it if they gained access to receivers.[82] The CA signal in early testing impressed everyone with an accuracy of 20–30 metres, exceeding the expectations of a 100m accuracy. To manage security concerns but still provide a useful service for commercial and private users—such as aircraft, railways, and trucks—Selective Availability (SA) was developed so that the U.S. military could 'activate' SA during a conflict or crisis, effectively degrading the accuracy of the CA signal on the entire GPS network. Commercial GPS receivers in the 2000s could achieve accuracies of 100 metres with SA on, but could achieve an impressive 15 metres accuracy with SA turned off. With local signal augmentation ground infrastructure, that could be further improved to 3–5 metres accuracy. This meant that in a time of conflict, or any time of their choosing, the U.S. could decide to degrade CA so that anyone with a commercial GPS receiver could not use it to guide munitions or any civilian vehicle with precise accuracy or synchronisation. It would also create havoc for civil or commercial users that needed accuracies greater than 100 metres.

There was a political snag to this. The U.S. promised only to switch SA on in the most dire of circumstances but left that undefined. The reduced accuracy when SA was turned on was simply not good enough for aircraft and a range of other important critical infrastructure uses, and therefore many countries did not take too kindly to the risk of the SA mechanism being turned on. This in part led to the development of the European Union's Galileo GNSS, which would ultimately provide Europe with guaranteed precise GNSS signals no matter the will of the United States in a crisis or war. In the 1990s, France decided not to implant commercial GPS receivers into its Apache cruise missiles due to fears they would be rendered unusable should Washington activate SA.[83] Though these fears turned to be unfounded when eventually France, along with other NATO members, gained access to the military P-Code receivers, it shows a distrust over the stated policies of the United States, and a French rationale for a 'European' GNSS independent of American control to guarantee PNT signals for the French state and armed forces. This also shows how a crucial part of large technological systems and infrastructures are the political institutions and governance mechanisms around them, not merely the actual technological devices themselves.[84] The EU wanted governance and control over its own critical infrastructure and emerging GNSS technology markets—not U.S. military-industrial domination. Today, Galileo and the European GNSS Agency (GSA) has formed the heart of a new EU Agency for the Space Programme because a higher level of 'European' space infrastructure has established itself and will impact all European industrial and commercial sectors, according to rules and standards set in Brussels, not Washington.

It is fitting that GPS's journey into commercial markets echoes that of inertial guidance systems: developed first for nuclear missiles in the 1950s which then found their way into civil aviation.[85] Civilian life has been changed by the information and navigation infrastructure from GNSS systems, not least with GPS which spearheaded the civilian applications of GNSS through the 1980s and fully into the 1990s. Indeed, 'what began as a classified Cold War military program had spawned a private sector GPS industry with a multitude of uses dwarfing those related to defense and a global market in the tens of billions of dollars.'[86] The success of GPS and other GNSS systems today is really the story of a convergence of many technologies and industrial infrastructures, such as the Internet, the web, the microprocessor, mobile phones, and remotely piloted or autonomous vehicles.[87]

Russia's GLONASS achieved full operational capability for the first time in 1995 with twenty-four operational satellites in MEO, but fell into a semi-operational status as the economic difficulties of the Russian Federation led to an inability to regularly replace GLONASS satellites at the end of their 3-year lifespans.[88] The collapse of the USSR and the financial crash of 1997 resulted in significant reduction in launch rates, hollowing out funding for launch facility maintenance and staff pay, and a specialist 'brain drain'.[89] By 1996, the launch rate fell to a quarter of what the USSR had achieved in the 1980s. The quality of Proton, a workhorse rocket, dipped with a series of failures, and the Mir space station almost prematurely ended its life in disaster were it not for the heroic actions of its cosmonauts.[90] Crucial elements of the Soviet space industry and launch facilities were distributed across the Commonwealth of

Independent States, most notably Ukraine and Kazakhstan. It would not be until well into the twenty-first century that Russia fully reconstituted and modernised the GLONASS system. Today, GLONASS accuracies are apparently comparable to those of GPS, unlike the first generation of GLONASS signals.[91] However, the rollout of the next generation GLONASS-K satellites seems to have stalled with only four launched since 2011. Given the high-technology import sanctions imposed on Russia since 2014 and exacerbated in 2022, it remains to be seen whether Russia will be able to keep upgrading GLONASS in the years to come.

China began its GNSS journey with the BeiDou navigation system as a research and development project in 2000, and after leaving the Galileo project it pursued its own full GNSS. BeiDou first began with a series of satellites in GEO, rather than the usual MEO, for initially (by GPS standards) inaccurate PNT services. The current generation of BeiDou, now fully deployed and operational, contains twenty satellites in GEO but also a more traditional constellation of thirty satellites in MEO. This combination of satellite locations and signal types allows for comparable accuracies for civilian and military users, with civilian BeiDou signals now matching the accuracy of augmented GPS signals from constellations such as Wide Area Augmentation System (WAAS), the European Geostationary Navigation Overlay Service (EGNOS), and Quasi-Zenith Satellite System (QZSS). Even when proven in concept and enjoying solid political and financial support, complex technologies and infrastructure take time to build, roll out, and adopt. WAAS, EGNOS, and QZSS generate their own signals to complement or 'overlay' the signals from the GPS constellation which makes civilian and commercial positioning data more accurate than the GPS satellite signals alone over a fixed region on Earth as they are usually located in GEO.

China's ability to field its own precision weapons through its own PNT infrastructure and its fleet of increasingly real-time ISR satellites is a far cry from the 'unforgettable humiliation' of the Taiwan Crisis of 1996, where missiles equipped with commercial GPS receivers fired by the PLA went astray after alleged U.S. GPS signal interference.[92] BeiDou-equipped munitions and vehicles will be able to accurately bombard detected targets in possible war scenarios across its periphery in line with 30 years of precision warfare as demonstrated by the United States and other NATO militaries.[93] However, BeiDou should not be reduced to its military uses, although it is an extremely important aspect of China's military modernisation. As early as 2007 the relevant Chinese government ministries 'had established [the] satellite navigation industry as a national strategic high-tech industry', with several strategies on BeiDou adoption in the civilian economy being developed in the 2010s. Yet this industry has struggled to take off on a commercial level outside of China.[94] Nevertheless, the infrastructure has been completed and Chinese industrial standards will drive an uptake in BeiDou devices in the domestic Chinese economy and can enjoy some 'latecomer' advantages in GNSS infrastructure and use development.[95] BeiDou is the subject of numerous agreements with Brunei, Laos, Pakistan, and Thailand which involve ground stations and the use of military and civilian receivers, improving the PLA's and friendly militaries' precision strike and power projection capabilities.[96]

India has deployed NAVIC (Navigation with Indian Constellation) which provides regional navigation services across South Asia and the Indian Ocean through eight satellites in GEO. Such a regional PNT service is not as accurate as a typical GNSS service, but it serves particular navigation and cartographic needs, as well as an evolutionary industrial and technological step towards a full GNSS capability, like the first generation of China's BeiDou. Additionally, India provides the GPS-Aided GEO Augmented Navigation (GAGAN) system that enhances the CA GPS signals in India, like the EU's EGNOS and Japan's QZSS.[97] Going further, India has also secured a position as a partner in GLONASS by launching GLONASS satellites on Indian rockets, fabricating low-weight versions of GLONASS augmentation satellites, sharing signals, and developing downstream GLONASS-specific applications in the Indian market.[98] This partnership should be taken seriously given the sensitivity of military-grade navigation technologies. It may be difficult to imagine the USA launching GPS satellites outside of its trusted military-industrial complex providers, let alone a foreign government provider.[99] As a result, India has secured a relatively secure place with two major GNSS infrastructures and its own regional variant, ensuring that in a time of crisis it is probably unlikely to lose access to all PNT services at the same time. This prevents the Indian economy and military from losing all twenty-first-century efficiency gains and provides a greater form of autonomy without developing a sovereign GNSS from scratch, with the associated costs.

Japan is expanding its QZSS which augments the CA GPS signal to work better in Japan's high-density, high-rise cities which block unboosted civilian GPS signals from reaching the narrow streets below.[100] Japan's close military relationship with the United States ensures a reliance on GPS and its military communications satellite network without much need for an alternate military PNT system. In recent years Japan's space policy has allowed military interests to invest in and use Japanese space capabilities and space industry, as well as directed data-gathering and sharing with other less-developed countries across Asia with Earth observation satellites to arguably 'compete' with 'practical' and hard Chinese spacepower in the region.[101] Whilst Japan will be reliant on GPS, India could enjoy a status as a more independent spacepower in terms of navigation services in a time of crisis or war. Like China and India, Japan shares economic interests in developing navigation technologies. The 'installed base' of the GNSS market—the number of civilian/commercial devices currently in use that use GNSS and augmentation systems—is presently dominated by Asia. In 2019, the EU estimated that 53% of Earth's entire GNSS device numbers were in Asia alone.[102]

With Galileo, the EU is making good on its long-term ambitions to develop a more autonomous Europe in the areas of technology, security, and defence.[103] Galileo is a GNSS constellation, fully operational today with thirty satellites in MEO. Like GPS, it broadcasts a number of PNT signals and services. Proposals emerged in the 1980s, and the system first began providing signals in 2016 following a development time that matched GPS.[104] An interesting political contrast between Galileo and GPS was that Galileo's troubled development was public and in the open due to its civilian and international nature, whilst GPS's fractious debates were kept behind the closed doors of the Pentagon. As with GPS, Galileo faced questions over cost, necessity in the face of alternatives, and interstate competition over resources and constraints (as opposed to inter-service fighting in the U.S.). A major sticking point on Galileo was that it was led by ESA at first, rather than the EU. ESA, being a mostly civilian-orientated agency, chafed at the fact that the most ardent drivers of Galileo were also NATO members that saw 'security' benefits to the system—a euphemism for military advantages.[105] Unlike GPS, Galileo is a civilian-controlled system, but nonetheless had a clear military utility, continuing the blurred civil-military distinctions in many satellite systems.[106]

A dispute between the EU and USA came to a head in 2002–4 over Galileo's signals and their ability to co-exist with GPS signals. Despite the provision of CA GPS signals to civilian users and P-Code user access to NATO militaries, and an EU GPS augmenter in the European Geostationary Navigation Overlay Service (EGNOS), many European states wished to not be solely reliant on the U.S. for GNSS infrastructure. The United States terminated the SA policy during the final months of the Clinton administration in May 2000, meaning that Washington promised to never degrade the civilian GPS signal across the whole network. Yet the new military approach to ensuring enemy forces did not have access to accurate GNSS signals was to jam all non-military GPS signals in a localised area with deployed forces in-theatre. This could still affect central, eastern, and southern Europe in a Balkan crisis, for example. Initially, the Galileo project had selected a frequency range that was extremely close to that of GPS, meaning that if the U.S. jammed Galileo they would also be jamming GPS.[107]

Sheng-Chih Wang summarises the dispute:

> the US asked Europe to continue depending on the GPS and opposed the Galileo system because Galileo would cause disastrous signal interference during US military operations, break the US monopoly in the market of satellite navigation, and invalidate the US efforts of preventing navigation data from being used against the US. However, the US would not guarantee the quality of GPS signals if the provision of signals undermined its interests. In this context, developing an independent satellite navigation system, rather than continuously depending on GPS services, was more cost-effective for Europe in protecting its own interests.[108]

Not only could a European GNSS provide degrees of strategic and operational autonomy for European military forces, but also provide the EU with a more powerful role in setting global industrial standards in the GNSS sector, something that had been dominated by U.S. industry. The EGNOS system, designed to augment GPS CA signals for use in civil infrastructure in Europe through GEO satellite signals, proved the initial navigation technology expertise in Europe, a pre-requisite to any ambitious GNSS project. The EU was keen to build from this toehold in satellite navigation technology and move into full-scale GNSS capability. Scott Beidleman, writing in 2004, strikes a trenchant and indignant tone in describing the situation:

> Considering that GPS has become a global public good, an international utility paid for by the United States and free for use by anyone, and that most of Western Europe has been a staunch American ally since the Second World War, Europe's pursuit of the Galileo GNSS could be taken as heresy from an American perspective. Europe has broken ranks... and is acquiring an independent space capability that seems sure to conflict with American national interests.[109]

A global public good or a large, global technological infrastructure becomes a tool of influence and a pathway of dependency if it is unilaterally controlled by one state. This was not lost on Brussels, especially for the EU's member states who remembered the Intelsat experience.

Early Chinese involvement in the Galileo project could be interpreted as a 'betrayal', too. This reflected Europe's role as a high-technology exporter to China at the time. China was also an attractive partner to have contribute to the project due to limited European funding for Galileo.[110] As soon as China became a participant in the design and building of Galileo in September 2003, 'US diplomats went on a full diplomatic offensive' and persuaded the EU to exclude China due to concerns over 'navigation warfare' with the People's Liberation Army in a future Taiwan Strait crisis or conflict.[111] The United States did not want Galileo to become a usable PNT service for the PLA.

The U.S. and the EU compromised in early 2004, despite Galileo being possibly the biggest discord in security and commercial interests between Europe and America since Ariane and the Shuttle.[112]

The bargain was struck: the U.S. managed to alter Galileo's signals so that they would not compete or interfere with GPS and ensured no further Chinese involvement, whilst the EU prevented a U.S. veto on future Galileo development, gained technical support, and secured a recognition of the military-grade Public Regulated Service (PRS) of Galileo as a peer to the GPS's military-grade P-Code signal. An important U.S. concession was that it agreed to give up complete autonomy in the management of GPS frequencies and signals, committing to consulting the EU and agreeing not to disturb Galileo with any future changes, as well as non-discrimination by the United States towards European companies in the GNSS marketplace, ending Washington's economic monopoly. Industries on both sides also pushed for compatibility between the two systems so IT systems, mobile devices, and their various components and chipsets could be easily sold and moved around the North American and European markets. As compatible and complementary systems, Galileo and GPS could become each other's backups for military services and critical infrastructure should one suffer massive failures through accident or hostile action.[113] In January 2021, the U.S. formalised such an approach in Space Policy Directive 7, which articulated a desire to pursue participation of the U.S. in appropriate GNSS systems as a resiliency measure, to avoid complete dependency on GPS for military PNT signals.[114] At present, Galileo is the only system that would meet such a goal.

The Galileo-GPS dispute highlights space as a 'political environment and so determining the distribution of economic, military and political power on Earth. Contrary to the assessments of several scholars' in International Relations there is 'evidence of significant and substantive attempts to "balance" against the power of the United States since the end of the Cold War.'[115] Discord is the norm, as opposed to wider expectations of cooperation within the same transatlantic security community; it often appears like a struggle for industrial and economic autonomy between Europe and the United States.[116]

As with launcher development and space access, Galileo can be seen as another area where a convergence of interests formed, or aligned, to realise an area of greater 'European' autonomy as stressed in multiple EU strategy and space policy documents throughout the

2000s. European industries', as well as the more 'strategic' goals of EU member states, were satisfied with the Galileo project, which led to the justification of the billions of euros spent on it. Indeed, as Sheng-Chih writes, 'after Airbus and the Ariane launcher, Galileo was the third large-scale aerospace program that aimed at relieving European overdependence.'[117] The economic and military benefits of such infrastructure were broad enough to satisfy many interests across the EU, and also contributed to the EU's 'balancing behaviour' against the United States' hegemony over many aspects of advanced economic and military technologies.[118]

Such an assessment directly clashes with those who claim that there is no balancing behaviour at all against the power preponderance of the United States in military and economic dimensions.[119] The EU and China's pursuit of GNSS may be substantive balancing activity against the United States. As GNSS systems provide dual-use capabilities, it is possible to interpret such balancing as hard and soft balancing against the U.S. Whilst the EU accepted various signals compromises, the U.S. lost complete autonomy in the management of its GPS signals. The EU's increasing dual-use space infrastructure may be at the vanguard of the devolution of military power,[120] where smaller forces can retain a high level of effectiveness against massed low-technology opponents, or fragmented powers abroad, thanks to space support and the qualitative advantages that it brings to high-technology military units capable of waging and resisting precision-strike warfare. Through common and shared space infrastructure, smaller economies can enhance their internal efficiencies and more readily convert their resource bases into further development or military power as needed, harking back to eighteenth-century Europe where calculations in power hinged on internal efficiencies rather than territorial and demographic gains.[121] An EU infrastructure provides advantages for the smaller economies of its member states in capitalising on global economic forces. Space integration in Europe should not be seen as merely an ancillary to the EU's attempts to seek autonomy and coherence. Spacepower and the EU's space policy seem to be at the heart of modernising the EU's economy, security, and intelligence capacities for the twenty-first centuries by pooling resources to compete with American, Chinese, and Russian spacepower.

Today, the GNSS sector is a multibillion dollar industry, particularly as commercial receivers and chips adapt to use the civilian signals of GPS, Galileo, GLONASS, and BeiDou through licensing and manufacturing rights for receiver technologies and the applications developed with them.[122] The feats and funds required to build GNSS constellations do stand out as significant because the space-based technology is not widely proliferated, and remains within the hands of four major entities: the USA, Russia, China, and the European Union. So far, regional and augmentation systems are restricted to the same entities, as well as India and Japan. The possession of a GNSS constellation may now be a prerequisite for joining the modern standards of the highest rungs of a global 'Space Club'. Whilst these systems provide a range of civil and military uses, they also require coordination and cooperation so that these different systems may coexist and complement each other as global utilities.[123] However, the terrestrial receivers of the satellites' signals are highly proliferated. The freely available civilian and commercial signals are increasingly used by multi-GNSS receiver chips, meaning a more accurate service from different systems is available around Earth, increasing resiliency and redundancy for civil infrastructure. However, the broadcasts of those signals are still controlled by the GNSS operator and are more easily jammed by local jammers than the restricted military-grade PNT signals they also broadcast, which are currently not interoperable.

Conclusion

Space technologies form the foundations of global, large technological infrastructures that modern states and societies rely on as critical infrastructure. Such infrastructure provides a wealth of services for military and intelligence benefits as well as commercial power, and therefore political institutions reflect their importance and become part of these technological systems, and vice versa. If they do not own such systems themselves, their economies, infrastructures, and militaries will depend on those owned by others. The efficiency gains of such services and capabilities described in Chapters 3 and 4 are difficult to resist today, though they each had to find political coalitions to be built. This is the heart of humanity's space age and its practical impact on international relations today: the exploitation of Earth orbit for war and development with mature, large technological systems, and less so for prestige and the exploration of the cosmos. Both in the development of panopticons and infrastructures, major powers and political blocs are trying to avoid total dependence on others in key areas of space technologies. Industrial capacity, economic competitiveness, and military power are being shaped by space technology's *maturation*—not its emergence as a novel trend.

Rooted in this maturation of competing technological infrastructures is a political reality of international anarchy where there is no forceful arbiter of right and wrong in the political world above the most powerful states of the international system that are able to enforce their will—the technopolitics of the Global Space Age. Therefore dependencies on critical infrastructure require high levels of trust and the acceptance that whichever state controls that infrastructure will enjoy a relatively stronger level of bargaining power in the relationship. Where that trust is not sufficient relative to the importance or opportunities provided by a particular infrastructure or capability, more sovereign or less dependent options may be pursued. Space technology, spacepower, and their infrastructures have therefore become an important factor in creating, enhancing, exploiting, and resisting technological power and political influence because they generate physical effects and opportunities that benefit some and disadvantage others. This makes satellites useful or dangerous enough to be potential targets in wartime and their industries' objectives in espionage and commercial competition.

The twists and turns of the maturation of space technologies as they met the needs of warfare was not a clear-cut path of technological 'progress' nor merely a story of rational policy making. Peter Hays echoes the work of Donald MacKenzie when he says that:

> cumulatively, these new and enhanced space systems... significantly increased the value of space systems in multiplying the combat effectiveness of terrestrial forces... These vastly improved force enhancement capabilities represented the beginning of a revolution in space and terrestrial military operations which was little under-

> stood in the late 1970s. Moreover, Air Force and general U.S. military space doctrine was not particularly clear or coherent in guiding the development of these force enhancement capabilities and was not the main driver behind the development of these very significant capabilities. Rather, relatively unfocused, bottom-up, and incremental technical improvements slowly established this growing capability. Unfortunately, the limited space doctrine discussions… did not provide any clear guidance on where these technological developments were or should be headed.[124]

Spacepower would have its first dramatic 'appearance' in the 1990s and the United States would show the way ahead for high-technology military power into the twenty-first century, and underpinned what some analysts called a Revolution in Military Affairs, or RMA. Space systems came to impact the course of military campaigns, battles, and tactics, expanding their influence from what they had already achieved in transforming 'strategic' intelligence, high-level diplomacy, and nuclear war plans. This makes spacepower an unavoidable but an unappreciated if not ignored factor in assessing and understanding global technological developments and military-economic power in the post-Cold War world. Military powers that could command space and use space technology with impunity became ever-more capable on the modern battlefield and upset methods of mechanised warfare that had been mastered in the 1940s and 1950s, destabilising and undermining the confidence of the large mechanised military powers of the late 1980s and 1990s. Deliberately using space for the 'full spectrum' of military combat and support missions would ensure that those satellites would be targeted in wartime and crisis situations, challenging the typical interpretation of space technology as 'stabilising' in international security.

PART III

STRATEGY IN THE GLOBAL SPACE AGE

5

SPACEPOWER AT WAR

Arfer yw mam pob meistrolaeth

Practice is the mother of all mastery

From the earliest years space technologies were intimately tied to the needs of fighting wars. SLVs, ICBMs, and the earliest military and intelligence satellites were to help fight a nuclear war should the order be given. As the Cold War progressed those technologies' later generations became more capable and relevant to non-nuclear wars, from conventional or major war fighting to counter-insurgency campaigns. The 1991 Gulf War, the so-called 'First Space War', was a watershed moment in demonstrating the potential of space-based communications, ISR, and navigation systems in influencing combat capabilities. The war and the U.S. military set a technological and doctrinal template for major military forces to follow or adapt against. As Washington led the way in the twilight years of the Cold War, the stunning battlefield successes of the U.S. military have led to imitations by other major military powers into the twenty-first century.

Not only was the 'strategic' level of war transformed by the space technology as it had brought about the Thermonuclear Revolution, but operations and tactics were now being shaped by satellite support services too. In U.S. military terminology, this is referred to as 'Force Enhancement', the 'tactical application of spacepower', or 'spacepower integration'. By the twenty-first century, U.S. satellites could network the individual trooper, vehicle, and short-range strike missile into a larger Command, Control, Communications, Computers, Intelligence, Surveillance, and Reconnaissance (C4ISR) network infrastructure, meaning more and more forward bases, vehicles, and weapons controlled by the military could be communicated with in almost real-time by the Pentagon and regional commanders. Deployed forces could draw upon communications and information from the centre as the needs of battle demanded.

Spacepower has made U.S. military forces far more efficient and effective on the ground, at sea, and in the air even as absolute force sizes have reduced since 1991. Space systems enhanced the combat effectiveness of aircraft, ships, ground vehicles, and munitions. They transformed cartography and coordination in unfamiliar environments and at much greater speeds. They guide bombs to their targets within metres or less, with almost a one-shot one-kill efficiency in peak conditions. They allow Special Forces higher levels of control and coordination with ever smaller, lighter equipment. They can update the airstrike missions of aircraft after they have taken off from their airfields and forewarn against range missile strikes to allow a military base's personnel to scatter or hunker down. Satellite imagery provides detailed post-strike damage images and can track heavy vehicle movements when shallow water tables are disturbed by a battalion of 40-ton tanks or spot the polluting exhausts of a convoy of pickup trucks. Radio chatter is always a risk as SIGINT satellites in addition to terrestrial SIGINT-equipped vehicles and installations are eavesdropping. As explained earlier, this does not make space-supported military forces omniscient as military intelligence needs to know to look at specific or likely areas to catch such things in the first place by 'tasking' space intelligence assets. During heightened tempos of military operations and channelling satellite resources to the right theatres, it increases the chances that such things will be caught and then forwarded to the necessary military units to make the most of that information on the operation or looming battle. Often, space systems will 'cue' aerial ISR assets that can take a closer and more persistent look after satellites have provided the first look or spotted something of potential interest.

This chapter traces U.S. military and intelligence space systems as they became 'integrated' with the requirements and platforms of the battlefield to create the modern ways of warfare that we are now accustomed to in the most advanced military forces. It then looks at how the RMA debates on military technology obscured what was going on in spacepower development and military space technologies throughout the 1990s and 2000s. Space is a distinct place—not simply an extension of airspace—and it needs dedicated strategic thought to be understood on its own terms. The RMA moniker drew attention towards IT systems and abstract concepts of 'information warfare', neglecting geostrategic and techno-strategic thought about outer space. At the backbone of such changes in military power in the RMA and 'defense transformation' years of the Pentagon was the continuing Global Space Age and the vanguard of efforts to use space for military and economic advantages in the anarchic international system. The chapter then moves to Russian and Chinese space-supported military modernisation, taking away the U.S. monopoly of spacepower-supported terrestrial forces. The chapter ends with a discussion of the risks involved with using space systems for nuclear war and non-nuclear war, and explores methods of resistance against space-supported military forces, as has been seen for a few decades.

The Road to the 'First Space War'

By the late 1970s, U.S. and Soviet military forces around the globe could better communicate with their political masters and other military units that were beyond their line of sight—or 'over the horizon'. They could also find, track, and target the enemy in far more efficient, timely, and accurate ways and began to track moving targets too. Satellites were weaving a web of connectivity through a diverse range of space-based platforms and terrestrial stations and receivers. Coupled with emerging PNT technologies and other navigation improvements, 'tactical' nuclear and conventional weapons were putting new premiums on the guidance, communications, and networking satellite systems underpinning it all. Soviet anti-ship operations were increasingly dependent on space-based sensors, helping to add to the rationales for the United States to develop anti-satellite missiles such as the Miniature Homing Vehicle.[1] By the early 1980s U.S. Navy satellite-cued missiles could hit targets beyond the horizon of their launching ships, again helping to justify Soviet anti-satellite weapons such as the Istrebitel Sputnikov (IS).[2]

Reagan's Strategic Defense Initiative (SDI) space-based missile defence proposals only increased Soviet desires to develop anti-satellite systems—it did not create the Soviet ASAT programme which long predated SDI. Satellites became more useful in the conduct of military operations, and therefore more of a potential target from the enemy to undermine over-the-horizon strike capabilities. The 'sensor-to-shooter' idea arrived long before the RMA debates of the 1990s, and was a mainstay concept in the U.S. Navy in the 1980s.[3] Anti-satellite weapons therefore should not be seen as relevant *only* in nuclear war fighting and in midcourse intercept ballistic missile defence functions, but also in conventional warfare that may not necessarily 'go nuclear' immediately, if at all. Silencing some satellites could meaningfully help tip the scales in some conventional operations in one's favour in a way that it does not for the mutual suicide pact of nuclear war.

Spacepower could help win battles (tactics) and enhance overall campaign planning and deployments that tie several battles together (operations). The U.S. called this the Tactical Exploitation of National Capabilities Program (TENCAP) which sought out ways to make 'strategic' space systems relevant in tactics and operations for non-nuclear forces. In 1986, TENCAP was an $86m project and described in one report as using space-derived 'intelligence' as a 'force multiplier'.[4] Technologies used for the 'strategic' level mean nuclear war fighting, top-level communications between government leaders and military headquarters and commands, and 'national intelligence'. The U.S. Army's work in TENCAP helped bring about many 1980s 'AirLand' battle concepts for fighting the Warsaw Pact: high degrees of flexibility, manoeuvre, and precise 'deep strikes' from long-range missiles, aircraft, and artillery against enemy positions, and military coordination across continents between air and land forces. It renewed Army interest in engaging with how the Air Force developed its space systems.[5] The Army's TENCAP was so successful in modernising its units and capabilities that it was emulated by the other services.[6]

Satellite-based intelligence is not singularly responsible for effective military intelligence for operational planning or 'national' level intelligence work. Satellite systems were used in conjunction with other terrestrial technical systems—such as undersea sensors—as well as HUMINT agents and subjective net analysis of Soviet naval doctrine to provide the U.S. Navy with a credible method of tracking the Soviet Navy. This 'fusion' of intelligence in the U.S. Navy:

> that allowed the Maritime Strategy's aggressive focus upon [anti-submarine warfare] in Soviet home waters was thus made possible in part by the US Navy's high-technology integration of command, control, computers and intelligence (C3I) systems to create a satellite-linked, highly-informed hunter-killer network.[7]

By the late 1980s the U.S. Navy, with NRO cooperation, was effective at tracking the Soviet Navy and ready to cue its long-range weapons against their vessels thanks to the mix of technological intelligence-gathering infrastructures and HUMINT assessing Soviet naval doctrine. Space was instrumental in the 1980s drive in the United States to develop more computer network-oriented weapons systems.[8] Dwayne Day explains that:

> for over half a century the United States has fielded increasingly capable systems for surveilling the oceans, bringing the data they collect directly down to the captains of the ships that most need it, and into the electronic brains of the missiles they fire.[9]

These new ways of making satellites useful for individual weapon systems and platforms for battles and skirmishes were novel and experimental in the Cold War. Though many of the satellites themselves were proven, the information they provided and the way they were tasked were not seamlessly integrated into a smooth sensor-to-shooter cycle. Superpower military and intelligence space infrastructure was in large part originally designed for nuclear war fighting and 'national intelligence' gathering with large, strategic perspectives and gathered less time-sensitive information. They were not built for extremely high-resolution, small-scale, real-time information for individual military deployments on conventional or other missions that they may conduct. Tactical ISR can be outdated even if it is a few minutes too slow reaching troops on the ground. Pat Norris writes that whilst:

> satellites are excellent at giving the big picture summary of… strategic [nuclear and missile] forces, they are less good at giving black and white answers in one particular scene at one particular time.

> Part of the process of gauging the strategic picture is to examine the changes in a facility or location over time—weeks and months, rather than hours or a day. The tactical requirements of forces engaged in regional wars are much more immediate and explicit: Is there a tank ahead of me? Is there an enemy soldier in that trench?[10]

In Operations Desert Shield and Storm, it was difficult to get the right data from space systems to the right users at the right time. Space reconnaissance systems were the major source of intelligence gathering on Iraq, particularly during Desert Shield.[11] The emphasis on space systems communications and data throughputs were focused on speed and bandwidth in transmitting raw data rather than thinking through who needs what specific information and 'analytical products' or the relevant kind of intelligence assessment the most. The sensitivity of IMINT and SIGINT also means that much cannot be placed on less secure but bigger bandwidths. Mann writes that 'once the specific need had been identified to the "right" people in the system… delivery was nearly immediate'. To use a simple parable, having a technically efficient email system is one thing, but knowing who to email and what information to include is more important to getting things done. Mann also explains that:

> intelligence delivered 'tons' of information as fast as possible… while operations wanted specific 'pounds' of it delivered much more quickly than the system was capable of. Operations planners, unable to get a satisfactory [imagery] resolution within the intelligence system, resorted to unofficial workarounds and informal arrangements outside the system.[12]

Mobile, miniaturised computers are useless if they have only a trickle of data to process and disseminate or they provide useless data to the wrong people.

Military bureaucracy shifted in Washington to better coordinate its maturing space infrastructure, especially after the Goldwater-Nichols Act of 1986. Yet efforts were underway before then. Set up in 1982, USAF Space Command focused a drive on 'operational' space support to military forces, and was learning the lessons on tactical space applications following the Vietnam, Falklands, Grenada, Libya, and Panama wars, conflicts, and interventions throughout the decade.[13] This led a small cadre of space enthusiasts and advocates within USAF to generate debate and ideas on space doctrine within the service into the late 1980s, with David Lupton's doctrinal treatise *On Space Warfare* remaining a useful product of that cadre to this day in military space education.[14] It was reported in U.S. media at the time that the United States had supplied space-based IMINT to Iraq during the Iran-Iraq War, including information that helped Iraq analyse the damage of their airstrikes and plan more rapidly for successive, more precise missions.[15] This not only provided more insight into the U.S. Intelligence Community about the impact of space systems on operations, but also showed how good U.S. spy satellites and their surrounding agencies had become and how they could impact fortunes on the battlefield. Today's apparent space-based intelligence support provided from NATO countries to Ukraine in their repulsion of the Russian invasion has a long precedent and is hardly novel in political or strategic terms. Not only does it allow Ukrainian forces to hide and avoid incoming Russian air and missile strikes, but also to see land force movements and build-ups and perhaps enable long-range Ukrainian air and missile targeting of Russian logistical targets across the border.

Despite such efforts, the 1991 Gulf War confirmed that spacepower was unprepared to support the theatre Commander in Chief beyond the well-established nuclear war requirements. There was a lack of space doctrine and space literacy across the entire U.S. military, including the bulk of USAF, despite the work of Lupton and associates.[16] The Gulf War is sometimes dubbed the 'First Information War' due to the importance of IT networks in the weapon and logistics systems involved. Naturally, both these labels imply that neither spacepower nor information were used or important in wars or military strategy before then, which is very much incorrect. Such labels are statements about relative effects and capability, as well as future promise, particularly in retrospect as space and IT systems took off in enhancing U.S. and allied military forces through the post-Cold War years. Spacepower's use in the Persian Gulf was a small-scale, real-world demonstration of technological potential that was known to some specialists on both sides of the Cold War for some time. Effective systems take many, many years to build and develop, and then integrate into other, larger existing systems and modus operandi. Often organisations want new technologies to help them do what they already do better, rather than completely change the way they do things. That is one way to understand how spacepower *intensified* existing military trends in more precise and longer-range strike or bombardment weapons. Spacepower led to more elaborate and capable C4ISR systems to improve existing trends towards longer-range weapon systems, more manoeuvrability, professionalisation, and reducing military force sizes.

In Desert Storm, satellite-guided precision munitions only accounted for 7% of all U.S. ordnance expended, but the potential of these now battle-tested experimental weapons and guidance systems was clear to USAF. In the Bosnian War in 1995 such munitions use rose to 60%, and in the Kosovo war it rose again to 90% as more ground equipment was modified to receive space-based data and services.[17] As the technology matured after 1991, any vehicle, building, or heavy equipment became a liability to the lives of its crew and anyone near it when the U.S. commanded the skies and Earth orbit. In Iraq 2003, many tank and artillery crews simply abandoned their vehicles to avoid being killed by the terrifying precision strikes such obvious equipment invites. Importantly, not only did an emerging one-shot one-kill capability make military forces more efficient, but also allowed for more discrimination in targeting, especially in urban environments. Radio and other electronic emissions from the Iraqi leadership allowed the U.S. military to identify and track the movements of the Iraqi leadership, and also strike critical targets. Specific buildings could be struck with one missile without needing to carpet-bomb an entire district. Being precise and more efficient with sorties and bombing missions meant that mission ratios turned into how many targets could be hit per sortie, rather than how many sorties it took to hit the target.[18] Airstrikes became less likely to commit war crimes through indiscriminate bombing, though precision bombing has not eliminated inadvertently killing innocent bystanders and 'non-combatants', or what is euphemistically called 'collateral damage'. In the Al Firdos bombing in Baghdad in 1991, 400 civilians or non-combatants were killed, leading to the requirement for top-level military approval for 'precision' strikes in Baghdad.[19] Civilian casualty estimates for the Gulf War number in the hundreds of thousands, whilst Iraqi military casualties ranged from 60,000 to 200,000.[20] In the Kosovo War a

GPS-guided Joint Direct Attack Munition dropped from a B-2 bomber hit the Chinese embassy in Belgrade, triggering a significant diplomatic incident.

Despite the United States' continuous development and deployment of successive SATCOMs for government agencies and the military services through the Cold War, the satellite infrastructure barely coped with the information demands of deployed forces. Much additional communications bandwidth had to be requisitioned from commercial and civilian SATCOM constellations.[21] The lack of electronic warfare attacks by the Iraqi Army was perhaps a critical error on their part, as Coalition forces relied heavily on a specific tactical communication system which had particular vulnerabilities to jamming techniques that were not beyond the capabilities of Iraqi forces to exploit.[22] With radiofrequency tethers connecting military forces and C4ISR systems together, the premium on electronic warfare against space systems has only increased since then. Between the 1991 and 2003 Gulf Wars, the U.S. military's dependency on SATCOMs for bandwidth increased by 30%. For Operation Enduring Freedom (2001) and Operation Iraqi Freedom (2003) 80% of the bandwidth was provided by commercial satellites. Military SATCOM, contrary to commercial bandwidth, provides an essential or minimum guaranteed communications service for the military with the highest degrees of anti-jamming technologies, encryption, and survivable hardware designed to work in a nuclear war (short of being directly caught in the fireball of an orbital nuclear detonation).[23]

The U.S. Navy was the single-heaviest user of satellite systems for navigation, targeting, reconnaissance, and communications in the Pentagon for much of the Cold War and the 1990s. Their communications satellites were mostly in LEO, and more vulnerable to Soviet anti-satellite weapons.[24] The tactical and operational benefits for deployed forces became only more apparent as miniaturised computer processing power proceeded along Moore's Law of doubling the number of transistors fitted onto a microchip every 18 months. This meant that the computers needed to plan missile strike missions could be placed on ships, rather than left at headquarters an ocean or two away. This reduced the bandwidth consumption between HQ and deployed forces if heavy processing could be done in-theatre, and also increased redundancies should long-haul communications be cut. Still, demands on bandwidth were high. Secure and high bandwidth SATCOM was the only method to reliably get the raw data needed for local computers to process and provide target and mission profile data for cruise missiles and airstrike sorties of U.S. Navy vessels, wherever they were. Cruise missiles such as the Tomahawk could rely on the Terrain Comparison (TERCOM) system that relied on imagery databases from the NRO loaded onto the cruise missile's on-board computer which matched the terrain the missile was flying over with what was in the computer memory. 1980s computers had become smaller and more powerful to the point that ship-board computers could effectively programme such missiles in less than an hour as opposed to a day or two. This dramatically sped up the process of acquiring targets and launching missiles and aircraft at them, making weapons more rapid and lethal, but increasing the burden on dispersed space communications networks.[25] Today, this problem may only be getting more acute as the raw data available, even for land forces which tend to require less data and IT systems, continually exceeds available communications bandwidth in U.S. and UK forces.[26]

Much of spacepower's integration in 1990–91 had been ad hoc, as 'much of what they did from August through February had not even been dreamed of in July'.[27] The communications system supporting the Coalition could barely keep up with demand as jury-rigged terrestrial gateways for SATCOMs were hastily set up across the Middle East in the latter half of 1990.[28] The U.S. military was fortunate that it had six months in a relatively safe and benign environment to set up its expeditionary infrastructure in Saudi Arabia and the wider region. Iraq chose not to roll into Saudi Arabia at this moment of vulnerability for the first U.S. deployments. This included altering some satellite orbits and antenna directions, and leasing the necessary commercial platforms and terminals, as well as grabbing as many commercial GPS receivers as could be built and shipped out. Crucially, spacepower sped up cartography for military engineers, as Cynthia McKinley writes:

> We have no maps! When coalition forces were deployed to the Persian Gulf region, the maps of Kuwait, Iraq, and Saudi Arabia

> were old and out of date... Desert Shield and Desert Storm engineers had valuable data that enabled plans for military airfield construction; Marines knew which areas were best for amphibious assault; land forces could monitor enemy operations; and air attackers could examine attack routes, verify targeting coordinates, and identify potential landing zones.[29]

The Soviet Union was rather dismayed at the ease with which Coalition forces could detect and strike Iraq's mostly Soviet-equipped mechanised military force, but they were pleased at their achievements in camouflaging techniques.[30] The Iraqi Army scored notable deception and manoeuvre successes with the Scud missile force.

Iraqi forces were able to 'shoot and scoot' their Scud missiles and escape destruction from the 1,500 USAF air sorties sent after them. Spacepower allowed sufficient warning times of their launches to be provided.[31] Many ground, space, and aerial ISR advances since 1991 may make mobile missile launchers easier to detect for the United States today than in 1991. However, it is important to not see future conditions as the same as 1991:

> The Scud hunt is both a misleading analogy and a distant data point from a technology perspective. As an analogy, the Scud hunt bears little to no resemblance to either the effort to track Soviet mobile ICBMs during the Cold War or future mobile missile scenarios.[32]

Unlike Iraq, China and Russia have much more ground to use with their mobile nuclear forces, making the work of ISR systems far more taxing. Concealment and deception against enemy sensors from above are only increasing in importance for the survivability of modern military forces from long-range precision strikes against nuclear arsenals, and may help explain the 'shell game' of the recent expansion of China's ICBM silo numbers. A 'shell game' refers to using a large number of missile silos, many of which may be 'empty', to give the enemy more targets to hit and therefore decrease the chances that all actual nuclear missiles will be destroyed before launch. Still, deception is extremely important when the enemy is desperate to find any technical trace of your activities. That in turn can be exploited: Serbian forces built wooden MiG-29s as decoys which NATO aircraft 'dutifully bombed', showing significant limitations in finding appropriate targets from technical ISR networks.[33]

Once a sensor's workings are figured out, they can be fooled with deceptive measures or given the slip. The exact details of methods of the cat-and-mouse game of sensors and fooling sensors will change, but intelligence opponents will play that game and try to win it. Complacency by any side will cause problems for themselves.

The implications of spacepower for modern warfare were not restricted to precision munitions. In 1991, GPS guided the U.S. Army's VII Corps to storm through the Saudi-Iraqi desert in a daring 'left hook' manoeuvre into Iraq's exposed western flanks—without getting lost. GPS supported a Special Forces and U.S. Army helicopter gunship sneak attack on Iraq's two air-defence radars which were 40 miles apart but destroyed within 20 seconds of each other.[34] Iraq's command and control network was systematically picked apart, so that the Iraqi Army could not see, hear and talk whilst Coalition forces could.[35] Space infrastructure helped USAF bombers and tactical support aircraft to take off without needing to pre-determine their targets and could be 'tasked' whilst *en route* with up-to-date information. This led to a far more intense and concentrated bombing campaign across Iraq, a level of intensity and scope that simply was not expected. One Iraqi prisoner claimed that 'thirty minutes of coalition bombing was worse than what he experienced in eight years of war with Iran.'[36] Within 100 hours the entire Iraqi air-defence system had been destroyed, and coalition airpower could begin to strike obvious targets with impunity.

'Tank plinking' had shown how lethal PGMs could be towards tanks, trucks, and artillery pieces with an almost one-shot one-kill capability.[37] As Iraqi forces retreated from Kuwait, the main road back became a deathtrap thanks to the rapidity with which coalition forces could detect them and send in airstrikes. This infamous 'Highway of Death' and the Bubiyan 'turkey shoot' rout of the Iraqi Navy episodes showed how vulnerable massed enemy forces can be when the enemy has sufficient ISR and air attack systems to hand. A notion of 'overkill' crept into operations as the main road between Kuwait and Iraq became:

> littered with the burnt-out vehicles and charred corpses of those who had been trying desperately to get out of the Kuwaiti theatre of operations. Footage from Apache helicopters [showed them engaging] individual soldiers running away from burnt vehicles

> with 30mm cannon designed to destroy light armour. These factors persuaded President Bush to take the decision to suspend the fighting… at the 100-hour mark, to prevent excessive Iraqi casualties, which would inevitably have generated international condemnation and turned the memory of a famous victory into a seriously unequal contest.[38]

The communications infrastructure established years beforehand allowed the White House and Pentagon unprecedented access to U.S. Central Command (USCENTCOM) leaders, and Cable News Network (CNN) broadcast the war as it happened to the living rooms of millions around the world to a degree never seen in the Vietnam War. Media management was also a concern, especially as there was the risk that the press could potentially access commercial or foreign imagery of the theatre of operations, and publish them. Indeed, French and Soviet imagery at the time could have revealed the 'Hail Mary' flank in the desert, the linchpin of the main offensive.[39] U.S. imagery revealed the true disposition of Iraqi troops and allowed CENTCOM to conduct the flank through the desert with high degrees of confidence.[40]

Fortunately for the United States, the Soviet Union and France condemned the Iraqi invasion of Kuwait and cut off their space information to Iraq or sources that may pass on such information to Iraq, and 'left the United States in possession of a temporary monopoly on the ability to routinely and unobtrusively probe the enemy's battlefield with highly accurate reconnaissance satellites.'[41] However, Gorbachev did try to prevent U.S. intervention, but failed and decided the relationship with the U.S. was more important than that with Iraq.[42] Doing so challenged some 'open access' policies in some civil satellite systems and organisations. Cynthia McKinley commented that:

> SPOT officials have repeatedly reminded the world of the corporation's open access policy and refusal to censor its imagery products. Rather, it was the unique circumstances surrounding the Gulf War that caused the French corporation to temporarily modify its policy. When SPOT has viewed a conflict situation as an opportunity to provide newsworthy imagery, it has readily offered to do so… Interestingly, their altruism in this situation would have quickly

> disintegrated if any other imagery agency had decided to provide similar data.[43]

Whether or not SPOT would have sold images in the event of commercial competition remains speculation, but it highlights political-economic tensions that are relevant today as commercial imagery companies proliferate. In the event, the Iraqi information blockade was not complete as mentioned previously regarding weather satellite data, and CNN satellite news broadcasts could be received in Iraq with simple dishes. It is not unreasonable to presume from this vantage point that similar issues have occurred in the Russian-Ukrainian War.

The combination of spacepower systems, their syncretic effects on years of preparation by the U.S. Army, Navy, and Air Force in preparing for 'AirLand Battle' in central Europe through the 1980s unleashed an insurmountable conventional war machine against the Iraqi Army in 1991. However, it is important to remember that though it was experienced, the Iraqi Army had been fighting an exhaustive attritional war for most of the 1980s against Iran. Still, 1991 was the first time a series of space and IT systems came together in battle at scale. Shimko writes that:

> this combination of technologies marked the emergence of a 'guided-munitions battle network' or 'reconnaissance-strike complex' with three basic components: sensors locating and tracking targets; platforms, weapons systems, and munitions able to attack with precision, often from great distances; and command, control, and communications assets linking sensors and 'shooters.'[44]

The 1990s and 2000s witnessed a continuous refinement and expansion of the tactical application of space infrastructure for the needs of deployed military units. Perhaps most notable was that Joint Direct Attack Munition modification could be bolted onto 'dumb' bombs at $16,000 a piece, a fraction of purpose-built PGMs which were priced at above a million dollars each.[45] Space-enhanced combat capabilities were no longer experimental but became mainstream and reliable. Through the wars in the Balkans, Afghanistan, and again Iraq bandwidth and connectivity improved, ISR tasking became more sensitive to the needs of local commanders, and precision weapons became more precise, reliable, and numerous—not to mention gaining an increasing ability to continue working in bad weather and at night. However, in many circumstances aerial ISR platforms—though linked up and networked via SATCOMs—are more directly useful in a specific battle than satellite ISR once a reliable command of the air is won.

Space ISR cannot do every ISR task, especially not as well as some other platforms depending on the specific situation at hand. High-altitude loitering drones are complementary ISR capabilities and sometimes unique and useful missile strike platforms that satellites and crewed aircraft cannot be. Such is the nature of joint and combined arms. Joint warfare refers to coordinating assets from many 'domains', such as land, sea, air, and space—like getting an air squadron to provide close air support to a landing marine amphibious force. Combined arms refers to different kinds of weapons, vehicles, and equipment being used in concert to take advantage of their strengths but compensate each other for their weaknesses and limitations. An example is the use of infantry units with anti-air and anti-tank sub-units to follow up on a tank advance to seize and hold territory and protect armoured columns' weak points on the flanks and rear. These are complicated things to do and require lots of training and communications equipment, as well as professional forces.

It is important, then, not to over-state the importance of spacepower in these wars or claim tactical successes as the product of any single form of power. As in war in general, success is usually the result of a combination and coordination of factors and capabilities, tangible and intangible (including luck), rather than singular events, individuals, or technologies. Other technologies working as they should also mattered. Even the basic fact that British and U.S. tanks could move and shoot accurately at a range of 2.4km, whilst Iraqi tanks could only effectively fire when stationary and at shorter distances is profoundly important in understanding the success of the 'Hail Mary' flanking move through the western Iraqi desert. Like airpower, spacepower was an increasingly important part of a larger whole of military-technological systems, and all technology and equipment is only ever one part of warfare and understanding war and strategy.

Writing in 1995, the U.S. analyst Steven Lambakis pointed out that:

> the U.S.-led coalition came to the desert battlefield with a near-total dominance of Earth orbits and marshalled [sic] unprecedented space-dependent military capabilities to help it achieve decisive victory… Perhaps this unchallenged, lopsided advantage in space has persuaded American military analysts that such orbital hegemony may be taken for granted. If so, that is a dangerous error.[46]

That is not being taken for granted today. It was not taken for granted as far back as the 1950s, but it certainly became a focused issue for superpower militaries in the 1970s. Today, the command of space and who gets to use space in a time of war is the concern of every military power because it underpins so much of the advantages gained during the RMA years of military transformation in the post-Cold War NATO militaries.

A Revolution in Military Affairs?

When planning for a range of possible wars with large and technologically competent enemy forces, there are strong incentives to attack the satellites providing battlefield C4ISR capabilities, an edge to terrestrial military forces in networking, coordination, and long-range precision bombardment capabilities. These satellites, their related infrastructure, and their effects underpinned what many analysts and practitioners dubbed a 'Revolution in Military Affairs' (RMA) through the 1990s and 2000s. One way to attack the strengths of the American military machine was to threaten the space systems underpinning that RMA in the first place. This is the primary military rationale for exploring ASAT techniques, not merely because one's own satellites are threatened or a possible foe has explored such technologies. ASATs need worthy targets to hit—the so-called RMA, the tactical integration of space systems into conventional military power, provided that justification. The original sin of space technology creates even more momentum for military satellite constellations and the weapons needed to attack them.

Occasionally novel technologies and techniques can combine with remarkable successes on the battlefield to define and transform a generation of study in contemporary warfare and international security. This occurred in the 1990s and the early 2000s with the so-called RMA following sweeping American military successes in the post-Cold War era. The RMA emphasised precision munitions, rapid offensives, joint warfare, sophisticated C4ISR technology, and professional forces with relatively little casualties for the U.S. military and its allies. Rather than seeing it as a revolutionary break it is time to see this period of military modernisation as evolutionary, making more extensive use of space and information systems in waging conventional wars and military campaigns. The RMA never settled in its definitions or core claims, and went through several evolutions through the 20 years following Desert Storm. Network-Centric Warfare, Dominant Battlespace Knowledge, Defense Transformation—all were grasping at the effects of IT systems and space communications in enabling more widely dispersed military forces that could coordinate over vast distances, conduct long-range precision strike operations, detect large-scale enemy force concentrations or missile attacks, and reduce casualties. These terms were merely buzzwords of the day and disguised the incremental and sometimes troubled development of space and IT systems as they were integrated into more of the U.S. military's day-to-day activities and combat operations.

The RMA's origins can be traced back to Soviet Marshal Nikolai Ogarkov in the mid-1980s and Soviet discussion of a Military-Technical Revolution (MTR).[47] The Red Army anticipated the effects of long-range precision strike capabilities afforded by new sensors, missiles, and aircraft NATO planned to use in its AirLand Battle concept, to strike with air-launched and ground-launched non-nuclear and lower-yield 'tactical' nuclear weapons deep behind the Soviet front to slow down the second and third waves of Soviet and Warsaw Pact forces were they to invade central and western Europe. The 'emergent technologies' of the 'reconnaissance-strike complex' such as advanced multispectral sensors, mobile peripherals and receivers proliferated—partly developed for the Strategic Defense Initiative's missile defence purposes—were seen by the Soviet Union's military as ways for NATO to attempt to shift the 'correlation of forces' in their favour through the 'deep attack' of Soviet forces across Europe.[48] Contemporary commentary on 'emergent military space technologies' are really referring to base technologies and systems first tested or developed several decades ago. These systems are now mature and essential, not emergent and exotic.

A U.S. analyst, Andrew Krepinevich, wrote of this MTR in influential studies in 1992 and noted that the Gulf War was perhaps the start of a new revolution in combat effectiveness, based on precision weapons hooked into surveillance and reconnaissance systems, alongside effective tactical command and control systems. Importantly, Krepinevich did not restrict the changes to technology, as new doctrines and organisational structures contributed to a revolution and were also needed to take advantage of technological changes.[49] The Director of the Office of Net Assessment, Andrew Marshall, commented in 1993 that:

> referring to it as a military-technical revolution, is to be avoided because of the emphasis it puts on technology. Technology makes possible the revolution, but the revolution itself takes place only when new concepts of operation develop and, in many cases, new military organizations are created.[50]

In the face of an information technology and space supported military force which could be smaller, 'leaner', more mobile, more efficient, more precise, and more survivable, large mechanised militaries made of thousands of heavy tanks and artillery were vulnerable without the necessary C4ISR systems to protect themselves and attack their enemies' systems.

Changes were afoot in combat effectiveness. But the methods of war are always changing whilst the socio-political phenomenon of war itself remains the same. In Strategic Studies, this is how the 'character of war' changes but the 'nature of war' stays the same.[51] With information perceived as being the key to strategic as well as tactical success, the charge levelled by the RMA evangelists was that the Clausewitzian concept of the 'Fog of War' was now moot. Friction, chaos, and uncertainty in military campaigns and wars could surely be practically eliminated with modern information technologies. With ever-greater sensors and computing power, everything could be known on a transparent battlefield, and then targeted.[52] Omniscience was seemingly within reach as 'strategy' became synonymous with 'targeting' for many analysts. War had become a battle of networks held together by IT systems. Admiral Cebrowski claimed in 2002 that the changes were evident of a shift from the industrial age into an information age.[53] Without 'dominant' information capabilities, heavy, massed, cumbersome forces would always be on the back foot, vulnerable to devastating enemy fire that could impose unacceptable attrition and losses. Others dubbed this a 'system of systems' approach, which Shimko explains as the:

> seamless linkage of commanders, forces, platforms, and sensors, with everyone sharing the most complete picture of the battlespace possible so that tasks can be allocated to whatever elements of the architecture are best positioned to carry them out.[54]

Network-Centric Warfare and the RMA made it into *Joint Vision 2010* and *Joint Vision 2020* capstone documents which outlined the George W. Bush administration's long-term vision for the U.S. military. They sought to take advantage of precision warfare and networked capabilities to increase the mobility and precision of the U.S. military whilst decreasing its size. To survive under the then-Secretary of Defense Donald Rumsfeld, everything in the DoD was couched as 'transformational' by bureaucrats and think-tank wonks. As a result, the Pentagon and the military-academic-industrial complex had drunk the 'RMA Kool-Aid'.[55] They were in for a rude awakening when the occupations of Afghanistan and Iraq descended into insurgency and civil war in the 'Global War on Terror'. U.S. conventional superiority amounted to little in winning counter-insurgency wars when winning battles meant little against a persistent, popular resistance and a half-hearted attempt at state-building.

The 'RMA Kool-Aid' reduced the practice of war and the study of strategy to merely questions of targeting and battle. In other words, focusing on the means of destruction rather than the ends of strategy: how those methods are meant to help achieve the ultimate political goals which justify that killing and destruction in the first place. Two specific examples are the Observe, Orient, Decide, and Act (OODA) loop coined by USAF Colonel John Boyd, emphasising the need to act more quickly than the opponent based on faster information-reaction times as well as a better understanding of when is the best time to act.[56] Warden's Concentric or Five Rings approach, which sought to outline efficient methods of attacking the enemy's war-making capacities via airpower, with each inner ring holding more critical targets leading to the enemy's capitulation, suffered similar problems. Simply having more information and acting quickly is not sufficient as unconventional or irregular forces can also take advantage of information, wait things out, and as previously mentioned, Clausewitz warned of false information and analysis paralysis following information overload.[57] Both the OODA loop and Concentric Rings approaches were concerned with tactics and operations, not strategy, but many began applying these ideas concerned with targeting and destruction to the strategic level, where violence meets ultimate political purpose and is only one method among many of meeting political objectives. Acting quickly is not the same as acting prudently, especially when violence of any kind in one situation is inappropriate due to the nature of the political objective sought or the political conditions at the time.

The strategist Colin Gray argued 'the basic problem is that the OODA loop does not capture conceptually the key to victory for a country as it might arguably do for a fighter pilot engaged in aerial combat'. The 'bullseye' nature of the Concentric Rings is too mechanistic and makes flawed assumptions that the effects of strategic bombing on the enemy can be calculable and predictable.[58] These are theories about specific strategies and tactics that could be effective in *some* situations. But in some wars, a bombing campaign or simply acting faster and with better timing than the other side is simply not enough or may even backfire. As discussed below with coercive airpower, bombing campaigns may do little to persuade the targeted population to cooperate. The OODA loop's encouragement of standing on tenterhooks ready to pounce makes you more vulnerable to being played by an enemy goading you to respond quickly to any sign of activity without thinking, falling for deceptions, and creating bigger second-order problems. Such is the terror of 'agile' command and leadership trying to match violent tools with political goals in haste without thinking things through beyond the immediate heat of the moment and the chaos of battle.

Taking these targeting concepts too far when thinking of war—which is about politics, emotions, uncertainty, and logistical matters—results in the 'tacticisation of strategy'. It reduces the profound, existential questions of war, peace, and the ultimate rationales and purpose of using political violence to questions of how to best kill people and assuming 'victory' simply follows destruction. Michael Handel noted the tendency in professional military circles to allow lower-level operational considerations to define the higher levels of strategy in war. In other words, relying on tactical and operational success (reliably winning many battles) as a substitute for comprehensive, calculated strategy (winning the war and getting a desirable peace by stopping the ability or will of the enemy to take up arms). A simple fact of strategy is that it is about winning wars, not battles. The RMA was obsessed with the latter at the expense of the former.

The precision targeting trends of the 1990s and the *Joint Vision* documents 'has specifically created a situation in which targeting has de facto become a substitute for proper strategic planning'.[59] The lower levels of war—tactics and operations—will of course matter and influence what strategies are available, but the point here is that tactics and operations, and specific weapons technologies, are but one part of the whole when thinking about strategy and war. For Clausewitz, in war planning the onus was on the overall military strategy to create the conditions within which tactical victories would be likely and relevant in meeting the political goals set by leaders.[60] These general observations are useful regardless of the precise methods and equipment used in warfare or the buzzwords of the day in military communities. This is not just a Western concern. Mao Zedong was also preoccupied by the problems caused by those who believe success in war boils down to tactical or operational success on the battlefield alone.[61]

Colin Gray emphatically argued that:

> better weapons always are preferable to worse weapons—provided large opportunity costs are not incurred—but tactical and operational military prowess is easily squandered if battles are ill-chosen, campaigns are wrongly pointed, and war is ill-conceived.[62]

Also warning against blinkered tactical-technological thinking, Andrew Marshall warned in 1992 that:

> It goes almost without saying that one of the things you would want to do is to develop some longer term strategy for conducting yourself in this situation. We need to be clear about our long-term goals, clear about what we wish to achieve.[63]

Sadly, these fundamental points do not go without saying given that the recent years have seen any number of new 'wonder weapons' that disproportionately draw attention and whose evangelists make wild promises of easy victories or certain defeat without them—drone swarms, artificial intelligence, and 'hypersonic missiles' to name a few. Wars that feature guerrilla campaigns and counterinsurgencies continually demonstrate that winning battles does not automatically win wars. For the weaker side, simply surviving and staying in the fight at all counts as 'winning'. Yet it seems to be a lesson that constantly requires teaching for professionals who ought to know better.

The 1991 Gulf War was remarkable in that one of the largest armies on Earth was defeated with so few casualties and so quickly once ground offensives began. Alan Campen remarks that 'it was a war where an ounce of silicon in a computer may have had more effect than a ton of depleted uranium'.[64] When the focus on the material changes in military equipment did occur, it tended to ignore spacepower in favour of IT and missile systems. Humans can see computers and wires. They cannot see radio waves. The important technological trends that were happening were sometimes missed perhaps due to the more subtle, infrastructural, and logistical nature of satellites, which were in space and out of sight, and then perhaps out of mind.[65] The classified nature of much military space infrastructure and intelligence products exacerbated this and made it less accessible.

A physical environment 'up there' was being used to facilitate the RMA 'down here' but discussion often spoke in highly abstracted terms about 'information' and 'knowledge' rather than basing thought on geographic and technological realities. This held back the education of military and civilian analysts about the geopolitics of a strategically crucial environment which is equivalent to understanding the use of the seas and the air in wartime and in meeting any number of state goals in many different situations, with or without the use of force. Recognising space as its own distinct 'geographic' environment and theatre of operations and conflict like the air and the sea was restricted to a cadre of niche experts in military doctrine, academia, and the defence policy community. That may only now, 40 years later, be changing.

The RMA debate, as well as the SDI of the 1980s, unduly mischaracterised what spacepower was and could be. It was not defined by 'information' in warfare, as supposed in the RMA debate. It was not restricted to missile defence systems as portrayed in the Reagan-era missile defence debates of SDI. Instead, spacepower became just another environment for warfare as it became tactically relevant in wartime. Gray and Sheldon argued in 1999 that 'over the better part of 10 years, space power has changed its status in the US armed forces from one of typically "useful and important adjunct" to terrestrial forces to, at the least, "indispensable adjunct."'[66] Unlike information in warfare, which has always been present and important in war, the systematic exploitation of Earth orbit for warfare—spacepower—is relatively newer and is geographically specific in its meaning.[67] This requires geographically orientated expertise if spacepower is to be understood, exploited, and resisted in its fullest sense.

The important parts of those discussions—whether or not one believes in a 'revolution' in the ways in which wars are fought—is a recognition that spacepower has played a necessary role in bringing about some of the biggest technological changes in the methods or 'character' of warfare. This has raised the pressure on other military forces to follow suit or adapt to undermine space-supported enemy forces, or both. Whilst methods of space warfare are important, it is even more important to understand why more countries have greater interests in engaging in space warfare and examining how global military forces are evolving as spacepower makes its influence felt beyond the U.S. and allied militaries today.

Russian and Chinese Modernisation

The Russian military was only too conscious of the relative inferiority of its conventional military forces as the dust of the Soviet empire's collapse settled in the 1990s. Nuclear weapons came to be viewed in Russian doctrine as a way to compensate for such weakness on the battlefield. Among the documentation on Russia's threat perceptions and high on the list of perceived threats was a massive air and missile attack from the United States and NATO, combined with 'information operations'—or political subversion—to bring about regime collapse. To counter this, modern conventional forces are needed as nuclear weapons are not useful for every contingency, especially in the early stages of conventional military operations as the weaker party.[68] Russia has—on paper—modernised conventional forces which has involved a resuscitation of Russian spacepower. However, as we have seen in the Ukraine-Russia War in 2022, such technological modernisations in certain parts of its space infrastructure and missile systems cannot make up for seemingly major and elementary problems in the Russian Army in waging a major war of conquest, even on its doorstep.

Following the collapse of the Soviet Union in 1991, the troubles of the Russian economy and 'shock therapy' throughout the 1990s, Russia took great pains to rebuild its military space infrastructure through the late 2000s to today. Foundational spacepower capabilities, such as GLONASS, a range of general IMINT and SIGINT satellites, and a suite of military SATCOMs under the banner of the Integrated Satellite Communication System (ESSS) are now online. Not only have old Soviet capabilities been reconstituted, but in some areas improved upon, such as with the GLONASS-K generation and improving civil and military IMINT satellite resolutions.[69] Yet future and continuing modernisation will be harder with increased sanctions and high-technology import restrictions. After fits and starts in trying to reform the Russian military, it was not until the Russo-Georgian War of 2008 that the 'New Look' military reforms took root and put the armed forces on the path to modernisation and professionalisation. Though ultimately victorious over the skies of Georgia, the losses of the sluggish Russian Air Force to such a small and inexperienced enemy triggered alarm bells. Such a performance simply would not do against any properly equipped and trained opponents. Backed up by large funds and determined political will, the Russian military set out to modernise its equipment and technology, transform its command structures, and push for professionalisation, doing away with a reliance on conscription.[70] Spacepower was very much a part of, if not essential to, this modernisation. Yet aircraft, naval, and armoured losses, not to mention the reported military death tolls in Ukraine today seem to hint that despite years of effort, much remains to be achieved and corruption may have siphoned much of the funds.

To counter or blunt modernised NATO forces, Russia needs sophisticated air-defence systems and long-range precision strike capabilities to strike at their C4ISR systems. Whilst Russian concepts of a massed air and missile attack on Russia vary, they often include hypersonic gliding missiles, cruise missiles, ballistic missiles, UAVs, combat aircraft, combat helicopters, strategic aviation, offensive space operations, and the use of electronic warfare.[71] These complex weapon systems require an extensive space-based infrastructure to function as has been described above with the United States. Since Russian forces will not accept a guerrilla-style approach of accepting enemy aerial dominance and an upper hand in every land engagement, they had to modernise enough and be able to put up a 'good enough' level of conventional military resistance and significantly increase projected kills and attrition on NATO forces. Russian planning has gone down the road of mounting a feasible defence and counterattack capability in conventional terms and with nuclear weapons as a last resort,[72] as opposed to retreating into a guerrilla force.

In the 2010s the Russian enclave of Kaliningrad saw a build-up in missile strike and anti-aircraft systems. With an emphasis on the early phases of a major conflict driving such thinking, long-range missile and airstrike capabilities, as well as robust air-defence systems, were seen as a way to blunt NATO precision attacks and naval power projection by attacking deep into NATO territory to hit vulnerable infrastructure, airbases, missile sites, and naval vessels in an early or pre-emptive moment. Though coined by Western analysts on China's long-range missile strike and air-defence systems, the concept often used in the English-speaking world to define this is anti-access and area denial (A2/AD).[73] Jargon aside, the reality is that both Russia and China have improved their ability to identify, detect, and track U.S. and allied military units on land, sea, and air, then launch precise missiles, and lots of them, at ever-greater distances. This threatens potentially significant losses that may enhance deterrence and also mount a credible conventional defence, counterattack, or pre-emptive strike in the actual business of fighting a war. As we have seen in 2022, however, military equipment and units are only properly tested in real war and it is hard to judge true military effectiveness before a war begins.

Such capabilities fortified in Kaliningrad include guided rockets, ballistic missiles, electronic warfare platforms, submarines, 'man'-portable air defence systems (MANPADS) and integrated air defences, and potentially in the future high speed intercontinental cruise missiles and glider vehicles dropped from rocket boosters. The last two missiles are often misleadingly referred to as 'hypersonic' due to the speeds exceeding five times the speed of sound as they fly in-atmosphere—but ballistic missiles are faster than many of the boost-glide or high-altitude cruise missiles systems labelled 'hypersonic' that have stolen much attention in military circles in recent years. Since 2014, the build-up of long-range strike systems and defences has increased significantly and include Iskander-M SRBMs, S-400 air-defence batteries, coastal missile complexes with 600km ranges, the Tsirkon cruise missile, and theatre missile defences which intercept very short-range missiles. Today there is—on paper—a credible depth and number to Russian electronic warfare and missile platforms in Kaliningrad.[74] These rely on an ever-maturing and expanding military space infrastructure to command and control, to identify targets, and make battle damage assessments following any bombing action. As an enclave separated from Russia proper, Kaliningrad's weapons systems will also rely on space systems once ground and marine cables are cut.

Russian conventional capability improvements have led to an apparent decline in the expectation that nuclear weapons may need to be used to repel a major 'aerospace attack' from the outside, but it is not gone as a weapon of last resort. Conventional and nuclear military options may be seen as complementary.[75] Yet this is declaratory nuclear strategy written in documentation and stated in speeches—these are no guarantees of how a leader, organisation, or local commanders may behave in an actual crisis. The Russian Strategic Rocket Forces enjoyed high degrees of funding and prioritisation even in the 1990s as the rest of the military degraded, showing a clear priority on nuclear weapons within Russia's perceptions of what matters in military power. However, conventional military forces have regained standing and some sections of the conventional forces now approach the same level of quality as the nuclear forces.[76]

Specific parts of Russia's modernised armed forces have been demonstrated in Crimea and Syria. Crimea saw competent work by air and naval units that had enjoyed funds and attention in their reforms and training in the years prior. Although it was an important success for the Russian military, they had not concocted some new war-winning stratagem or formula, and nor did the Russian Army conduct any major ground warfare operations. Many western analysts thought some 'new way of war' was afoot. Some officials and doctrinaires in the west call this 'hybrid' or 'grey zone' warfare.[77] Russia had unique advantages in Crimea that may be hard to replicate elsewhere, such as a major naval base in Sebastopol, a receptive audience to its information operations and propaganda, and surprise. Nevertheless, it showed an improved strategic and political ability to fine-tune specific military means to specific political ends.[78] Hyperbole over Russian 'hybrid' or 'grey zone' warfare conveniently ignores the war in the Donbas in east Ukraine where there was much more popular resistance to Russian and Russian-supported forces. Large, heavy Russian infantry and artillery units had to be deployed to counter the successes of the Ukrainian Army's infantry, armour, and artillery capabilities which successfully checked the initial militias and proxies that seized towns in the early phases of the conflict in 2014–15.

The war in the Donbas did show an allegedly effective use of electronic warfare by the Russian military, in particular in the Battle of Debaltseve in 2015, among other capabilities such as jamming cell phone networks and detecting and firing upon moving targets regardless of ground and weather conditions.[79] Yet in the 2022 invasion of Ukraine it has been noted that Russian electronic warfare has been rather muted or small scale, but given the secretive nature of it we should not draw hasty conclusions at this time based on partial observations that are only available in the public domain.[80] In 2015, Russia demonstrated its own advances in similar battlefield applications of its own comprehensive satellite and space infrastructure. At least ten IMINT satellites were used, including civilian imagery satellites, to support Russia's military campaign in support of the Assad regime against Syrian rebels, Islamic State, the Kurds, and Turkish proxies. Russia also used such imagery to demonstrate the extent of Turkey's own intervention, and GLONASS has been used to guide precision Russian bombs. Combined, these capabilities increased the efficiency at which Russian aircraft and other strike forces could identify and accurately destroy targets.[81] Before 2022, Russia's Syrian intervention was arguably the most extensive battlefield test of a Russian reconnaissance-strike capability and relied on the radiofrequency tethers between space systems and terrestrial forces. Not only that, but Syrian forces learned to integrate with Russian military space systems, enhancing their own efficiencies and killing power.[82] In many ways, the Syrian campaigns showed the tangible, broad results of technological renewal and power projection in the Russian forces, in particular the Russian Aerospace Forces.[83] Around Syria and on NATO exercises, civilian GPS signals have been persistently jammed by Russian ground jammers. However, their impact on the military GPS signals, if any, remains unknown.

Renewed sealift and airlift capabilities add new weight to Russian interests and expeditionary potential beyond its immediate neighbourhood. Without space systems, such expeditions would not have been possible with lower costs and lower numbers of personnel, where UAVs and precision strike weapons have been employed to compensate for relatively low Russian personnel deployments. Bettina Renz is of the view that Syria has demonstrated that the Russian Aerospace Forces have arrived at where the U.S. Air Force was in 1991, and have downsized but retained heightened levels of permanent readiness to deploy, like U.S. and UK forces.[84]

Russian military modernisation is not without its shortcomings or problems, as with any military's transformation. Many shortcomings have apparently been worse than many may have anticipated given Russia's failure to quickly defeat Ukraine. These targeted advances did not give Russian forces a *fait accompli* in the attempt to storm Kyiv with paratroopers and a rapid armour advance from Belarus in February 2022. The paratroopers were encircled and failed to use Gostomel airport as a forward base, and the tank columns got bogged down by sticking to roads and facing organised Ukrainian artillery and anti-tank infantry tactics. The sinking of the flagship of the Russian Black Sea Fleet, *Moskva*, after being reportedly hit by a Ukrainian coastal missile, was a particularly damaging blow. Ukrainian air defences were not decisively annihilated by the Russian Air Force as Iraq's were in 1991 and 2003, and the Ukrainian Air Force have stayed in the fight several months into the war.

Before 2022, the Russian military was known to face major hurdles: personnel shortages, economic lethargy, corruption, and struggling defence innovation.[85] Whilst there are strengths in the Russian defence research sector, such as aviation, nuclear, and space, electronics for example remains in the words of John Engvall a 'problem child'.[86] This was a problem even before sanctions were imposed following Russia's annexation of Crimea in 2014, which blocked the imports of many electronic components. This has only worsened as of the war in 2022. In 2015, up to 95% of electronic components in Russian ships were foreign-made.[87] However, having 'the best' technology is not important in and of itself, but having 'good enough' technology is. What counts as 'good enough' requires marrying technological capabilities with specific plans, bureaucratic interests, military service cultures, strategies, threats, financial means, and political objectives, not a simple comparison of equipment specifications. Against lower-technology foes, Russian forces certainly have an upper hand in combat operations today, and what counts as good enough there is different to what is good enough against conventional and nuclear NATO forces.

When facing a more technologically sophisticated opponent, Russia will hope to make good on its C4ISR advances and long-range precision strike weapons and concentrate a lot of heavy mechanised forces in its immediate neighbourhood in Eastern Europe and Central Asia. The more high-technology elements may be temperamental and vulnerable as the enemy will be seeking to undermine their effectiveness and their enabling infrastructure, just as Russia will be seeking to do the same to them. Planning for the disruption to spacepower, or for a command of space to be seriously challenged, is important. As Grau and Bartles note:

> Although the Russians have precision-guided munitions for their artillery, they still believe in the effectiveness of massed artillery fire. Tactical EW systems may jam or prematurely detonate electronic VT fuses, but the mechanical fuses of conventional artillery rounds cannot be jammed by electronic signals. There are occasions that call for the use of surgical, precision fires, but massed artillery fires carry a mathematical probably of kill with which it is easier to predict tactical success.[88]

It is no surprise that sieges of Ukrainian cities and the use of 'low-tech' indiscriminate, imprecise bombardment weapons are following the established Russian pattern in Chechnya and Syria.

The Russian military view counterspace operations, or attacking enemy space systems, as an important part of overall 'aerospace defence', with systems like the S-500 being advertised as air defences, ballistic missile interceptors, and anti-satellite systems. Recently, Russia appears to have deployed the Peresvet anti-satellite laser system. This is a ground-based laser designed to dazzle the sensors of unfriendly satellites which accompanies mobile ICBM units so that the nuclear missiles can manoeuvre without being spotted.[89] Russian forces, like those of China and the United States, have to anticipate situations where their most sophisticated long-range precision-strike capabilities, as well as remotely piloted or autonomous systems, will suffer degradations to their communications and navigation systems due to the proliferation of anti-satellite or 'counterspace' weapons and techniques, as discussed in the next two chapters. PGMs are indeed more efficient than massed artillery when they work as they should. If the support or guidance systems of PGMs are broken or degraded, massed artillery becomes a 'sure bet' on levels of destruction that can be visited,[90] so long as those artillery formations are not subjected to the enemy's fully functioning PGM and reconnaissance-strike capabilities. Otherwise, Russian artillery and armour units may suffer the tank-plinking visited on the Iraqi Army. If the enemy fails to sabotage and harass Russian space infrastructure that supports the Russian military, then Russian forces will be able to capitalise on them.

What Russia has had to prepare for against the spacepower-supported NATO forces and plan to attack their space systems is now something that NATO states must reckon with in turn. Russian military successes since 2014 are not absolute or without problems, and the Russian military's conventional strike and C4ISR capabilities are supposedly better than they were a decade ago. This is not some dramatic, sudden turnaround—it follows long-term and consistent financial and political commitments.[91] Yet it seems the bulk of the Russian Army, Air Force, and Navy have struggled in Ukraine, perhaps echoing the worst moments of the First Chechen War, where 'there seemed no way out at all, and no ideology that could restore to the armed forces pride and faith in what they were doing'.[92] The spread of spacepower through the Russian military is increasing the potential payoffs for NATO and other potential Russian adversaries in attacking Russian space systems in the event of open hostilities. Left unchecked, Russian spacepower would allow Russia's increased military effectiveness to visit high costs, risks, and losses upon its foes. In the years to come the Russian military may learn and modernise from its shortcomings (again), and spacepower will be part and parcel of that modernisation.

Unlike Russian forces, China's People's Liberation Army (PLA) has not seen as much deployment and combat experience since its invasion of Vietnam in 1979 where three provincial capitals were levelled, Beijing showed a proclivity to using force against foreign threats for domestic purposes, and a bloated and outdated PLA Air Force and Navy performed poorly.[93] The PLA has spent the last forty or so years increasing the pace of modernisation and catching up to the United States with new C4ISR capabilities, within which spacepower plays a foundational infrastructural and logistical role. Plan 863, launched in 1986, set up the general modernisation plans that were accelerated following the U.S. wars of the 1990s and the Taiwan Strait crisis of 1996. Spacepower was one of the major planks in making the PLA a networked and more information-technology literate force.[94] Indeed, the extensive level of space technology cooperation and transfer between Washington and Beijing in the 1980s demonstrated how far ahead the United States was in terms of spacepower and information technologies for both economic infrastructure and military power.

The Sino-Vietnamese conflict from 1979 and into the 1980s also showed a need to downsize and modernise the PLA.[95] Alongside the space modernisation and civilianisation efforts that kicked off in the 1980s, Deng Xiaoping inaugurated the first attempt at modernising the PLA with joint, multi-service warfare which today dominates PLA reforms and force structure planning.[96] Like many others, the PLA was quick to notice the impact of 'network-centric' warfare which was 'so vividly demonstrated in the Gulf War' and recognised its own conventional forces were outdated, as well as potential weaknesses in U.S. dependencies on space and IT systems for battlefield capabilities.[97] China's military has pursued modernisation to fight what it calls 'local wars in informatised conditions', which can be traced back to Jiang Zemin's military reforms in the early 1990s.[98] Indeed, as much recent news and policy discourse has leapt on the U.S. Space Force's and the Trump administration's use of bullish doctrinaire language such as 'space dominance' and 'war-fighting domain', that language is echoed in Chinese military writings on space warfare, and has done for quite some time and itself echoes decades of USAF space doctrine documents and military graduate student writings.[99] Over the past 20 years, spacepower has been increasing in recognition and prominence within the PLA as evidenced across numerous analyses of military publications, doctrines, and official statements.[100]

Arguably, spacepower in military operations has been 'mainstreamed' in PLA thinking as in the central *Science of Military Strategy* text which claims that a 'nation's capability and levels in space military struggle often can objectively reflect that nation's comprehensive real strength.'[101] It goes on to explain that:

> Space information assisting support is now, and for quite a long period to come will be, the main mode for the application of space forces among the various nations. Stopping the friendly space information system from encountering jamming or sabotage, to ensure its normal effective operation, and at the same time when necessary jamming and sabotaging the adversary's space information system, to degrade its system operating effectiveness, are important content in military struggle in the space domain.[102]

In 2015, major reforms were pushed throughout the PLA from the top to modernise with improved IT and space systems infrastructures. Among many sweeping changes, the Second Artillery Force which was in charge of China's nuclear missile forces was renamed the Rocket Force (PLARF), and a new organisation was set up to develop and coordinate space, cyber, and electronic warfare assets and operations—the Strategic Support Force (PLASSF). The PLARF now oversees a range of ballistic, cruise, and emerging boost-glide hypersonic missiles, for nuclear and non-nuclear bombing missions. SRBMs, MRBMs, IRBMs, ICBMs, HGVs, and a mixture of cruise missiles are increasing the ability of PLA commanders to hold obvious targets at risk up to several thousand kilometres away from the Chinese mainland and well into rearward bases and logistics bases of the United States' Pacific forces. Fortified and militarised islands beyond China will extend such strike distances further. This drive towards modernising conventional precision strike capabilities, such as striking mobile naval targets, for instance, with the infamous DF-26 MRBM with MaRVed warhead (the so-called 'carrier killer'), is part of what many Western analysts have dubbed an 'A2/AD' suite of capabilities.

Long-range strike capabilities against fixed and mobile targets have become more prominent in recent exercises, reflecting PLARF's strike capabilities rather than the Navy keeping distant enemy maritime forces in check.[103] These capabilities require an elaborate space-based infrastructure, or C4ISR systems. Conventional missiles, particularly SRBMs, form a major plank of the PLA's power projection capabilities, especially with regard to a Taiwan War scenario. However, against obvious, fixed targets like government buildings and bridges, roads, and factories, space-based targeting is not needed. Ballistics and inertial guidance can be sufficient. That said, space-based infrastructure and other ISR and guidance systems have in concert increased the accuracies and hit probabilities of over 1,200 SRBMs across Taiwan, the Strait, and other potential stationary targets in the region. Many longer-range Chinese ballistic missiles can be equipped with MaRVed warheads, such as FD-15B, DF-21C/D, and the DF-16, and not only the DF-26.[104] Additionally, what might make these warheads even more dangerous could be more sophisticated terminal guidance systems that could take advantage of the newly operational BeiDou GNSS constellation. The PLASSF is directly involved in BeiDou's operation and engineering[105]—though this is not alarming by itself given the military nature of GPS and its operation by the USSF. As the PLARF enjoys an ever-growing arsenal of nuclear and conventional missiles, PLASSF is increasing the potential effectiveness of that arsenal by providing space, information, and other integration services. According to Elsa Kania and John Costello,

> the PLASSF can directly enable joint operations through serving as an essential conduit to integrate the disparate units and systems across the PLA's services. In any future conflict scenario, the PLASSF would be critical from the start, enabling the PLA to contest information dominance and contributing to power projection within and beyond the first island chain through bolstering the PLA's reconnaissance-strike complex.[106]

In a major conflict the PLA may be able to impose greater costs on any U.S. military intervention, both with and without nuclear weapons. Doing so without nuclear weapons requires a more sophisticated and robust C4ISR network to allow it to threaten U.S. carrier strike groups with missiles, airstrikes, cyber-attacks, anti-satellite attacks, and PLA Navy vessels.[107] Naturally, this opens up a new vulnerability for the PLA along the same lines as the U.S. military in 'navigation warfare' and interfering with the ISR and guidance systems the PLA rely on to make such long-range strike weapons work in the first place. Whilst imitation can be flattering, it also means the flaws and weaknesses might be copied as well. As such, when looking at 'new' weapons or capabilities in the news, it is important to put them in the wider context of countermeasures and weaknesses, much of which is not available in detail in the public domain, as well as the fact that until a war actually happens it is hard to know exactly how each military force will perform.

Chinese military modernisation is not only about the long-range strike systems and nuclear strategy, but rather the wholesale modernisation of the PLA with a range of support systems for more fluid, combined, and joint operations and shifting to more of a balanced force as opposed to an Army dominance. These 'new type forces' are meant to be more responsive and adaptive to the situation as it develops, and do so without needing to concentrate in any large mass formations outside of key geographic locations (such as landing sites and cities). PLASSF is the top-level manifestation of joint warfare reforms, but down to the brigade level information operations support units are being linked to tank, artillery, and missile strike units.[108] Moving away from the traditional PLA Army dominance of the PLA as a whole, joint operations emphasise the PLA's services cooperating to achieve common goals as needed in the campaign at hand, rather than a campaign being solely a PLA Navy or PLA Air Force mission. Joel Wuthnow explains that:

> commanders need to be sufficiently knowledgeable about subordinate forces and empowered to direct their activities, while also retaining the flexibility to delegate operational authority to lower levels when desired. A sound joint C2 structure also requires the ability to devise joint operational plans according to commonly

> understood terms of reference, maintain a resilient communications network through which information can be passed quickly between higher and lower echelons, and ensure interoperability of different service arms and branches.[109]

The PLASSF Space Systems Department has 'integrated and consolidated the Chinese military's space systems and capabilities in ways that could enhance intelligence and information support to the PLA's newly established theater commands'.[110] Technological and organisational modernisation cannot happen without space infrastructure, especially as the PLA moves towards operations beyond its mainland territory and the PLA Navy and Air Force expand their reach as well as ship, boat, and aircraft numbers.

As the PLA decreases in absolute size, it is not getting any cheaper. Not only does sophisticated equipment cost more, but so does the training needed for the skilled personnel operating them.[111] Since 2017, over 1,000 units at the regiment size or above have been disbanded, with the brigade becoming the standard unit size and focus of capability as opposed to the division.[112] Smaller yet higher-technology and more space-supported 'joint' operations planning means that the PLA is imitating the reforms seen in NATO countries and Russia. The combined arms battalion is now the basic combat unit of the PLA and is designed to call upon land, sea, air, information, and space support and capabilities as needed. This is not to say that Russian, Chinese, and American forces are identical, rather, there are large common trends borne of spacepower integration that cut across them.

Demonstrating the scale of these reforms, 'the entire PLA Army is in a major state of flux. As a result, many officers and enlisted personnel are undergoing significant stress and discomfort, so much so that many units are conducting psychological counselling sessions to help their personnel cope with the situation.'[113] Such disorganisation and bureaucratic trauma may pass in time, particularly if the reforms bring about the successes in the PLA that the U.S. military received from the Goldwater-Nichols Act in the 1980s.[114] Spacepower forms a crucial technological plank in these reforms as satellites are needed for the PLA to fight the kind of high-technology wars it may do in future, not only against the United States but also other military powers in Asia that are modernising with space systems and missile technologies such as India, Japan, South Korea, Taiwan, Australia, and Vietnam.

Entanglement and Responding to Spacepower Integration

The thermonuclear revolution has gotten no less acute in thinking through a war between many of the major space powers today. 'Entanglement' is how C4ISR systems and weapons can be dual-use: they are used for nuclear and non-nuclear military operations, and are vulnerable to nuclear and non-nuclear forms of attack. Trying to keep a war non-nuclear may be impossible due to the need to attack these systems to hamper the enemy's conventional military power, but will inevitably undermine the enemy's nuclear warfighting capabilities too, leading to a nuclear escalation. Entanglement can happen in three ways—geographically, operationally, and technologically. Geographic entanglement refers to the physical proximity of non-nuclear forces with nuclear ones. Operational entanglement refers to the organisations, units, and systems that command and control both nuclear and conventional systems. Technological entanglement refers to the ability of weapon systems and other technologies to be used for either nuclear or conventional missions without much, if any, modification.[115]

James Acton claims that compared to the U.S. military—citing SBIRS as a particular escalation and entanglement risk—Chinese and Russian risks of entanglement are not as great, but not negligible either.[116] Lewis and Xue noted in 2012 how Chinese missile forces represented a warhead ambiguity problem.[117] However, even with recent increases in China's planned nuclear missile silos it would not be able to build enough warheads for all its deployed missiles. Looking across these three kinds of entanglement, Logan cautions that entanglement is not as entrenched in Chinese military systems as many fear. Many conventional capabilities remain rather distinguished from nuclear systems, with the DF-26 remaining something of an exception.[118] As Chinese and Russian conventional forces continue a spacepower-enabled modernisation drive along American lines, Logan expects the entanglement of nuclear and conventional forces in the PLA to increase in future.[119]

The United States courts entanglement not only with systems like SBIRS, which are now conventionally important yet are first and foremost performing nuclear warfare duties, but also with several other capabilities it has fielded or contemplated over the years. Prompt Global Strike (PGS) was a capability ambition that dates back to the Bush Jr. administration, aiming to strike any target on Earth with a conventional bomb within an hour. Effectively, this could turn ICBMs and SSBNs into long-range conventional ballistic attack systems that could take out specific buildings or positions. However, their flight paths could still look like a nuclear attack. Such global strike capabilities could help U.S. forces respond more flexibly to an unforeseen military intervention if the number of local forces were low, and would enable 'deep strikes' against enemy systems regardless of air defence and other long-range missile strike systems without resorting to nuclear weapons.[120] Many capabilities can fit the PGS umbrella, including converting the Ohio-class SSBN Trident-II missiles to carry conventional explosives rather than nuclear warheads, as well as 'hypersonic' boost-glide systems such as the Hypersonic Test Vehicle 2 (HTV-2) programme from the late 2000s and early 2010s. Whilst PGS has struggled to secure large amounts of funding—by U.S. military spending standards—and has failed to become a tangible weapons system, the potential remains as R&D continues to this day on PGS options across many services in the Pentagon.

Of greater concern regarding nuclear-conventional entanglement are the 'strategic' or 'deterrent' roles of aircraft, such as the F-35 Joint Strike Fighter's A variant which is certified as 'nuclear capable'. Other aircraft such as the B-52 and B-2 bombers have also demonstrated a ubiquitous nuclear and conventional bombing capability. Another risk is the perception of 'lower yield' nuclear weapons as different to 'strategic' nuclear weapons in terms of their escalation risk, again harking back to NATO's 'Flexible Response' strategy. Using lower yield, 'tactical' nuclear devices may bring about the very nuclear exchange some planners are trying to avoid. Tomahawk cruise missiles were originally developed for nuclear use, but today have become a 'normal' conventional capability. However, the most troubling note is Acton's: nuclear and non-nuclear U.S. forces rely on overlapping space and IT networks to function, attacks on which undermine Washington's nuclear-warfighting capability for anything other than inaccurate city vaporisation should satellites be degraded in wartime. An important factor in dealing with escalation is the surrounding context and behaviour rather than the technical platforms. An all-out ICBM launch with hundreds firing at the same time is very different to a lone B-52 bomber, and their flight paths relative to ongoing military operations or geopolitical crises can determine how such technical devices and their use can be interpreted at any given moment. Having something is one thing, how it is used is another.

Risks of escalation are always present whilst many of these systems complement each other. Disrupting significant chunks of precision strike and C4ISR systems could undermine the readiness of nuclear weapons, not just conventional, regardless of the attacker's intent. It could push one towards nuclear use. Any instance of space warfare between several of the major powers is also a clash between nuclear-armed powers, only increasing the intimacy between the space-nuclear-missile complexes that have defined the Global Space Age's original sin from the very start. We simply do not know how leaders and commanders will react to a situation in advance of it happening, nuclear or not. Yet that very uncertainty plays a role in increasing deterrent effects and preventing deliberately escalatory actions.

It is not just Russia, China, and the United States that are fielding high-technology military forces, and missile forces in particular. Israel has long fielded a handful of ISR satellites and communications systems, and possesses a midcourse ballistic missile defence system—Arrow—that may possibly be used as an anti-satellite interceptor. The two Koreas have extensive and growing varieties and numbers of missiles, with North Korea's having some nuclear options and South Korea's being enhanced by U.S. spacepower integration. At present, North Korea has fielded or is developing SRBMs, IRBMs, MRBMs, SLBMs, ICBMs, HGVs, MaRVs, and cruise missiles, using a variety of liquid- and solid-fuelled engines.[121] South Korea's missile programme includes SRBMs and cruise missiles, submarine-launched missiles, and an increased emphasis in recent years on precision-strike missile capabilities thanks to the ROK's position as not only a U.S. ally but also a high-technology economy and powerhouse.[122] Increasing its emphasis on picking apart the DPRK's command and control networks (and potentially China's), a mounting dependency on space systems is on the cards. Therefore it is not surprising that the ROK is moving towards investing in more of its own space systems for communications and ISR, and the United States has lifted constraints on South Korea's SLV engine development.[123] South Korea successfully launched its first LEO-capable SLV in 2013 (Naro) and launched a new SLV in 2021 (Nuri), and a handful of imagery and communications satellites are being operated by South Korean companies and government agencies. If domestic demand for space systems increases—either through public or private actors—it is reasonable to expect ROK-based space-supported military modernisation and less reliance on the United States in select areas. Seoul currently controls two government communications satellites and eight imagery intelligence or Earth observation satellites, showing significant investments for a state of its size.

Taiwan has also developed a suite of land- and air-launched cruise missiles and SRBMs, with a 2,000km-range cruise missile in development.[124] These ranges of missiles will help Taiwan conduct more precise strikes against Chinese forces on the Chinese mainland, across the Strait against naval and amphibious forces, and on heavy PLA units that manage to land on Taiwanese territory.[125] Like Japan and South Korea, Taiwan's strike options can be expected to enjoy U.S. ISR and GNSS support from space. In cooperation with Washington, Taipei currently operates 11 IMINT or Earth Observation satellites which increases its surveillance capabilities. As previously examined, India also has a large missile industry and its missile development and testing regime is centred on SRBMs, MRBMs, IRBMs, and short-range cruise missiles, and Indian SATCOMs, ISR, and PNT technologies are being integrated into the military services. Pakistan meanwhile is also fielding SRBMs, MRBMs, and cruise missiles.[126] Iran has a particularly wide range of missiles in its portfolio, including SRBMs, MRBMs, Anti-Ship Ballistic Missiles (ASBMs), cruise missiles, and an SLV in testing and development.[127] Missiles of various kinds are spreading and importantly being deployed in greater numbers, and all stand to gain from improvements with accompanying spacepower developments and major space powers sharing their space services with allied or client military forces. These tethers to allied and friendly space systems broaden conventional military options and enhance the lethality and capability of many militaries that otherwise could not afford all the necessary parts of space infrastructure.

States and other political-military organisations that may conceivably end up fighting against space powers or space-supported military forces will have to learn and adapt to the influence of spacepower on modern warfare and plan their own operations accordingly. If attacking space systems is not possible then adaptations have to nullify the advantages spacepower brings to military forces on the ground, at sea, or in the air. In the Balkan wars of the 1990s, space systems continued to prove their worth as the U.S. military came to rely on precision missile strikes more, and in the case of Kosovo, almost entirely on airpower and its supporting space and information infrastructure.[128] The 1990s saw much debate on whether airpower could be an effective tool for policymakers to achieve political results through coercion without needing ground forces, with precision airpower and a dominant command of the air allowing for a wide range of decapitation, punishment, and paralysis operations.[129] Spacepower played an important part of the ensemble that enabled airpower to reach a higher level of flexibility, discrimination, and reliability, but it did not make it an unstoppable tool of the U.S. military, or a silver bullet for strategic problems.

Serbian forces adapted to blunt the effects of precision NATO airpower through the use of forested, mountainous terrain to hide away from sensors, and also built decoys. In Afghanistan in the final months of 2001, U.S. forces worked with Northern Alliance troops to overthrow the Taliban and Al-Qaeda. The so-called 'Afghan Model' of matching local light infantry and cavalry with space-supported PGMs, Special Forces with laser pointers, and close air support proved too much to resist:

> If massed to take on the Northern Alliance, Taliban forces became the target of devastating aerial bombardment. But when the Taliban responded logically by hiding and dispersing, 'they lost most of their ability to conduct conventional operations, rendering them all but ineffective in their main mission of fighting rebel forces'... The ability of SOF [Special Operations Forces] units to laser designate tar-

> gets or transmit coordinates using GPS to aircraft delivering munitions was essential to the success of the Afghan model.[130]

This was an echo of the Serbian military's punishing losses when they massed to fight the Kosovo Liberation Army: they presented themselves as an obvious target to NATO airpower.[131] At the Battle of Tora Bora in the White Mountains south of Jalalabad in late 2001, the U.S. military faced difficulties given the mountainous terrain and entrenched Al-Qaeda defenders buying time for bin Laden's escape into Pakistan. Precision airstrikes were less effective in dislodging Al-Qaeda forces from their high-ground positions and the cave network. U.S. and Afghan forces had to advance against hardened defensive positions, compounded by the difficulties imposed by the high altitude of the battlefield. The Afghan Model was in many ways simply the wrong approach for Tora Bora. Precision airpower, even when supported by cutting-edge spacepower, could not make up for the lack of local forces' willingness to fight such a difficult battle.[132]

In Iraq 2003, the Fedayeen paramilitary forces resorted to 'stay-behind' operations to hit vulnerable Coalition supply routes, and marked some of the biggest engagements in the invasion, dispersing when pursued and blending into the wider population, denying U.S. technological sensors an obvious target.[133] U.S. leaders had no idea that Baghdad had sent these paramilitary forces in to slow and stall the advance, with virtually little insight from HUMINT to give any warning. The assumption from 1991 that opposing forces would simply melt away in the face of Washington's precision war machine simply did not hold as Iraqi paramilitary forces set up limited but noteworthy resistance.[134] Emblematic of the responses to precision warfare and other high-technology developments in space-enabled military power was perhaps the Apache helicopter attack on the Medina Division, which Finlan discusses:

> The Apaches were more than just a helicopter, but the embodiment of a corpus of ideas about warfare that had been steadily developing with the Army in the aftermath of the Vietnam War. The attack on the Medina Division proved to be a disaster. Of the 32 Apaches that launched the assault, 31 had sustained serious battle damage with one shot down and captured intact by Iraqi forces.[135]

The Iraqi military had learned how to deal with and punish Apache helicopter gunships since the rout of 1991. Space-enabled forces today have many advantages—but regular and irregular, hi-tech and low-tech military forces can learn to blunt, resist, and overcome spacepower-supported military systems just the same. Indeed, with space systems being felt in more and more military forces and theatres, the premium of learning how to deal with them, or adapt to their use by enemies, is only increasing.

In 2006, Hezbollah showed an ability to deny easy targeting opportunities for the Israeli Air Force, and continued to launch rocket attacks with near-impunity.[136] In the Cormoran 21 exercise in 2021, French forces conducted helicopter raids according to the schedules of enemy spy satellites to conceal their movements.[137] Spreading out your forces, dispersing them, and concealing them from enemy sensors is now a well-established principle in responding to enemy forces linked to space-support precision strike systems. Dispersion is an important feature of spacepower on the modern battlefield as it allows military forces to retain the effects of mass against obvious enemy targets but remaining dispersed across vast distances.[138] Manipulating, exploiting, and disrupting that dispersing influence is an immediate tactical reason for engaging in space warfare.

Conclusion

Today, many military powers are considering space warfare among the many tools of modern warfare as everyone's most serious or likely state-based adversaries come to rely on spacepower to varying degrees for conventional military power. What we are witnessing as the current century progresses is a continuation of what was set in motion in the Cold War. The tactical application of space systems, and consequently the risk of anti-satellite actions by the adversary in a time of conventional warfare, was recognised as far back as the 1970s.[139] This is an unsurprising and 'normal' consequence of the original sin of space technology, and certainly a hard edge of the Global Space Age.

Spacepower underpinned the so-called RMA of the 1990s in the U.S. military. It is the maturation and acceleration of technological and strategic trends and choices that have been in play for many, many decades. The tactical exploitation of and dependency upon space systems challenges the notion of passive military and intelligence space systems as inherently stabilising when those systems gave massive lopsided conventional military advantages to those who wielded them. Becoming more dependent on space for military capabilities leads only further the impetus for space warfare and for the relevance of spacepower in understanding power politics on Earth. Whilst the debates on space weaponisation tend to equate weapons development as destabilising, this confuses the cause of supposed instability with a symptom. The cause of space weapons development is the instability generated by making military power ever more dependent on satellites that can be targeted and harassed in a time of war.

This is not a novel phenomenon. Indeed, 'anxiety about the protection of America's space reconnaissance assets plagued every Cold War administration from Eisenhower to Reagan.'[140] As the last few chapters have demonstrated, space infrastructure has not come from nowhere or 'out of the blue'. Many countries and companies have spent generations building these space infrastructures and today the number of states relying on them or building their own systems is only increasing from the established, major space powers of the USA, Russia, China, Europe, India, and Japan. There is naturally a strong military dimension and impact to these technopolitical developments in orbit and the maturation of spacepower.

These military uses of outer space for tactical purposes, that is, making spacepower relevant to the battlefield and not just nuclear war or 'strategic' intelligence, ensure acts of space warfare in future crises or wars of sufficient seriousness to try to tip the scales on Earth in one's own favour. As well as dispersing your own forces to avoid the enemy's precision strikes, or at least minimise the chances of posing as an obvious target and being seen, military forces can try to parry precision strikes by intercepting munitions with close-in weapon systems (CIWS) such as Phalanx, Pantsir, or Goalkeeper, or theatre missile defences such as Iron Dome and Patriot. Another option, and perhaps a more preferable one than point-defence as a last resort, is to attack the C4ISR infrastructure that underpins the most effective precision weapons, and not least the space segment.

Space weapons are designed to challenge these effects from space infrastructure by undermining the ability of the enemy to effectively disperse and retain mass effects, and their ability to impose dispersion on others with precision weapons. None of these are silver bullets or simple solutions, but they complicate space-supported strike and reconnaissance systems. They increase the challenge, reduce confidence in any attack, increase the number of forces needed to ensure success, and reduce flexibility for the attacker.

In any major conflict we can expect space systems to be attacked, degraded, or suffer errors through overuse, including space and terrestrial segments. Even without enemy attacks and sabotage, information systems cannot provide an 'all sensors for all shooters' capability, meaning military C4ISR systems need to prioritise certain kinds of information to priority users at the right time.[141] This would only get harder as a space infrastructure comes under strain and attack. As a major war drags on, whether it goes nuclear or not, the infrastructure that allows for precise, networked, 'systems of systems' kinds of warfare with long-range strike and extensive surveillance and C2 capabilities will begin to fall apart unless major replenishment and repair capabilities are developed. Therefore, the kinds of space support available as time goes on may degrade or collapse in the most excessive conflicts, leading military forces to revert back to older methods of concentration and massing firepower on the enemy as the threat of precision strikes from great distances diminishes. The methods of disrupting and attacking those space systems is where the next chapter goes, the culmination of exploiting Earth orbit for military advantages on Earth itself.

6

ARSENALS FOR SPACE WARFARE

Tafl garreg at fur, hi a neidia at dy dalcen

Throw a stone at a wall, it shall leap at your forehead

The spread of military infrastructures in orbit is rekindling a desire to threaten enemy satellites of major military powers to a level not seen since the 1980s. Today we are witnessing the resuscitation, maturation, and proliferation of space weapons because there are many targets worth hitting in orbit and relations between three of the biggest space powers have taken a turn for the worse. Knocking out C4ISR infrastructure has potentially systemic impacts on conventional military power that would disrupt the modern capabilities of military forces as explored in the last chapter. Silencing enough of the correct satellites at the right time can undermine the logistical systems of complex expeditionary operations, degrade the efficiency and accuracy of strike weapons, and cause problems for the civilian economy and global supply chains. Space warfare is ultimately about picking apart supporting infrastructures in orbit and preventing the enemy from doing the same in return, and not destroying space systems or other space weapons just for the sake of it.

The bulk of the everyday work of military space personnel is maintaining and operating these machines in outer space and ensuring that the radiofrequency tethers connecting these orbiting machines with hardware and people on Earth's surface remain intact. Major military forces seeking to enjoy the benefits of space technologies for themselves need to master space logistics and infrastructure before becoming preoccupied with combat and attacking space systems. Combat operations in space between satellites and

ASAT weapon systems are secondary to the needs of keeping space infrastructure going or denying the enemy's use of their own space infrastructure. This book has focused on the creation of the infrastructure and logistical systems supporting terrestrial warfare and state power projection first, so that space weapons and space warfare can be placed in their strategic and political context. Space warfare is not something waged for its own sake, but rather over the exploitation and denial of infrastructures that now orbit Earth. A simple strategic force propelling the desire for space warfare capabilities is the ever-increasing dependency on Earth orbit for direct military power and economic infrastructure. If both sides do not rely on spacepower, then space warfare tools are less relevant to the war at hand.

In space warfare as in aerial and naval combat, battle is only a means to an end and not an end in itself. How satellites and space systems can be struck is not as important as their effects, whether political, psychological, military, economic, or environmental. Those effects are important in how such actions may help achieve ultimate political goals, or not. Focusing on combat and the methods of destruction and killing—the 'means' of war—at the expense of all else can become an analytical blinker. What Richard Wyn Jones calls 'means-fetishism' in some areas of Strategic Studies must of course be avoided. Jones is correct to argue that means and ends in war are mutually implicated; they are not separated and therefore means and ends have to be discussed at the same time.[1] How a weapon works and what it can do is a pointless conversation without also asking why you want it and for what effects relative to other weapons and political objectives. Space warfare must not be reduced to enthusiastic 'space cadets' living out their *Moonraker*, *Star Trek*, *Eve Online* or *Babylon 5* fantasies based on weapons and propulsion capabilities that are beyond technological and logistical capabilities today. Claiming that a weapons or spaceflight test 'defies' the laws of physics shows either an ignorance of physics or mere alarmist commentary.[2] Given the prevailing influence of science fiction tropes in perceptions of space activities, this is not an idle problem in trying to conceptualise the realities of space warfare in Earth orbit in a technopolitical system that is never envisioned in popular space operas.

Without relating weapons and violence to political purposes and achieving a political goal, the violence becomes senseless, wanton destruction.[3] The political objective of the war can shape what methods and weapons are deemed acceptable, something that will become apparent throughout this and the next chapter. There is nothing special in military or political terms about Earth orbit. Since it is already a place that is used for critical military support tasks that facilitate political violence on Earth, targeting and attacking satellites and space infrastructure is a logical progression. Earth orbit is merely an extension of the 'terrestrial' environment beyond the atmosphere, something that resembles more of a coastline or a new flank rather than a 'new ocean', as argued in the next chapter. However, having a practical understanding of how weapons actually work (or not) is still important. Strategic and political analysis needs to be grounded in the realm of material reality—how specific technologies actually work—and not science fiction. Otherwise, political theorising and strategic thinking can lose all connection to reality and lead to political critique based on a fantasy, such as the belief in a space weapons-based U.S. empire of the future, or the domination of Earth by the United States using space-based weapons to impose its will with impunity.[4] The technical and economic limitations of space-based weapons, discussed below, show how dominating Earth with space weapons becomes something of an unrealistic problem resembling the plot line of the 1971 film *Diamonds Are Forever* and not a useful basis for political theory or something to be seriously concerned about in the decades to come.

The previous chapters have given us a sense of what major powers stand to lose and gain from spacepower and what kind of targets space weapons might be aimed at. This is not a comprehensive survey of every ASAT system ever designed, tested, or deployed, or a summary of every effort at controlling them. Comprehensive and freely available technical primers on various counterspace weapons, their proliferation, and their finer points of operation and capability exist elsewhere.[5] This chapter looks at what space weapons are, as well as their limitations and strengths based on whether they are Earth-to-space, space-to-space, or space-to-Earth weapons, by using examples from the Cold War and present day, and how they support the metaphor or analogy of Earth orbit as a cosmic coastline rather than a high ground. This chapter gives a general sense of the classes of space weapons available today and what they mean for the bigger picture of space warfare, strategy, and military power, rather than providing a detailed, technical 'bestiary'.

Space Weapons

During the Cold War, the increasing influence of spacepower on the battlefield shifted arguments and perceptions within military space communities, putting a strain on the notion that Earth orbit should be a 'sanctuary' from weapons fire and hostile actions in a time of war. As Peter Hays writes:

> given the vulnerable LEO location, few numbers, and exorbitant costs of most space-based intelligence gathering assets, the sanctuary school met the operational requirements of the NRO very well and the idea of ASAT weapons was an anathema to this group. Of course, other groups within the Air Force... were beginning to realize that they were being threatened or targeted by Soviet space systems and saw ASAT systems in a very different light.[6]

That space was a 'sanctuary' that could be free from conflict and warfare, should a war occur on Earth, was a matter of debate in the space security communities in the United States during the Cold War, with the recognition by the mid-1970s that space was not a sanctuary from conflict winning out.[7] Space was already part of battle systems and therefore susceptible to attacks of some kind, whether political, military, and intelligence elites were concerned about their expensive orbital hardware or not. Colin Gray argued at the time that 'no satellite system... enjoys assured survivability. Space is not a sanctuary'.[8] In the Ford and Carter administrations, it was privately accepted that an ASAT capability would have little deterrent value in preventing Soviet attacks on American military space systems given the value of satellites to conventional U.S. military capabilities. In turn, a U.S. ASAT capability would likewise only have value in hampering the Soviet military's conventional capabilities against the U.S. military.[9] This was particularly the case at sea where satellites 'were vital to each side's plans to destroy the other's fleets, and they were therefore well worth destroying.'[10]

Parallels to the U.S.-USSR logics of destroying satellites to deny each other's ability to accurately threaten enemy forces are visible in debates over a U.S.-China war today.[11] Since the late 2000s the major space powers have resumed relatively small-scale, but persistent and incrementally improving anti-satellite weapons programmes. The fact that this follows the ever-increasing dependence of the major military powers on space systems for 'force enhancement' is no coincidence. Anything of great importance to one side is worth attacking by the other in some way.[12] This is the rationale for space weapons and counterspace operations for any sufficiently equipped and resourced military power, not necessarily because the 'other side' may have space weapons themselves. It is no surprise, therefore, that Russia and China have more matured kinetic ASAT test flight programmes than the United States today—they have had more reason to attack satellites in their war plans than the United States since 1991. With Chinese and Russian space infrastructure now expanding and mimicking what the U.S. has had since 1991, potential adversaries of Beijing and Moscow today have more reason to pursue their own ASAT capabilities, which may trigger ASAT developments in many other states that may be involved in regional wars against them.

Much of the debate in U.S. space policy and military communities for decades has vacillated on the issue of whether Washington should deploy 'space weapons' on Earth and in space. Discussion on military space in the 1980s was shaped by the Reagan-era Strategic Defense Initiative (SDI) or 'Star Wars' programme.[13] Whilst SDI was mostly about terrestrial and space-based interceptors targeting ballistic missile warheads as they lunged above the atmosphere into a suborbital flightpath, the technologies SDI required would be better suited towards kinetic ASAT missions and a range of other military targeting uses on Earth itself. Reagan's SDI relied on using pre-existing ASAT concepts and technologies from the Carter and Ford years and turning them to missile defence uses. Reagan was reluctant to pursue ASAT arms control because doing so would limit SDI and missile defence development.[14] These weapons had supportive constituencies in parts of the U.S. military given the increasing spacepower integration in the Soviet military in the 1980s and a general fear of a Soviet 'space resurgence'.[15] The 2000s saw a renewed debate among U.S. military space and national security communities following the Bush administration's withdrawal from the ABM Treaty which loosened restrictions on testing technologies related to missile defence and ASAT systems, alongside Secretary of Defense Donald Rumsfeld's 'Space Commission' of 2001.

Arguments over an 'American Space Empire' drove debate and academic publications at the height of U.S. unipolarity in the international system. Questions in this literature surrounded the maintenance and even expansion of U.S. domination (or a *Pax Americana*) on Earth.[16] The question of whether the U.S. should be 'first' to deploy weapons in space was truncated by the lack of obvious targets and concerns over whether the U.S. being 'first' to 'weaponise' outer space would then encourage other states to copy them. Given that most satellites in space were providing services to the United States and its allies in the early twenty-first century, stimulating space-weapons development would only have harmed U.S. interests. Such debates were overtaken by events in the 2010s—specifically the proliferation of tactically relevant military space infrastructure beyond the United States. Regardless of U.S. inaction, the U.S. and others have continued to put very important military and economic infrastructures in space, providing important targets that justify space-weapons development and deployment.

Colin Gray trenchantly argued that the 'arguments pro and contra space weaponization'—placing weapons in space—'are simply irrelevant'. His reasoning is instructive:

> Many people believe that the now longstanding absence of weapon deployment in space can be made permanent, both by treaty and culturally by taboo. This is a nonsense. It is foolish on two lethal counts. First, it disregards the strength of the strategic incentive to contest space control, at least in pursuit of some prowess in space denial. While second, it is technically absurd because space weaponization is an unduly general conflation of all sorts of means of waging space warfare.[17]

This does not mean military forces should do whatever they want or that building more space weapons is inherently a good thing. This is not a *carte blanche* for the USSF or any military space service to have their space weapons wish-list fulfilled without question—far from it. Some ASAT investments will be more feasible and suitable than others on political, economic, environmental, and military grounds. Gray also outlined useful questions when considering the benefits and drawbacks of various space weapons:

1. Would space weapons fulfil a strategic need?
2. Will the weapons work against (a) unhardened satellites and missiles? (b) hardened satellites and missiles? (c) manoeuvring satellites? (d) satellites that can deploy decoys or create false electromagnetic profiles?
3. Are there other more efficient and affordable methods of attacking satellites and missiles?
4. Can the weapons survive countermeasures?[18]

The answers will differ for everyone according to their needs, means, ends, and enemies. Some counterspace weapons are better suited for some tasks than others, and much depends on what effects are sought in which situations. Sometimes ground-based systems may be better alternatives or just as effective as space-based systems, which does away with the need for space-based weapons.

There is no universal definition of what makes an ASAT, counterspace, or space weapon. They refer to a large variety of methods, techniques, and devices, including specially designed weapons and innocuous space technologies that could be turned towards weapons use—a 'latent' or ad-hoc weapons capability. Much discussion of space warfare in general, and arms control in particular, stumbles over the issue of defining what a space weapon is. 'Weapon' implies a device specifically designed to cause destruction or harm. Many analysts and practitioners cite a difficulty in defining a space weapon because of the dual-use nature of satellites, space vehicles, and missiles.[19] Rendezvous operations (spacecraft docking), on-orbit repair and refuelling (orbital maintenance), and even lasers for debris removal can be used for space warfare as well as their more benign purposes. Any controllable satellite is a potential ramming weapon. But there are limits to such categorical, abstract technological threats. Those satellites are designed and needed for other things and would not be sacrificed as weapons so easily, and they may not be able to target everything of value to the enemy or in a timely fashion. As importantly, the people that operate these systems need to know how to use them as weapons or as targets of weapons, something people outside dedicated military units or intelligence agencies will not be trained for. Like in other environments, specialised platforms specifically designed as weapons will be better placed to target enemy space systems than satellites designed for infrastructure and C4ISR services.

On the positive side, as more 'space weapons' are developed and deployed, identifying specific vehicles and platforms will become easier for treaty purposes, if political will is there to pursue any form of regulations or curbs on space weapons developments beyond the existing ban on nuclear weapons in space, as codified in the Outer Space Treaty of 1967. However defined, determining the intent of a weapon's possessor is a completely different problem because the 'dilemma of interpretation' around weapons means they are always ambiguous in practice.[20] Even though weapons have objective physical characteristics and limits to what they can do, they still create uncertainties because humans can use them, or not at all, in different ways within their physical boundaries. Weapons deployment does not simply present hard facts about intentions. This 'security dilemma' of not knowing what the other side is really thinking and what they might do with what they have pervades much of international relations, and extends to astropolitics. Possession of something is one thing, but knowing under what conditions it may be used, if at all, and in what specific way, is another. Thinking those possible actions through is the purview of strategy and war planning, not 'deterrence', as seen in the next chapter.

The previous chapters have provided a broad outline of what kind of targets there are in orbit for ASAT systems to harass and destroy. Small constellations that provide essential C4ISR functions resemble 'chokepoints' for enemy attacks compared to large constellations that provide non-military services that have no bearing on a military operation. It is important to consider the sensitivities and vulnerabilities of specific platforms as not all space systems are equal. Different parties to a conflict can have very different degrees of dependency on space systems. Likewise, just because one state develops a particular space weapon system does not mean it is well-suited for another state to follow suit, especially if the latter state has no targets to aim at.[21] On the utility of ASAT technologies, Paul Stares wrote in 1987 that:

> the value of military space systems to each superpower depends largely on contextual factors, so the effect of their loss or threatened loss is likely to vary in the same way. This basic fact is rarely acknowledged, however. Instead, both sides of the anti-satellite debate rely on generalities to argue the merits of their cases.[22]

Space weapons and their targets can be based on Earth with targets in space, based in space with targets in space, and based in space but with targets on Earth.[23] In addition to these basing and target locations, further varieties can be drawn on whether they use 'soft-kill' or 'hard-kill' methods of attack. Hard-kill refers to physically destructive attacks that damage satellites, usually through kinetic impacts or ramming, co-orbital grappling, explosive damage, or intentional radioactive damage from a nuclear blast. Hard-kills can also include attacking ground stations and terrestrial infrastructure which supports or relies on space systems, and can be attacked with terrestrial weapons, meaning methods of space warfare and influencing the command of space in wartime need not necessarily involve actions in space itself. Special Forces or covert sabotage units being sent behind enemy lines to kill the people staffing a satellite control station and their equipment is one such method of attacking space *systems*, but not necessarily attacking the satellite itself. The focus on kinetic-kill interceptors of the arms control community was a problem back in the 1970s and 1980s, and risked confusing the regulation of one kind of ASAT weapon system for guarantees of peace in their time.[24] Banning one kind of space weapon does nothing to ban the rest, or make space warfare an impossibility. The same problem should be avoided today.

On the more extreme end of hard-kill capabilities are terrestrially-based nuclear weapons, which were among the first ASAT weapons developed. High-altitude nuclear detonations (HAND) would destroy any satellites caught in the fireball and degrade other satellites over time by enhancing orbital radiation from the bomb. Kinetic hard-kill systems can create large clouds of debris or enhanced levels of radiation in vast swathes of orbit by ramming into satellites at such high speeds. Debris in orbit can pose an indiscriminate hazard to other satellites, depending on the exact location and spread of the debris generated. Such acts will increase the probabilities of a cascade of debris collision events, where the population of debris pieces increases as more pieces collide into each other or functioning satellites, creating yet more pieces of lethal debris. A runaway growth in that population is referred to as the Kessler Syndrome.[25] Physical destruction accumulates to a 'scorched orbit' strategy when the long-lived orbital debris effects and the potential of the Kessler syndrome are taken into account. Given the high velocities and the lack of atmosphere in orbit, small pieces of debris even a centimetre in diameter can fatally damage space vehicles. Debris will stay in orbit for longer the higher up their orbits are. Over long periods of time, the debris 'cloud' will spread out and threaten more orbits mostly in the proximate altitudes, but some pieces can be thrown into much higher orbits. Therefore the consequences of hard-kill systems may raise the political threshold at which the use of such weapons may be thought to be worth using given their indiscriminate environmental consequences.

We should not assume that the political threshold for using such environmentally destructive weapons may never be crossed. As Clausewitz argued, the political objective of the war and the condition of the fight may be serious and important enough to justify especially desperate and destructive efforts.[26] One side might think that fouling swathes of Earth orbit may be worth it to win a war over what matters on Earth. The security of machines in space will be secondary to political control on Earth. Military victory and political control on whatever the dispute is on Earth is what matters most, not necessarily machines in orbit. How badly does one side want to win the war on the ground and determine the immediate political future there, and simply accept the consequences for machines in the lifeless volumes of space?

Soft-kill refers to methods of attack that may neutralise or harass a satellite's normal operation, such as electronic warfare or computer network operations, with effects that are often reversible or temporary and cause no lasting or only minor physical damage. These are included as space weapons due to their potential usefulness in attacking and disrupting space systems, effectively struggling over the command of space and the control and denial of space infrastructure. The effectiveness of electronic warfare can vary from an everyday nuisance, such as unintentional jamming which occurs frequently for commercial SATCOMs and can be reversed or undone easily, to system-spamming communications which shut out controllers and users from the space system for a certain time period. Electronic warfare (radio jamming) is 'an endless game of point and counterpoint as threats are identified and evaluated, and countermeasures are developed and deployed in response'.[27] Under electronic warfare or computer network (cyber) attack, satellite command and control could be lost, sending the satellite drifting into an orbit which puts it on a collision course or a useless orbit without fuel to return its original useful orbit. This could lead to physical destruction. In skilled hands, and with some luck, jamming and cyber-attacks can be just as permanent and destructive as kinetic ASATs. Another scenario could be the disabling of a satellite control centre in advance of a planned debris avoidance manoeuvre, causing damage or destruction to a satellite. A jamming incident could have similar political effects to a kinetic-kill attack if it is catastrophically successful against very sensitive space systems, and invite a serious response. The USAF has long referred to lasers and electromagnetic pulse (EMP) devices as 'weapons', and cites radio frequency jammers as threats to U.S. systems. EMP should not be seen as a weapon in its own right as depicted in the 1995 film *GoldenEye*. Generating a massive EMP strike requires multiple nuclear detonations in low orbit and could very well start a nuclear war, at which point EMP effects are the least of many concerns. Although radio jamming devices may not be classed as 'space weapons' by everyone, deliberate and serious interference may be responded to with force regardless.[28]

Electronic warfare is a mainstay of military operations on Earth, and space is no different. If anything, electronic warfare is essential for defenders and attackers in space because everything is remotely operated through the radiofrequency spectrum. Without the radiofrequency tethers connecting orbiting machines to users and networks on the ground, satellites are useless pieces of junk in the making. Any military service neglecting its electronic warfare capabilities in space courts disaster. Without people on orbital spacecraft, they must be remotely controlled through radio communications (or fully autonomous) to keep operating. If these radio tethers are cut or compromised, satellites have no other way of relaying their data or gathering it in the first place, autonomous or not. Electronic warfare is about using and denying the use of the electromagnetic radiofrequency spectrum, which is used for a variety of purposes such as communications, surveillance, navigation, and weapons guidance. Electronic warfare against space systems is the art and science of broadcasting a signal at the frequency your enemy is trying to communicate through, drowning out the frequency with white noise or garbage communications, preventing two ends from communicating meaningful information between them. In effect, it would be like trying to scream nonsense at a high volume so that the original senders and receivers could not hear anything else—only loud noises. Jamming an uplink can also be so 'loud' as to convince a relay satellite to retransmit the jammer's garbage signal noise to the receivers instead of the authorised sender, meaning the intended recipient of the relay satellite's downlink receives unintelligible noise.[29]

The radiofrequency spectrum is a particularly busy place—all sorts of machines and equipment rely on radio links on specific frequencies to work. It is not just communications systems that are susceptible to radio jamming, but precision guidance and SIGINT collection systems too. Fooling these systems is an important method of reducing the confidence behind precision bombing campaigns and tracking systems, increasing attrition, and exhausting the enemy's munitions supplies. SAR satellites can be fooled with materials or emitters that produce false radar telemetry back to the SAR antennae. Electronic warfare is also something that is done throughout peacetime, including industrial espionage which can be conducted in the radiofrequency spectrum.[30] Electronic warfare is a vast subject in its own right, and in terms of space-centric views of jamming we must remember that they are often part of wider modernisation drives and capability programmes in multiple environments. For example, Russia's electronic warfare can be interpreted as part of a 'wider effort by Moscow to adopt and strengthen its network-centric capability, which focuses upon C4ISR integration'.[31] Terrestrial jammers are ubiquitous and can be placed in aerial and maritime vehicles and platforms, as well as ground vehicles and carried by individuals as requirements dictate. Contrary to hard-kill methods, in soft-kill space warfare methods many military forces will have significant levels of experience to call upon in wartime. Various Russian satellite jammers have been found in the Ukrainian War and many commercial and civilian satellite communications and navigation signals in the region have been disrupted.

Electronic warfare, like cyber operations or computer network attacks, have many techniques and methods exploiting various weaknesses and procedures. But this variety of attack method can be met with a variety of defensive measures or counterattacks. Electronic warfare is a highly classified and sensitive area of activity, and it is therefore more difficult for open-source assessments to gauge the 'strengths' of various military forces in these areas. Cyber warfare, or computer network operations,[32] are another option open to states and non-state actors in terms of attacking space systems. Space systems are a mix of machines relying on hardware and software, piloted remotely, to complete their tasks. Like any other high-technology system, cyber warfare is another 'front', but like electronic warfare is another discipline of warfare in its own right and only the most basic of overviews can be provided here. A useful way to view cyber operations in space is through the lenses of espionage and sabotage against infrastructure, rather than like missile strikes or artillery firing on military units.[33] Cyber warfare and spacepower will be more like stealing data, commandeering certain satellite control systems, disabling some systems, or 'spoofing' the signals and information satellites receive and transmit. Depending on the desired effect and software installed, cyber intrusions could cause hard kills as well as soft kills, such as commanding a satellite to de-orbit or mismanage its heat systems and fail.

The boundaries between electronic warfare and cyber warfare can blur if a radio emitter is used to transmit malicious code or false data to the target system wirelessly. This is the plot device of the 1997 film *Tomorrow Never Dies*, one of the few plausible representations of space warfare in film: a military GPS signal is 'spoofed' with incorrect data to send a British warship off course and into Chinese territorial waters, engineering a major nuclear crisis that a diabolical media baron wishes to use for his own purposes. The complexity and diversity of software, control systems, and intranets means that it is impossible outside of practitioner communities to know how vulnerable any particular space system is to intrusion via software methods. Like electronic warfare, it is important to recognise it for the cat-and-mouse game that it is, where vulnerabilities or 'exploits' are continuously detected and countered with patches and updates. Attacking space systems will involve cyber warfare as well as space warfare, but cyber warfare is not the focus of this chapter. In the same sense, maritime or aerial warfare today remain geographic specialisms, but also have to be aware of their interactions with cyber warfare and computer systems, just as space warfare will be impacted by cyber operations. In the same way geographically oriented military services have all had to adapt to electronic warfare for the past century.

Electronic warfare is a technical activity that space powers must master and counter to ensure their command of space and the integrity of their space systems, as well as to threaten those of adversaries. Electronic warfare can pursue two general kinds of attack: wide-area jamming or narrow beam jamming. Wide-area jammers are blunt tools designed to broadcast a powerful signal in all directions for localised jamming. For instance, a civilian GPS signal jammer on a vehicle can 'drown out' the relatively weak CA code of GPS that non-military receivers use by simply broadcasting a stronger local signal. Such jammers require more power to be effective against stronger signals, and the stronger the source of the jamming, the more likely it can be found and targeted by military forces and destroyed with anti-radiation missiles—missiles designed to home in on specific signal sources. Contemporary anti-radiation missiles can work at great speeds so that they can destroy the jamming emitter before it has a chance to react to the incoming missile by switching off and moving or changing frequency. If known to be in play, anti-radiation munitions and their platforms can loiter so as to discourage the use of radar systems or jamming equipment.[34]

Narrow beam jamming is a more precise electronic warfare method that is tailored to a specific target. It is a directed signal beam that is pointed at the sender or the receiver, or both, of a radio communications link and casts a very specific beam of radio signals that those specific senders and receivers use. Outside of this narrow beam, it is hard to detect that jamming is going on, let alone find the source of the jammer, unlike wide-area jamming which broadcasts over a large area in all directions. Narrow beam jamming, however, poses its own restrictions: it requires more detailed knowledge of the targets including communications specifications and locations. To respond to this the targets can move to alternative links or 'frequency hop' to different frequencies that may be unknown to the attacker, rendering the jamming ineffective. Jamming satellites in GEO with a narrow beam is more feasible than those in LEO because a GEO satellite appears stationary from a ground-based jammer, whilst a satellite in LEO will clear the horizon within minutes. Jamming mobile targets with narrow beams can be extremely difficult. In addition to this cat-and-mouse dynamic of electronic warfare, radio broadcasters can engage in deception activities by broadcasting fake electronic signatures to fool, misdirect, or overload enemy sensors, masking other activities in the process.

Lasers today generate soft-kill effects and are not powerful enough to physically destroy satellites. Proven anti-satellite lasers on the ground are known to 'dazzle' the optics of an imagery satellite, much like a laser pointer can temporarily blind a person. The power required can be as low as 10 watts due to the sensitivity of the optics or sensors, but it does require a high degree of accuracy to reach them. This blinding effect may be temporary, or if sustained can cause permanent damage. Some lasers could overload sensors—if they are accurate enough—by focusing a powerful signal on it that it is too powerful for the processor to handle, leading to a permanent degradation in the system's capabilities.[35] Like electronic warfare, the speed of its propagation is at light speed. However, a laser's damage effects take longer because it is additive or cumulative in nature. Lasers need time, focus, and several aligned emitters to focus enough energy for physically destructive effects such as burning or melting satellites. Physically destructive lasers require a lot more power:

> a laser welding machine in a factory typically uses a laser with a few hundred to a few thousand watts of power directed by optics with a diameter less than 0.1m. A space-based laser intended for targets on or near the Earth requires millions of watts of power and optics with a diameter of about 10m... Weapons capable of destructive effects will be large and expensive.[36]

Even in the vacuum of space where there is no atmosphere to dilute a laser's effects, space-to-space lasers would still need a lot of wattage to burn through outer layers and materials, especially if it wishes to do so quickly without having to sustain a focused beam for an extended period of time. Laser anti-satellite weapons have been the subject of intense debate in the past, in particular in the pages of *Aviation Week* in the U.S. in the 1970s where several alarmist stories about Soviet laser technologies were run.[37] Satellites are more vulnerable targets for lasers than targets inside an atmosphere. Satellites could suffer heat damage or imbalances in their heat distribution leading to failures if enough laser energy is effected upon them, without necessarily needing to directly melt them.[38]

Strategies of space warfare must also include other weapons systems or techniques on Earth that can strike at terrestrial elements of space systems such as ground stations and spaceports. Just as attacking naval facilities—even from the ground—is part of maritime strategy and has a direct impact on naval warfare, attacking ground-based facilities from Earth itself has an impact on space strategy and wider efforts in a space warfare campaign. If the necessary number of sufficient satellites or up/downlinks are silenced, the service is denied, regardless of how or from where they were silenced. What is 'necessary' will vary depending on the effect sought, the space infrastructure and satellite constellation targeted, what is politically permissible, and what is likely to work. Jamming and cyber intrusions always raise the possibility that service might be resumed if political demands are met, or the source of the interference is eliminated. Space strategists must be open to all weapons basing locations and methods of attack, and targets that are on Earth and in space, with the greatest interest in the effects of those weapons for the political goals sought.

A grounding in the actual past and present of ASAT programmes is more important than monitoring hyperbolic chatter over 'emerging' technologies or futuristic concepts that may be 'just around the corner' or the 'threat of the week' promoted by government briefings and news cycles. This is not a novel problem as it plagued 1980s discussions of space warfare, with fantastical technologies under the all-encompassing brackets of SDI seemingly promising an escape from nuclear destruction.[39] Hedging against 'potentially effective' ballistic missile defences is a strategy as old as missile defence programmes themselves, and the 1980s saw no shortage of promises over effective and efficient laser or particle-beam weapons that can physically destroy satellites. These have yet to materialise, therefore it is important to consider what proven space warfare methods actually exist or are being developed before discussing their wider strategic and global power political implications in the next chapter.

Terrestrially-based ASAT Weapons

Earth-to-space weapons are the most highly proliferated type of space weapons capability today. They are also the longest-established in terms of their origins. Nuclear weapons as ASAT systems emerged within the United States military in the 1950s and became operational into the 1960s. The Nike-Zeus system, or Program 505, was a repurposed air-launched terminal-phase ballistic missile defence system. It was designed to 'nuke a nuke' as the enemy warhead was reaching the final phases of atmospheric ballistic flight. Zeus was deemed ineffective as a terminal missile defence system against nuclear attack. As an ASAT weapon, however, it held greater potential. Regardless, it was reduced to a research and development project by the final years of the Eisenhower administration. At around the same time, Weapon System 496L was developed to check any Soviet effort at space-based or orbital weapons platforms—which the Soviets duly achieved with the Fractional Orbital Bombardment System (FOBS) in the 1960s, as seen below. Safeguard was an operational ABM system fielded by the United States but was only active for a short time. Bold Orion and High Virgo air-launched systems were also geared towards detonating a nuclear warhead in the desired region of orbital space (HAND), and Bold Orion managed to fly close enough to the Explorer 6 satellite that it would have destroyed the satellite had a nuclear bomb been detonated.[40] Whilst the thermal and blast effects of a nuclear warhead in space are minimal in the vacuum of outer space, the radiation it would release would spread to various parts of orbit and along the patterns of Earth's magnetosphere and seriously degrade the electronics of other satellites over time. The Starfish Prime nuclear weapon test in orbit in 1962 generated an electromagnetic pulse which knocked out a significant fraction of the orbiting satellites at the time, and disrupted Hawaii's power grid. The lack of military gain from such weapons and a desire to not accidentally destroy so many satellites through testing helped bring about the 1963 Limited Test Ban Treaty which forbade the testing of nuclear weapons in space, the atmosphere, and on Earth's surface.

Program 437 used a repurposed Thor MRBM missile and proved the feasibility of a direct-ascent satellite-targeting nuclear weapon, carrying a 1.4 megaton warhead up to 1,300km altitude. In September 1962 the Air Force Secretary Eugene Zuckert said that the U.S. 'had no choice but to extend our defences as far as they need to be extended to save freedom on Earth', intended to counter Khrushchev's earlier remarks that the Soviet Union could 'hit a fly in space'.[41] Project Mudflap in June 1963 reached test readiness, and a Nike-Zeus system successfully intercepted an old rocket body in orbit above the Kwajalein Atoll. There was a fair amount of mirror-imaging going on in terms of the U.S. military justifying the pace of the military space programme by implying the Soviet Union had started on equivalent systems. As Parker Temple writes:

> undoubtedly many of these capabilities the Soviets were working on… For every weapon system the Air Force started, the Soviets would be postulated as also having started their own version, for which the Air Force then became committed to developing a counter. Such an upward spiral of postulated threat and response could not continue indefinitely. Eventually, reality would demonstrate the inflated nature of the postulated threat.[42]

Almost all Soviet space systems have U.S. equivalents, but Washington's commitment to a specific capability was not usually decisive in Moscow's decision-making, though often part of the mix.[43] That may provide some hope to the U.S.-China threat perceptions today regarding ASAT developments. U.S. ASAT development in the Cold War was not done so in a single drive towards space dominance, rather it was a piecemeal response to potential threats posed by enemy space systems to U.S. nuclear capability, and later, to conventional terrestrial military forces. 'Threat hyping' to justify various weapons development has a shelf life, as seen in the Cold War, but behind such hypes there is often genuine concern over specific activities and threats that hype embellishes or misdirects attention on.

In the larger balances of nuclear weaponry in the 1960s, ASAT systems remained a rather marginal and exotic capability. National Security Memorandum 345 in January 1977, in the Ford administration, spelled out the clear logic of a need to attack Soviet radar satellites with hard- and soft-kill ASAT weapons because they were enhancing Soviet military forces in conventional combat capabilities.[44] The Carter administration sought to use ASAT development as a bargaining chip in nuclear arms control, although not necessarily as interested in ASAT capabilities in themselves.[45] With SDI in the Reagan administration, ASAT weapons were back on the agenda for space warfare purposes, not arms control bargaining.

In 1983, the CIA estimated that the USSR had twenty-two Tsiklon-2 boosters stored ready for short-notice launch in a kinetic-kill ASAT attack role into LEO altitudes. Seven RORSAT/EORSAT were possibly stored to be launched at very short notice too, which would be launched by the Tsiklon-2, to replace any losses caused by U.S. ASAT operations.[46] In 1985 the U.S. air-launched Miniature Homing Vehicle was tested with an altitude range of 560km, still the lower reaches of LEO. However, only fifteen of the missiles were created and very few F-15s were modified to equip them. The programme was mothballed in 1988, but not before the system had been considered for modification for launch from aircraft carrier-launched F-14s, SM-2 missiles from Aegis destroyers, or Poseidon missiles launched from SSBNs.[47] Today, the United States fields an ASAT-capable SM-3 missile aboard its Aegis missile defence systems housed on U.S. Navy destroyers that can reach lower altitudes in LEO.[48]

As well as the Aegis missile systems, another ASAT-capable 'missile defence' system the U.S. has deployed is the Ground-Based Interceptor (GBI). Forty-four are currently deployed in Alaska and California, and may have greater effectiveness at ASAT missions than the SM-3. However, GBIs have not been used in any ASAT intercept test.[49] Given the vastly expanding nature of Chinese space infrastructure, and the revival of Russian military space infrastructure, considerably more than forty-four GBIs would be needed to pose a significant ASAT threat to the useful Russian or Chinese LEO-located space systems at the same time, leaving MEO and GEO assets free from harm.

These terrestrially-based systems emerged from the SDI programme, which themselves were drawn from earlier missile defence and ASAT concepts from the 1950s, 1960s, and 1970s. Paul Stares claimed that SDI promised to be a 'breeding ground for new generations of ASAT weapons if nothing else' because techniques of missile defence were so similar to ASAT techniques. Intercepting a nuclear warhead is more difficult than intercepting a satellite. Therefore striving for the more difficult goal of warhead intercepts would produce systems capable of ASAT attacks.[50] Satellites tend to be easier to detect through various sensors, travel along well-established orbits, and emit radio signals. Warheads are fleeting, small, extremely rapid, and give off no radio emissions. Stares was proven correct with the U.S. Aegis SM-3 and GBI 'missile defence' interceptors that are capable ASAT weapons. In 2021 a U.S. Navy Vice Admiral referred to the SM-3 as a 'space weapon', clearly referring to its proven ASAT capability and vindicating arguments that many kinetic interceptor BMD systems are functionally ASAT systems, and specifically the SM-3 missile and the Aegis system.[51] Ballistic missile defence weapons with 'midcourse' interception windows should be seen mostly as an ASAT capability that have many impacts on conventional warfare as more military forces become more dependent on spacepower for their vital conventional warfighting functions, rather than missile defence and nuclear warfare alone.

SDI may have impacted Soviet policies in some ways as different bureaus and departments used it to justify their own policies and actions or were genuinely concerned at the eventual potential of the technologies involved. The USSR's Kontakt system was developed not necessarily because of SDI, but it was later 'branded' as an 'anti-SDI' weapon like many others.[52] This echoes the multiple re-brandings of the SBIRS-Low idea at the U.S. Missile Defense Agency. Kontakt relied on the A-135 missile which achieved operational status in late 1989, but had first begun development in 1978.[53] Six MiG aircraft were modified to equip the Kontakt air-launched ASAT in the 1980s, and may have been resuscitated in the late 2010s, or perhaps replaced by an air-launched co-orbital ASAT, Burevestnik.[54] Unlike the U.S., the USSR as far as is known had never explicitly developed ground-based nuclear-tipped anti-satellite weapons, but some nuclear weapons could have been used in such a way.[55] The

USSR eventually concluded that the SDI programme would never present a major threat to the Soviet nuclear arsenal as plentiful countermeasures against space-based assets and missile interceptors were available and could be developed.[56] Today, Russia fields the Nudol ground-launched kinetic ASAT system, whilst the ground-based S-500 air-defence system may have some ASAT potential.[57] According to Weeden and Samson, the Nudol is a road-going transport-erector-launched ballistic missile that can reach LEO, with several flight tests being conducted since 2014.[58] Until November 2021, these tests were merely flight tests without intercepts. The November 2021 test included an interception, creating more debris in orbit and resulting in evasive manoeuvres being taken by the International Space Station. Many details are lacking on Nudol in terms of its targeting potential and altitude ceilings, but importantly its development is accompanied by extensive tracking and targeting systems, including ground-based lasers and mobile radars, as well as space-based tracking and measurement systems alongside SSA infrastructure.[59]

China's ground-based kinetic ASAT efforts can be traced back to their first dedicated ASAT programme in the 1970s (within Programme 640) which was spun out of an anti-ballistic missile programme, as well as a new dedicated ASAT system in the 1980s (within Programme 863).[60] Programme 863 was a wide-ranging investment-in-national-technology programme, focusing on high technologies including space, for a range of civilian and military applications.[61] The hit-to-kill weapons system received a boost after the U.S. bombing of the Chinese Embassy in Belgrade in 1999.[62] Flight testing proceeded through the 1990s and 2000s. China caused a diplomatic stir in 2007 when it successfully conducted its own kinetic-kill ASAT test on a defunct Fengyun satellite at 800km altitude, creating over 3,000 trackable pieces of debris in LEO that will take decades to deorbit. There was a lack of coordination between the military conducting the test and the Ministry of Foreign Affairs having to deal with the political fallout, and perhaps the military leaders in charge of the test were eager to demonstrate they had achieved a mature capability, not thinking of the political impact of the test. This was viewed as a 'net negative' event in terms of security interests by the Chinese leadership.[63] This was the SC-19 system, which is currently deployed, and is seemingly based on the DF-21C MRBM which is also road-mobile using transport-erector launchers. That said, known kinetic ASAT test launches have come from prepared launch pads. Though China has proven a kinetic ASAT capability that can reach LEO, it is possible that China has tested similar direct-ascent weapons that could reach altitudes in MEO and GEO.[64]

India may have conducted a failed test in February 2019, but Mission Shakti in March 2019 launched an indigenous missile and successfully intercepted an Indian target satellite at around 300km altitude, shortly followed by Indian statements against the weaponisation of space. The target satellite was much lower than the altitude of China's test in 2007, so most debris will be short-lived. However, some pieces have been detected up to 1,000km altitude due to collision effects.[65] India joined the 'club' of space powers that have tested a kinetic-kill ASAT system. Like Russia, China, and the USA, India's kinetic-kill ASAT capability intertwines with ballistic missile defence programmes and their tracking systems. India's 2019 kinetic ASAT test, therefore, has as many implications for conventional war planning and border response capabilities between China and India as it does for long-term nuclear missile procurement, particularly as satellites become essential C4ISR infrastructure for both border forces. The Indian ASAT test should be seen first and foremost as an ASAT test, and not an emerging 'midcourse' ballistic missile defence demonstration. In line with India's general approach to spacepower, their military space infrastructure development involves a diversity of partners and providers. India's Green Pine radar was constructed by Israel (using its experience with its Arrow missile defence system) and the country purchased a handful of S-400 surface-to-air missile systems from Russia.

It is important not to over-emphasise the threat of hard-kill ASATs, despite the risks that these weapons pose. Offensive actions do not determine everything in space. There are limits and countermeasures to what any ASAT can do. These technologies are proven and deployed, but not at a massive scale for prompt and successive attacks against all potentially important enemy targets. Their impact on military power and therefore strategy by themselves will be marginal if their numbers and feasible targets remain limited. As Jaganath Sankaran argues, many crucial U.S. military and intelligence satellites will be out of reach of China's SC-19 ASAT weapons (and by extension, Russia's initial LEO kinetic ASAT capability). Any large-scale operation, even to make a serious dent in U.S. ISR assets in LEO, may put significant strain on Chinese logistics.[66] In LEO, depending on the exact altitude, which can range from 300–2,000km, the kill vehicle could take only a matter of 5–15 minutes.[67] Direct-ascent kinetic ASAT weapons will take hours to reach MEO and GEO. Reaching MEO and GEO, where more sensitive military infrastructure lies, requires another leap in the quantity of available rockets and fuel to reach a significant number of targets.

Forrest Morgan argues that 'satellites are difficult to defend against adversaries with capabilities to attack them. As a result, space, like the nuclear realm, is an offense-dominant environment with substantial incentives for striking first should war appear probable.'[68] This is an isolated view of one engagement. A one-satellite one-interceptor scenario is not the totality of a space warfare campaign and military space strategy. It is not so simple when entire infrastructures and large technological systems and backups, as well as the number of available enemy interceptors that can survive prolonged terrestrial hostilities, are taken into account. Some systems might be more vulnerable than others, and the devil is naturally in the detail. As Stares commented above, the value of any ASAT capability changes a lot depending on the specifics of the case at hand and we should avoid speaking in totally abstract terms. For example, the SBIRS constellation is small and important—only a handful of satellites in HEO and GEO provide crucial infrared spotting capabilities to the U.S. military and Intelligence Community. Silencing the handful of U.S. AEHF satellites will silence the most reliable and major channels between the President and nuclear weapons command and control. Disruptions to AEHF will affect the British, Canadian, Dutch, and Australian militaries too. A coordinated series of kinetic ASAT strikes against constellations of less than ten satellites is a different kind of task to taking out larger constellations. GPS is a critical infrastructure due to its entrenched and ubiquitous military and civil PNT services. The GPS constellation today has around thirty satellites in orbit. It would take half a dozen successful intercepts reaching out to MEO to begin to undermine the 24-hour global coverage of 3D positioning data provided by the constellation, and even then the degradation would be a matter of degree and not absolute because it is a system, not a single platform.

Serious depth in ASAT capabilities only begins when dozens of interceptors, ready to fire, are deployed at any given time over a variety of launch points. Localised terrestrial GPS signal jamming, therefore, may look like a better alternative for navigation warfare. With even larger and more distributed, decentralised constellations, such as Iridium, Planet, and Starlink, a higher number of interceptors and attacking launches would be required to significantly degrade the system. Though these systems are not used for essential C4ISR in the U.S. military, they provide auxiliary intelligence sources or less secure military communications services instead. However, distributing space infrastructure among many more platforms is recognised by Morgan, but is not squared with claims of the supposed offence dominance of the space environment.[69] Such a view is a focus on tactics of a single engagement rather than an entire campaign. There are many ways to reduce the significance of attacks or defend against them. Sankaran is correct to argue that direct-ascent ASAT systems will probably not be decisive in future conflicts. Rarely are single-weapon systems so decisive by themselves. Even the Maxim Gun of the late nineteenth century, arguably one of the most significant automatic-weapons innovations of the time, required skilled crews and extensive logistical networks to sustain its use and would not prove decisive—though important—in many wars.[70]

At present, ASAT weapons, particularly kinetic kill variants, are not deployed at scales of hundreds of interceptors. Yet they are proven technologies that can be rolled out on a much larger scale if required. Space-weapons testing and development programmes that span a decade or more should not be confused for an arms race, which requires haste and mass deployment in comparable terms by more than one power. These are not 'crash' programmes and therefore there is no need for alarmist reporting on such developments, regardless of whether one supports or opposes them. Beyond the kinetic options, electronic warfare and cyber operations may provide more immediate effects and results, potentially catastrophically so should a computer virus dismantle GPS software. However, there are many more defensive options available to the defender in electronic warfare and computer network operations than a kinetic interceptor strike on a satellite. Launching a major attack on GPS ground stations, including known backup facilities, is an example of a non-space option of undermining a space system, though it might have far more escalatory potential than attacking the satellites themselves.

Both Russia and China have been jamming civilian and perhaps military GPS signals with localised downlink jammers, meaning GPS in other parts of the world are unaffected but a certain theatre of operations may have degraded some signals of GPS. Russian counterspace electronic warfare units have been deployed into eastern Ukraine, with the region suffering from many civilian and commercial GPS signal issues.[71] As with their precision strike weapons and C4ISR systems, the Russian military has been able to practise counterspace jamming in Syria as well. This does not mean GPS will be useless or a complete write-off in a conflict as the U.S. military naturally devotes resources to defending the integrity of the military signal and attacking others' methods of doing so. The United States has fielded the Counter Communications System (CCS) since the mid-2000s: units that can be deployed in-theatre to jam enemy satellite systems and satellite jammers. Formerly USAF, these are now USSF units. At present there may be over a dozen deployable CCS units. The U.S. military has other jamming capabilities targeted towards GNSS, such as GLONASS and BeiDou under the NAVWAR (Navigation Warfare) programme—which is used to locally degrade many other GNSS signals since Selective Availability for GPS was shunned in 2000. With the United States forgoing the option of flipping the 'off' switch for civilian GPS signals, it developed localised military jamming units as a way to secure U.S. forces from enemy forces using accurate civilian GPS signals.[72]

We do not yet know what impacts such efforts have had. It is important to note that:

> conducting operationally-useful, dependable, and reliable jamming of highly-used military space capabilities, such as GNSS, is more difficult than most commentators suggest. Military GNSS signals are much more resilient to jamming than civil GNSS sig-

> nals, and a wide variety of tactics, techniques, and procedures exist to mitigate attacks. It is much more likely that an EW counterspace weapon would degrade military space capabilities rather than completely deny them.[73]

None should take unfettered access to GNSS for granted—whether civilian or military. Iran may have been able to jam the military GPS signal in conjunction with other interference techniques to bring down a U.S. UAV, which further enhances the profile of Iran's antisatellite jamming capabilities beyond commercial SATCOMs and television.[74] Iraq attempted to jam coalition satellite signals in the 2003 Iraq War.[75] These examples serve as reminders that some methods of waging 'space warfare' or disrupting space systems are proliferated beyond Earth's richest or largest military powers. But countermeasures and adaptations are possible. In the civil sector, maritime shipping continued to work despite many GNSS disruptions and there are several terrestrial backup opportunities.[76] In civil infrastructure as well as defensive operations, friendly local infrastructure can provide augmentations and backups for GNSS services, meaning some space infrastructure coupled with friendly terrestrial backup systems can assist strategic defence on Earth.[77]

China and Russia may be conducting research and development into various kinds of directed energy weapons, including microwave emitters (or radiofrequency weapons) which project a beam of energy that can damage electronic circuitry. Both states have a long history of interest and development in laser weaponry and technologies, but Russia has more legacy programmes to fall back on.[78] Some reports of dazzling by Chinese ground lasers on reconnaissance satellites have been circulating as well. The United States and the Soviet Union pursued various research and development projects through the Cold War, but advanced little beyond the laboratory as weapon systems.[79] In 1996 the Clinton administration pursued the Airborne Laser (ABL) system, which was a megawatt-range chemical laser fitted onto a Boeing 747–400F in order to fire on missile boosters as they launched. A successful test firing against a solid- and a liquid-fuelled rocket in boost phase did not save it from cancellation during the Obama administration.[80] One of the major limitations of the capability being that laser aircraft would have to be on-station and in close proximity to all potential launch sites at all times. Whilst the U.S. maintains a widespread laser research and development ecosystem, there is little direct evidence of turning laser equipment to counterspace roles, such as dazzling, though the U.S. faces no technological roadblocks to doing so if it wished.[81] A Russian airborne laser developed and reportedly successfully tested against a Japanese satellite in 2009 has been restarted in 2012. Details on the 1LK222 airborne laser, the Krona ground-based laser, and the Peresvet mobile laser (designed to accompany ICBM units) remain scant.[82] As in terrestrial military operations, riot control, or aggressive and forceful responses to protests, lasers are useful for blinding and irritation, but not yet proven as destructive weapons on a large scale. Weeden and Samson prefer to refer to 'dazzling' lasers as a countermeasure rather than a weapon as their effects are mostly reversible.[83] With regard to anti-satellite weapons, 'it is one thing to be able to damage a satellite that obligingly floats by through the laser beam when beam-transmission conditions are just right, and quite another to be able to zap satellites on demand.'[84]

Ground lasers firing up at targets in orbit suffer penalties the higher their targets are. A beam that focuses energy on a 1-metre diameter at a range of 360km will reach 100 metres diameter at GEO altitude of 36,000km, whilst containing the same energy over that surface area, thereby diluting its thermal effect to 1:10,000 of the intensity at 360km altitude. Ground lasers suffer further penalties from only viewing what is within their horizon or what may be within reach of a relay mirror in space. That dilution increases the lower the satellite is in the terrestrial laser's horizon due to the increased thickness of the atmosphere at low angles. Fixed ground lasers will also be known to capable adversaries and measures can be taken to anticipate their use when satellites overfly them. Against targets in LEO the amount of time a terrestrial laser will have to fire its beam on the target satellite will decrease because of its relative movement to Earth's surface. At 400km altitude, a satellite would only be within effective firing range of a single ground-based laser on one point of Earth for 2% of its orbit—a few minutes.[85] As well as the limitations of weapons-grade lasers, satellites can be hardened to reflect particular light wavelengths, absorb them better, or ablate away any heat. Hardening satellites against many forms of laser attack can be achieved. Powerful lasers are therefore big and bulky because of power requirements, making ground lasers cheaper to supply, maintain, and upgrade, and their aim is far more stable from the ground than in the air or in space.

Continent-sized states have many opportunities for kinetic satellite intercepts, lasing, and jamming with ground-based systems. Russia, China, and the USA have larger territories to work with in satellite overflights for targeting, but also in the mobility of ground systems that have many places to hide. Ship-borne ASAT capabilities would only increase the flexibility further based on naval logistics, but may be harder to hide. States at extreme latitudes may have more options for ground-based counterspace operations against LEO due to the convergence of polar orbits at the poles, which can form something like a chokepoint. Large states can ensure a higher number of different points to increase the detection fields and fields of view of terrestrial anti-satellite weapons fire—whether kinetic-, radio-, or laser-based. Relay mirrors in orbit could enhance the target coverage of ground-based lasers, but restrictions would still apply if the ground-based laser was obscured by inclement weather. Deploying mirrors fit for the task is also no small or cheap proposition. As seen below, many space-based lasers would be needed to maintain high chances of being within the line of sight, or reducing the 'absentee ratio', and so there is no easy solution to the limitations of either terrestrial or space-based beam weapons.

There are no easy, cheap, or simple 'wonder weapons' for space warfare, just as on Earth. Depending on the targets, weapons effects, and the resources available, it is not immediately obvious when a space-based weapons platform may be superior to a terrestrially based one, or other terrestrial methods of attacking space systems. However, terrestrial ASAT weapons do enjoy logistical advantages over space-based systems—mainly through reduced cost and the ease of access to maintain, resupply, upgrade, replace, and decommission. Terrestrial systems, especially those on land, can be hidden and fortified as well, unlike aerial, naval, and space-based weapons which may be less inconspicuous and more fragile, especially in space. Whilst ground-based systems need to 'wait' for targets to appear overhead, many polar-orbiting satellites in LEO will appear in range sooner or later. This makes fully synchronised attacks difficult for any state without globally deployed weapons platforms and SSA networks, or a massive launch of heavy ASAT interceptors would need to be conducted in sequence to ensure a simultaneous strike of all targets at once from the ground. Synchronisation at such a scale will usually require an elaborate space-dependent C4ISR system, further increasing dependencies on space and the payoff for attacking space systems for the other side. However, for continent-sized states, or for states fighting in their neighbourhood or defending their home territories, this may be less of a concern.

Space-based Weapons

Space-based weapon systems can offer some advantages in orbit. This is not necessarily due to mobility—satellites are restricted to particular orbits based on how much fuel they have to conduct manoeuvres, and any changes take time, and they can only alter slowly from their original trajectory due to inertia. Advantages for space-based weapons come from the available range and time provided to land the desired effect on the target. Some space-based weapons can be useful against higher-orbiting systems given the relatively vast distances involved for kinetic weapons effects to travel, or provide more accessible weapons coverage on another part of the globe well beyond sight from the ground of the state in question. A missile or intercept vehicle only needs minutes to reach LEO from Earth's surface—but that extends to hours for targets in MEO and GEO, even for the most direct routes. Fuel-efficient routes to the GEO belt may take a day or two along what are called Hohmann transfer orbits, where a craft moves from LEO to GEO along the most efficient but predictable path. Weapons already placed in MEO or GEO will have a reduced time to destroy the targets when the order is given. Kinetic, interceptors, or grappling weapons may enjoy some advantages if they are already in closer altitudes and orbits, but not without their own drawbacks, mainly the cost of deployment at the necessary scale and numbers to be a significant asset or threat.

Spacecraft designed to remain in orbit, stalk, or advance on another target in orbit are sometimes called 'co-orbital' and reach orbital velocities which means they can stay in orbit until ordered to de-orbit or they hit a target. This contrasts with direct-ascent Earth-to-space hard-kill weapons which are usually on a ballistic suborbital trajectory and cannot loiter in space—a 'pop up and hit' system. Space-based or co-orbital weapons are designed to maintain orbital flight until otherwise instructed to de-orbit. One co-orbital ASAT system was the IS deployed by the Soviet Union. It was designed to operate in LEO up to 1,600km altitude and carry interceptors that would detach, home in on target satellites, and detonate, hurling lethal shrapnel towards the target satellite's path. It also carried a significant amount of fuel to conduct several orbital manoeuvres to get into an interception orbital path. The Soviet military included space in the 1968 Soviet Military Strategy which envisioned using space for weapon systems on Earth, preventing enemy use of space, and the development of space-based weapons.[86] First conceived in 1959, IS began development after the go-ahead from Khrushchev in the early 1960s. Though he did not support Chelomey's more outlandish plans for orbital battle stations, the UR-200 ICBM and IS programmes were approved. The first IS test vehicle was launched in November 1963, and was the first satellite to conduct manoeuvres in space and change orbits.[87] With twenty-three launches and seven tests until 1971, the IS system was declared operational in 1973. The ABM Treaty of 1972 led to a pause in testing that lasted until 1975, when Soviet officials blamed the military utility of the U.S. Space Shuttle as a reason to resume IS development and improvements, until a moratorium prohibiting further flights was announced in 1983.[88] Paranoia over the military utility of the Shuttles was not preposterous at the time given the role of the DoD in the Shuttle's development and its potential as a nuclear-bomb-carrying vehicle. Orbiting between 230km and 1,000km altitude, the IS tests created around 900 pieces of trackable debris. A successor system called Naryad was in development and continued until the collapse of the Soviet Union in 1991. It was designed to be more manoeuvrable and target any orbit between 150km and 40,000km—all the useful orbital altitudes around Earth—at inclinations of 0° to 130°. At one point, Soviet planners considered pursuing the ability to launch 100 such systems at the same time, though nothing beyond small numbers of demonstrators were deployed.[89]

Today, Russia fields other systems including the air-launched Burevestnik as well as other satellites designed for close-orbit inspection and manoeuvre capabilities as demonstrated in the GEO belt in recent years, which have a potential weapons utility. The Olymp-Luch satellite conducted manoeuvres in the GEO belt and came uncomfortably close to target satellites, including commercial COMSATs. In the 2010s and into the 2020s Russia has conducted several rendezvous and proximity operations (RPO) in GEO and LEO.[90] Such loitering and close inspection capabilities could be comparable to the modern U.S. GSSAP system designed to build SSA and orbital inspection capabilities in GEO, or something akin to the unconfirmed Prowler satellite deployed by the Space Shuttle Atlantis in 1990. Some Russian satellites conducting RPOs may have released smaller satellites, vehicles, or perhaps interceptors of some kind designed for possibly aggressive uses given the relatively high velocities at which those smaller objects were released. Such a co-orbital or space-based ASAT capability could be effective against GEO satellites in range, and hold at risk the relatively small number of crucial command and control (C2) and missile early warning satellites used by the United States there, and with potentially some element of surprise should any weapons payload be concealed. However, collateral damage in the GEO belt could be quite spectacular as the debris would eventually spread to the entire GEO belt and threaten the safe operations of every state and company with assets in GEO. GEO satellites are often the most expensive and important satellites in use, therefore such military options as this may be desperate and held in reserve if the situation on Earth deteriorates enough to warrant it.

The United States, Russia, and China have been conducting 'cat-and-mouse' manoeuvres in GEO. In August 2020, USA 271, a U.S. space surveillance satellite and part of GSSAP, moved close to the Chinese SJ-20 satellite in order to inspect it and perhaps eavesdrop on its communications. Chinese satellite operators detected what at first were supposedly subtle manoeuvres by their U.S. counterparts, then moved SJ-20 away from USA 271 and into a different location in GEO. In another incident in 2018, a Chinese satellite, TJS-3, apparently conducted synchronous manoeuvres with its old rocket body—a piece of junk—in such a way so that SSA sensors would mistake TJS-3 for a piece of junk and not an active satellite.[91] Whilst these are not necessarily dedicated weapons systems, the technologies, techniques, and sensor systems used would be needed in any space-based kinetic-kill GEO ASAT capability. This also demonstrates deception attempts at greater operational freedom for military assets in space. Such behaviour increases the premium on satellites such as Canada's Sapphire and the U.S. GSSAP system to better determine what exactly is deployed in GEO and monitor behaviour there on a persistent basis. Seeing what is on a satellite is useful to determine whether a satellite is carrying anything that could be construed as a weapon or projectile, particularly when international launch site and spacecraft assembly inspections are not possible. SSA infrastructure has considerably improved since the Cold War years, meaning there is a greater capacity to conduct more precise operations in GEO, and more closely monitor activities there.

One of the best-known examples of space-based weapon concepts was part of the SDI programme. The Brilliant Pebbles system envisioned hundreds, if not thousands, of satellites (Pebbles) equipped with infrared-seeking missiles in LEO to intercept ICBM launches from the Soviet Union, attacking them early on before the boosters finished launching. This boost-phase interception capability required thousands of platforms to be deployed in LEO to ensure a sufficiently low absentee ratio—or enough interceptors within effective range of the USSR at any given time. The boost phase of an ICBM only lasts a couple of minutes, meaning any infrared-homing interceptor needs to be very close to any launch site. Once the boost phase ends, the warhead is in its midcourse phase, 'coasting' in a sub-orbital, ballistic flightpath at extremely high speeds and not giving off any heat emissions until it starts re-entering the atmosphere towards the end of its flight.

The number of required space-based weapons can be rather large when considering the use of SBIs. For round-the-clock coverage to intercept an ICBM launch from North Korea, at least 1,012 SBIs at a 45° inclination orbit may be needed.[92] At present, even with the emergence of megaconstellations, the total number of active satellites in orbit is approaching 6,000. For a state the size of a continent with hundreds to over a thousand nuclear missiles, the number of required SBIs for constant interceptor coverage becomes staggering. The cost of building a megaconstellation of thousands of SBIs would be astronomical given the expense of building any satellite system that often reaches into the billions of dollars to set up, let alone maintain and replace. It would be cheaper to build decoys and greater numbers of terrestrial-based nuclear missiles to overwhelm any defence. Additionally, such weapons based in space will be just as vulnerable as any satellites they may wish to target at other ASAT weapons, including ground-based direct-ascent ASATs using kinetic-kill and HANDs. Though space-basing may provide advantages in providing a 'global' coverage for certain orbits and inclinations, their logistical expense and complexity severely undermines that advantage. One estimate is that the fuel required to deploy and later safely de-orbit a kinetic energy or interceptor weapon in orbit may be fifty times the mass of the weapon itself. Conversely, ground-based systems are cheaper, but have to be moved around Earth if they are not deployed where they should be at the outset of a conflict.[93]

As well as kinetic SBIs, SDI included space-based laser concepts. Space-based lasers that can reliably and effectively target others in space will need to be 'a combination of a next-generation space telescope with a large rocket engine and its propellant tanks'.[94] Some SDI plans envisioned eighteen space stations in polar orbits. With an 8,000km operational range each station would be able to direct 1,000 laser pulses on as many targets.[95] Space-based laser ASAT constellations would be large and cumbersome due to power and other logistical requirements, and also pose as obvious targets for multiple ASAT barrages from Earth. They may not 'provide leak-proof defences... because their effectiveness falls off with the square of the range to the target, they will likely have lower orbits—in easier reach of terrestrial weapons. Such orbits designed for missile intercepts also mean that the absentee ratios for engaging time-critical targets on or near the earth will be high.'[96] The relative speed and velocity of orbital vehicles does not automatically translate into 'mobility' or 'flexibility' in weapons reach, coverage, and firing options compared to ground systems. Additionally, friendly fire and collateral damage may well be an issue when trying to get a clear firing line between two objects in orbit separated by 8,000km across congested orbits. Despite these issues, U.S. weapons concepts and proposals such as neutral-particle-beam weapon research still draw on SDI.[97]

Solid-state as opposed to chemical lasers are a logistically easier burden, but are not without their own challenges. Chemical lasers require large amounts of liquid fuel that must be manufactured and processed, then stored in the hostile environment of Earth orbit. Such fuel is heavy and increases the costs of launch. Solid-state lasers use a more robust battery that can be recharged, depending on the type of battery used, but that will degrade with time, particularly if unhardened against solar radiation. Regardless of the energy source, more rapid kills from lasers require more sophisticated focusing lenses and greater amounts of power, leading to greater expenditure of fuel reserves or stored energy and larger batteries. The constraints on power generation for space-based lasers cannot be overstated. Many satellites operate with a power consumption of around one kilowatt hour. For context, boiling a kettle consumes three kilowatt hours. The International Space Station's four solar arrays can produce between 84 and 120 kilowatts of power.[98] For a space-based laser targeting other satellites, a few hundred kilowatts would be needed at the very least.[99]

In effect, each laser weapon platform requires several times more power generation than the ISS at peak production to sustain fire, and many times more in stored energy to sustain a beam for long enough to start causing heat damage or start firing at subsequent targets. Even with modern improvements to solar panel efficiency, it is still a daunting task. Like any kind of general capability, more than one actual weapon will be needed to prevent the system from being saturated with targets and decoys, and to provide continuous coverage or reduce the 'absentee ratio'. Preston et al. calculated that a constellation of twenty-four lasers at an altitude of 1,248km would be needed to provide enough coverage of a minimal ICBM launch from a 'rogue' state, not accounting for saturation attacks from dozens of ICBMs launched at the same time. Other configurations included constellations of 120 laser platforms.[100] Such a constellation, however, would have more success as an anti-satellite weapons constellation, though it may struggle to sustain a strong enough laser effect on satellites in GEO given the distances involved and the relative motion of platforms in LEO. Trade-offs abound in the configuration of such constellations. In the 1970s and 1980s, there was plenty of unfounded speculation and alarm at a Soviet space-based particle beam weapon that could entirely neutralise the U.S. nuclear arsenal.[101]

Though the Cold War and SDI ended, technological development continued under the guise of National Missile Defense (NMD), Ballistic Missile Defense (BMD), or Global Protection Against Limited Strikes (GPALS). Echoing Paul Stares, Peter Hays comments that SDI paved the way for 'high-ground' space systems like space-based ASATs, even when attention in the 1990s turned towards ground-based systems. The Bush administration's withdrawal from the ABM Treaty in 2002 also freed up testing and deployment of both ballistic missile defence and ASAT technologies. Indeed, any space-based missile defence interception capability would struggle to have its ASAT capabilities 'engineered out' of them.[102] This wounds the argument that ballistic missile defences are inherently 'defensive' weapons—they can be used in offensive manners or in ways that support general offensives on Earth.

Space-based jammers may be useful as part of a range of counter-space options and they share many of the same benefits and drawbacks with electronic warfare from the ground, as discussed above. Satellites are more restricted in terms of the power they can generate to overcome powerful jammers, but proximity can make up for it, especially when used in tightbeam jamming. However, relative motion and fields of view may be an issue. As well as jamming, such proximity can enable eavesdropping on tightbeam communications as has been alleged with some proximity operations in GEO in recent years.[103] However, space-based jammers are themselves vulnerable to jamming because their human operators have to have a radio link to use them in the first place. Autonomous or automatic systems requiring no human intervention or monitoring are the only way around that but conjure a range of new problems regarding control, targeting, accountability, and oversight. Unless space vehicles become fully autonomous, space-based weapons will be entirely dependent on secure radio links for effective operations, making electronic warfare an even more desirable method of space warfare and disrupting space operations.

Weapons placed in space that are designed to strike targets on Earth are not separate from the same constraints as space-based weapons targeting other targets in space. Space-to-Earth weapons, or orbital bombardment, is another class of space weapon that has a long history, though very little operational utility relative to their cost, complexity, and alternative terrestrial weapon systems. A 'classic' example of orbital bombardment systems is the Soviet Union's FOBS. FOBS was declared operational in the 1960s but was deactivated in 1971.[104] It was only useful to circumvent the USA's ground-based early warning radar systems by hurling nuclear bomb-carrying satellites into orbital (not suborbital) flightpaths that could drop their bombs and enter the atmosphere from a southerly direction into the United States. In such a situation, where most radar stations were pointed towards the North Pole, it was a possible 'surprise attack' weapon as it would reach U.S. targets without detection until the last moments. However, once new radars were built to fill these gaps, and with the emergence of U.S. space-based early warning systems such as MIDAS and DSP, the 'surprise attack' gains of FOBS diminished, and were outweighed by the costs of deploying nuclear weapons in orbit.

The Soviet Union closed the gap with the United States in terms of ICBM numbers and SLBM deployments by the 1970s—often termed 'nuclear parity'—meaning the Soviet Union had a second-strike capability against the United States and further reduced the need for FOBS. The Soviet Union had only tested and flown the FOBS without nuclear weapons on board, therefore it could claim that the system as deployed never violated the OST which only prohibits nuclear weapons or 'weapons of mass destruction' from being placed in orbit.[105] Additionally, if a nuclear warhead was placed on one of the space vehicles, it would be intended to re-enter the atmosphere by firing its retrorockets to slow down before completing an orbit, therefore avoiding the 'placement' or 'stationing' of a nuclear weapon 'in space' and violating the Outer Space Treaty. This would be the fractional orbit. However, FOBS was capable of longer orbital flights because an orbital velocity is achieved by the weapon-carrying space vehicle, making de-orbiting before completing the first orbit an operational choice rather than a technological restraint.

FOBS returned to the limelight in 2021 due to an apparent test of a similar system by China. From what information is available, China appears to have launched a spacecraft or vehicle of some kind into an orbital flight path, and then fired retrorockets to de-orbit it back into the atmosphere before completing that orbit. However, contrary to the ballistic warheads 'dropped' by the Soviet FOBS, it appears that the Chinese space vehicle glided at hypersonic speeds (Mach 5 or above) within the atmosphere before striking a point a few dozen miles away from the aim point. This marrying of 1960s Soviet FOBS technology with 1980s and 2010s MaRVs and hypersonic glider technologies does not overcome the same limitations of Soviet FOBS in terms of its potential as a nuclear weapons delivery method. The gliding aspect is a detail that may be more useful for other spacecraft technology developments, rather than a transformative method of vaporising U.S. cities. Chinese officials claim they flight-tested reusable space vehicle technologies.[106] Zhou Enlai commissioned a feasibility study on developing a FOBS capability back in the 1960s for the Chinese military, based on a three-stage DF-6 as the launcher. The idea was dropped due to significant technical issues, improvements in the U.S.-China relationship from the early 1970s onwards, and the priority of developing the DF-4 and DF-5 missiles into a reliable nuclear weapons arsenal against the Soviet Union.[107]

Flying into orbit and re-entering the atmosphere and gliding at hypersonic speeds is also what the U.S. Space Shuttle and the X-37b Orbital Test Vehicle are capable of. The X-37b looks like a small version of the Shuttle, but is uncrewed and its cargo bay tests various ISR sensors that can be brought back from space after the X-37b lands on a runway. It may well be that China tested something similar to these spacecraft rather than a FOBS. Neither the Shuttle nor X-37b are known to have been designed to carry weapons. This did not stop Soviet fears over the potential of the STS as a method of launching a surprise nuclear attack on Moscow, leading to their own Buran space shuttle-type system. According to Asif Siddiqi, the Buran shuttle was 'probably the most expensive space project in the history of the Soviet space program', and flew only once, in 1988.[108]

The USAF considered the use of space vehicles for orbital bombardment. The Dyna-Soar project which ran in the 1950s and into the early 1960s was a rocket-launched, winged, crewed vehicle designed for orbital flight, hypersonic gliding and powered hypersonic flight through the lowest depths of orbit and the highest reaches of the tangible atmosphere. The USAF saw it is as a 'strategic deterrent' by designing it for nuclear bombardment, space interception or ASAT roles, affordable and routine logistics into orbit, and a reconnaissance vehicle. In the 1950s, the idea for crewed military space vehicles had some potential over remote systems due to the bandwidth restrictions in communication and computation systems. There was not enough communications bandwidth and computer processing power needed for machines to 'point sensors accurately and interpret instantly what was seen, acting upon it at very high speed.'[109] It was ultimately cancelled due to its inability to compete with rapidly increasing uncrewed satellite photo reconnaissance capabilities, which was a primary mission of one of the earlier phases of Dyna-Soar development. More seriously, however, was the persistent failure to achieve lower-cost reusable and more reliable methods of accessing space than conventional, expendable rockets. In that sense Dyna-Soar foreshadowed a fundamental flaw in the STS programme.[110] Dyna-Soar was apparently projected to cost two and a half times the cost of the Mercury astronaut programme.[111] Dyna-Soar had proven its uses in gathering data on manoeuvrable hypersonic flight in the upper reaches of the atmosphere, but the costs associated with the project would not be sustained in light of the NRO's Corona successes.[112]

The USSR also experimented with an armed space station. Almaz, approved in 1967 and led again by Chelomey, was a multi-module space station designed for reconnaissance. The Almaz project (disguised as a Saluyt 'civilian' space station) proceeded further than the Spiral spaceplane project, but the Soviet military never became interested in such things because of their continued preference for uncrewed military reconnaissance systems, like their U.S. counterparts.[113] The Almaz space station featured a modified MiG cannon which was remotely test-fired after the cosmonauts had left and before de-orbiting the station.[114]

A widely-known concept among military space communities for orbital bombardment is 'Rods from God': orbiting satellites that release high-velocity impactors (such as tungsten rods) from orbit and strike targets on Earth with enough kinetic impact energy to cause significant damage. These kinetic impactors would, in theory, be difficult to defend against once jettisoned due to their high speed and relatively brief travel time from LEO to the planet's surface. However, the orbiting weapons platforms would be vulnerable and obvious targets to ASAT barrages, especially in LEO. The jettisoned 'rods' would face difficulties in maintaining velocity to create enough of an impacting effect on arrival so that they could destroy fortified buildings and kill people, whilst also remaining accurate enough and without losing too much mass by burning up or banking in the atmosphere.[115] Developing the materials needed to create such 'rods' without dramatically increasing their mass (and therefore cost to manufacture and deploy) in orbit remains a theoretical proposition at this stage. Rods from God would be an extremely unsubtle approach to bombing and lethal combat support, as well as expensive given the required mass required for a number of orbiting bombardment platforms needed to provide meaningful mass and depth to the capability to meet practical military demands on Earth. Dropping non-nuclear weapons against surface targets on Earth from orbit is feasible only for precise strikes on fixed targets on Earth.[116] Beyond kinetic impactors, space-to-Earth lasers are unfeasible today because the atmosphere dilutes any laser effect. Space-based lasers targeting anything in an atmosphere need to be measured in *megawatts* of energy, and even then such wattage would only burn and melt 'soft' targets such as aircraft and missiles that tend to have thin metal skin and no thick armour due to weight considerations.[117] Like kinetic orbital bombardment, space-based lasers melting and burning people and objects on Earth remain the purview of science fiction, perhaps most vividly portrayed in the doomed 'Jayhawk Wars' campaign of the *Footfall* novel, where alien close-support space-based lasers rout a joint air-land U.S. military attack on an alien beachhead in Kansas.

'Far-out' programmes in U.S. military space, such as orbital weapons, Dyna-Soar, and lunar missile bases, were consistently watered down in favour of satellite infrastructure and space-tracking systems.[118] Therefore, other than nuclear or conventional warhead delivery that replicates the terminal stages of flights of various kinds of ballistic missiles and re-entry vehicles, space-to-Earth weapons remain something of a niche or marginal consideration in space warfare. Even should breakthroughs in material sciences or an economic revolution be achieved in space access that make various kinds of orbital weapons more affordable and feasible, they will not escape strategic logic, terrestrial precedents in warfare, and physical and economic limitations. Such weapons may well be in a 'high ground' of sorts (as discussed in the next chapter), but those advantages and bombing capabilities only go so far even if they are not subject to a range of anti-satellite attacks, which they certainly would be in a time of war. Weapon-fire from space onto Earth would generate similar strategic effects to air superiority and not lead to simple or easy victories on the ground. These could range from nuclear annihilation and 'strategic bombing' to close weapons support for tactical, conventional engagements.[119] Indeed, it would begin to resemble the same debates over the uses and limits of airpower in delivering victories, advantages, or coercion, which themselves were preceded by debates over gunboat diplomacy, just delivered from higher altitudes.[120]

Conclusion

'Space is hard', as the truism goes in the U.S. space community. The most technologically advanced space powers are only now competent in sustaining the most essential orbital space infrastructure *en masse* that are routine and reliable, rather than exotic and exceptional. The 'space warfare' that could happen in the next few decades will remain warfare in Earth orbit and on Earth's surface, but not beyond. The methods of waging space warfare are diverse and range from nuisance or reversible soft-kill methods to massively destructive and often indiscriminate hard-kill techniques. The purpose of space warfare is to threaten important infrastructures and prevent the enemy from doing the same in return, to undermine the support military forces receive from C4ISR systems and to ensure the safety of your own, and to impose economic costs if the enemy's civilian economy relies on space systems and have no good alternatives. In short, space warfare needs to help achieve the goals of the war on Earth by creating strategic effects in space or with space systems. Therefore, commanding space is a normal corollary in strategy to commanding the sea and commanding the air.

Space-weapon basing locations have benefits and drawbacks. Space-based weapons do not simply 'overcome' terrestrial constraints because they are in orbit. Terrestrial systems may have some limitations in terms of time to target for higher orbits and fields of view, but continent-sized states have a great degree of flexibility in terms of launch opportunities in any direction. Terrestrial systems are easier to deploy, maintain, upgrade, decommission, and hide. Redeployment on the ground is easier as well, because once a vehicle is set into a particular orbit in space, drastically changing its inclination and orbital height consume significant amounts of its fuel reserves. Orbital physics and economic-logistical constraints mitigate most of the benefits of the greater views and global reaches provided by orbiting weapons platforms. Space weapons do not usher in an era of certain death from above, even if orbital bombardment systems are re-deployed. They have their drawbacks and very limited usefulness, as the Soviet Union, China, and the United States learned in the Cold War. Like any individual weapons system, space weapons provide benefits that are marginal and specific, and hardly ever 'game-changing' or sweeping. The sweeping change in terms of weapons capability in the Global Space Age has already happened—ICBMs with nuclear warheads on them.

Electronic warfare is crucially important to sustain and maintain a command of space as everything in orbit is a machine that relies on radiofrequencies to gather and relay their data as well as to receive commands and provide their services. Additionally, as these machines are a combination of hardware and software systems, information technology infrastructure and security should be expected to come under cyber-attacks. These soft-kill methods, whether based in space or on the ground, have more defensive options and countermeasures to meet them, making planners less certain of 'kills'. Hard-kill methods are more reliable in destroying their targets, but their environmental, economic, and political consequences are potentially greater. However, this is not an inherent bar on their use—the political objective of the war on Earth may be serious enough to consider a scorched orbit approach to space warfare. This could be the case in a war over a core political issue for one side that also does not rely as much on space for military and economic power as the other. The PLASSF, like any intelligent military force planning for space warfare, would want to keep a mix of hard- and soft-kill systems.[121] Russia, India, and the U.S. are each fielding a mix of capabilities, especially in terms of terrestrially based systems, to keep options open. A diversity of capability adds to strategic depth and flexibilities in war planning and responding to situations as they develop, which can never be completely foreseen due to the inherent chaos and uncertainty of war.

More diverse space systems with larger constellations may lead to more defensive options and resilience if states have access to more of them. Blowing up one satellite in a constellation of a hundred may not matter much, but single points of failure in their IT infrastructure will still be a concern. This means that a serious ASAT capability requires an extensive range and depth of systems and weapons, not just a few demonstrators. This does not mean the threat is negligible, but neither is it catastrophic by itself. The impact of space warfare can be varied and 'patchy' depending on what is hit, how, and when. Big satellites in GEO are still in use and do not appear to be going away anytime soon based on current modernisation plans for military SATCOM systems, meaning there may be limits to how much a satellite capability can be spread out over many satellites. Bigger satellites can do more things, partly because of the bigger power plant they can carry, the amount of hardware they can host, their propulsion systems and fuel reserves, and reduced debris risk. Bigger platforms also provide more opportunities and flexibility to update and place new equipment or modules in future years without needing to alter the core platform.[122]

Space warfare is a real prospect, as real as any major war on Earth, because there are many targets worth hitting in orbit. Understanding space warfare and the use of any space technology in conflict needs to be put in a larger context that takes all geographies of war, a myriad of technological capabilities, and the unquantifiable political, emotional, and chaotic elements of war into account. Spacepower does not exist in a political and strategic vacuum; space warfare is merely the continuation of terrestrial politics by other means. The next chapter discusses space strategy in the cosmic coastline.

7

WAR ON THE COSMIC COASTLINE

Bedd a wna bawb yn gydradd

The grave makes everyone equal

Not only has the original sin of space technology so far proven inescapable, Earth orbit stands to become a warzone if a major conflict occurs involving major space powers tomorrow. Though trends have been going in this direction for well over 40 years, this has gone unnoticed outside of expert military and intelligence communities. Until, perhaps, now. In the wake of the USSF's creation, there is a mainstreaming of 'space deterrence' and space arms control discussions on the impacts of spacepower and space weapons proliferation in the United States. Whilst more interest and awareness of contemporary military and intelligence space activities is welcome, deterrence and arms control perspectives and approaches are not always useful to think through the roles, possibilities, and dangers of space warfare. Deterrence and arms control are narrow subjects within the larger sweep of what the sum of space warfare is, what it is doing and can do, and why many military powers are pursuing it at all. Deterrence and space arms control place an undue emphasis on nuclear strategy, missile defence, and space weaponisation and do not capture the main purpose of space warfare and spacepower today: a supporting infrastructure from a littoral, adjunct environment like a coastline that tends to impose dispersion on military forces.

The original sin of space technology as military infrastructure means space warfare is a part of modern military strategy. The concepts of strategy and war apply in space as they do in terrestrial environments. To understand the impacts of military space technologies on the world, the feasible possibilities and impacts provided by space systems and anti-satellite weapons and techniques should be the starting point for analysis. However, less useful concepts about them are currently in vogue. Space warfare is a specialism in its own right and should be treated as an equivalent to land, maritime, and aerial warfare. It is not merely a subtopic that belongs to nuclear deterrence, missile defence, or space arms control.

We should not view Earth orbit as 'the ultimate high ground' of modern warfare, as so many officials and analysts do. It is a vacuous concept that leads to mechanistic and limited strategic thought. It is also a banal and overly generic statement that commanding space is advantageous in modern warfare. The limitations of holding the high ground are unaddressed to the detriment of strategic thought about space. Instead, Earth orbit should be viewed as a celestial coastline, a cosmic shoreline, in that it shares many of the same aspects of littoral and flanking environments rather than a high ground. It is a secondary, supporting, adjunct environment that provides support to terrestrial environments. Its usefulness as a force for logistical and other kinds of support to military forces makes it a target for weapons fire within space and from Earth's surface. This analogy captures the intimate connections between Earth and Earth orbit. Earth orbit begins only 100km or so above Earth, and is susceptible to multiple kinds of effects and power projections from the surface. In turn, events and infrastructure in space can directly impact activities on the ground. The impacts of the orbital infrastructure that many states have placed in space can be approached through the idea of dispersion on the modern battlefield. Space warfare must be used to allow terrestrial forces to exploit or deny the effects of dispersion on Earth. This practical utility of spacepower in modern warfare captures an important aspect of the Global Space Age, and why so many states are investing in their own large technological systems and infrastructures in orbit.

Normalising space warfare as a necessary part of modern strategy in the same way as land, aerial, and naval combat has not been achieved yet. Normalising space warfare does not mean that attacking space systems is a 'good' in itself. Rather, if we can accept land warfare, aerial warfare, and maritime warfare as 'normal' aspects of modern warfare, then space warfare should be a 'normal' addition to the way we analyse and study war in general. Strategy is a 'bridge' between military power and political purpose.[1] Space warfare is just another part of modern strategy. Rejecting space warfare alongside rejecting war as a whole is a morally coherent, pacifist position. Refusing to contemplate the conduct of space warfare whilst being comfortable with killing and destroying in the name of the state on Earth is a morally dubious position. Such is the flawed foundation of many policy and strategy discussions in U.S. literature on space weaponisation and deterrence, where there is more apparent discomfort at the prospect of destroying expensive machines in space than killing people on Earth. Keeping Earth orbit a 'sanctuary' from space warfare or ASAT operations only to ensure unimpeded access to orbital infrastructures that make their state better at killing people and destroying things on Earth shows a higher value placed on machines and property than people, their political futures, and terms of life.[2]

Deterrence and Space Arms Control

Two obstacles stand in the way of accepting space warfare as merely the continuation of terrestrial politics by other means. First, space warfare is sometimes reduced to subsets of nuclear strategy, ballistic missile defence, and deterrence, rather than 'warfare' and 'normal' strategic activity. Second, space warfare is often dominated by discussion over the 'weaponisation' of space and space arms control.

Until the 1980s, space warfare dovetailed with the needs of nuclear warfare on Earth. Since then, space warfare has increased its relevance for non-nuclear terrestrial warfare, given the ubiquity of spacepower and orbital infrastructures for a wide range of conventional military capabilities and lines of communication, as Chapter 4 and 5 have shown. Reagan's SDI displaced attention from space as 'just another' environment for warfare and supplanted it with a focus on nuclear strategy and space-based missile defence, rather than as using and denying space systems to influence terrestrial warfare of all kinds.[3] SDI and ballistic missile defence persists as a distorting lens on military space activities. It can be seen in one wrong description of the militarisation of outer space 'as an outgrowth of missile defense'.[4] This is at odds with the original sin of space technology which militarised space from the very start, and how ballistic missile defence (at least in the U.S.) was drawn from ASAT and other military space technologies. As with nuclear war, there is fortunately very little practical experience to hand in waging a large-scale shooting war in orbit or a major conventional war between space powers. However, as ASAT technologies spread horizontally across more states into the twenty-first century, military space strategists and operators must think through their potential use, since they may indeed be used. Military experts must take the impacts of hostile space warfare actions into account. The spread of ASAT techniques and technologies, such as satellite jamming, laser dazzling, and computer network operations into satellite control systems, should be seen in part as investments in counterspace or space warfare technologies that can impact warfare in all terrestrial environments in all kinds of operations—not only things that impact nuclear strategy or ballistic missile defences. The space assets of the major powers could be targeted by smaller powers, even in proxy wars, as many ASAT options are open to smaller powers.

As with the 'Wizard of Armageddon'[5] Herman Kahn's educational intentions in discussing nuclear strategy, it is important that space warfare is thought through to make individuals 'aware of the complexities, ambiguities, and terrifying choices that may be faced by decision makers'.[6] Simply deciding to possess a weapon without thinking through how you would use it is irresponsible. Advocates of any weapon system must be able to explain why and to what practical ends they should be possessed. Rather than abstractly dwelling on how to prevent space warfare through deterrent effects, it is more responsible and prudent to think through space warfare's possible conduct should deterrence fail, and work 'back' from there.[7] Destruction, violence, and actual political control cannot be 'faked', meaning that the perceived likely outcomes of actual combat and action should threats be carried out is the pivot upon which many decisions should swing. Death and destruction have objective, material effects that cannot be as easily ignored as declarations and statements. Clausewitz argued 'whatever number of things may introduce themselves [into strategic calculations] which are not actual combat, still it is always implied in the conception of war that all the effects manifested in it must have their origin in combat.'[8] How planners think about how actual combat may turn out will impact what they think is prudent to do before combat starts in the first place.

Deterrence, in very simple terms, is about preventing someone from doing something you do not want them to do. Actions that risk bad consequences should be refrained from to avoid punishment.[9] Deterrence is about promising punishment for bad behaviour or making such bad behaviour ineffective by denying its effects. Snyder usefully defines deterrence as a kind of political power that does not have to depend on military force alone.[10] Schelling referred to the 'power to hurt' as a bargaining tool, an ability designed to impose pain and suffering to coerce the other side into accepting a certain outcome over others.[11] Showing a credible power to hurt may require some performance of it.[12] As such it is important not to equate the study of space warfare and military space strategy with that of studying or imposing deterrence. It is by definition a practically non-violent and political communications activity. Yet effective deterrence requires a preparedness to carry out bloodshed and violence should threats fail to get the desired effect, which then becomes more the business of war.

This is not to say that deterrence is not important. It is simply not the same thing as strategy and waging war. Useful deterrence analysis focuses on applying imagined thresholds of political sensitivity and reaction based on the types of satellites attacked, how they are attacked, and when they are attacked in a crisis. For example, a Planet Labs commercial imaging satellite being jammed outside of a crisis is a different incident to a Keyhole imagery satellite being destroyed during a Taiwan crisis. The sensitivity and greater escalatory reactions Washington might hold for a Keyhole satellite as opposed to a Planet Labs imaging satellite may create a greater 'deterrent' effect. There are many Planet Labs satellites that can fill a gap in coverage, but they are not used for the same kind of high-resolution IMINT as Keyhole, which is not a highly distributed system and instead has a small number of platforms that could be attacked with relatively small numbers of ASAT platforms. Attacking a SBIRS satellite which provides a fundamental nuclear early warning service and is not highly distributed could be considered a particularly daring and dangerous move for the attacker because of the severe response it may cause. Silencing the entire Iridium network, though difficult, may provide more of an inconvenience for military forces than a debilitating effect, though its economic and civil consequences could be more severe. In contrast, disabling the GPS network in its entirety (again a tall order) will have systemic consequences for U.S. and allied military forces as they scramble to bring backups online, which will not be as effective as GPS. The critical infrastructure and civilian consequences of attacks on GPS may well be reducing as the twenty-first century proceeds because more civilian and commercial PNT systems can use multiple GNSS signals, such as Galileo, GLONASS, and BeiDou. The science of space engineering becomes the military and political art of deterring unwanted behaviour.

The fact that many western militaries rely on commercial space systems also ensures that they become valuable targets in wartime. The hand-in-glove relationship between 'private' space systems such as Iridium with the Pentagon as a customer should leave no-one in a naïve position that just because a system is 'civilian' or 'privately owned' it is above consideration for attack. Civilian and private property is constantly attacked in warfare, and civilian shipping and merchants have traditionally been targeted in war. There is no reason to presume that civilian and commercial space traffic should be spared, particularly when it is so useful to the military forces of the states they are often resident within.[13] If a business or profit model centres on war and supporting a war effort, there is a fair ethical question as to why that business should be spared from the horrors of war by the enemy.

A warfare approach embraces the fact that uncertainties make perceived consequences of 'bad behaviour' educated guesses at best and a real conundrum for imposing deterrence. Competent strategists seek to allow their side to 'get away with' bad behaviour through meaningful practical actions and ultimately winning the ensuing conflict or putting up enough of a fight to drag the other side to the bargaining table. Deterrent effects flow from war-fighting capabilities and those capabilities will be put into use should deterrence 'fail'. This paradox is central to the idea of deterrence—securing peace by preparing for war. That said, the thermonuclear revolution still imposes its logic on all would-be 'space warriors'. Like planning for major state-on-state warfare on Earth, things could always escalate to a nuclear exchange which would see cities cast into the fire of nuclear war. Non-nuclear war may just be the prelude to nuclear Armageddon.[14] War is a serious means to a serious end and we should have no illusion over the gravity of the subject matter.[15] Those in the business of deterrence are ultimately in the business of war, in one way or another, albeit the initial stages of it only.

As Sun Tzu the Chinese military philosopher taught, how can one turn warfare to its best use—that is, winning without fighting at all—without awareness of and planning through the horrors of war itself?[16] Without a belief that threats will be made good upon, the actors issuing such threats lose credibility, influence, and moral authority. Anything that might be termed 'deterrence in space' requires some level of ability and preparedness to engage in space warfare or pursue violence on Earth in the cause of space systems or against them. As in abstract discussions of nuclear strategy, we must avoid overly sterile, clean, euphemistic language which threatens to distance our minds and understanding from the brutality of the subject matter, whether threatened or real, as Carol Cohn argued.[17] Behind the rather sterile and code words of 'costly signalling', 'escalation thresholds', and 'mutual vulnerability' are threats to visit increasing levels of destruction and death, heightening the stakes and the costs, and daring others to respond in kind only to invite more punishment, revenge, and vigorous efforts at resistance and retaliation. 'Re-establishing deterrence' can become a euphemism for 'winning a war'. Clausewitz wrote of this as the emotional forces of war and the 'escalatory tendency' of violence to invite even more effort and resources when you suffer major losses and want to make prior sacrifices mean something.[18] Deterrence language disguises the emotional aspects of space warfare.

Spacepower's contributions to contemporary deterrence cannot be fully grasped without looking at space warfare on its own terms as a part of just another tool of modern military strategy. Deterrence theory relies on rationality and first moves, for if action happens then 'deterrence fails'. Yet thinking through and conducting war, whether nuclear or not, is rife with contradictions and irrationalities.[19] Actions and reactions, attack and counter-attack, the uncertainty and non-linearity of what may or may not be done or in what order, are complicating factors in any 'deterrent' posture.[20] There is always an after-pressure following the first move, the risk of counterattack, missed targets, and attacks coming up short through general friction.[21] There is always a second move after a first strike. So long as there is a political will to resist, there need not be a 'final move' in sight. Clausewitz warned that war cannot be decided like some equation of actions and consequences, that 'rational' interpretations of costs versus benefits cannot be known before war actually begins:

> how far from the truth we should be if we ascribed war among civilized men to a purely rational act of the governments and conceived it as continually freeing itself more and more from all passion, that at last there was no longer need of the physical existence of armies, but only of the theoretical relations between them—a sort of algebra of action.[22]

War is a phenomenon fraught with uncertainty, emotions, and friction as things go wrong when plans are put into action.[23] Threatened combat is important and has an effect because at its root is some belief of how *actual* combat would play out.[24] It is impossible to know exactly how any military campaign will play out before it is put into motion. Thinking through space warfare campaigns as best we can is useful to better understand pre-war decision-making and calculations over probable actions, whether merely to create deterrent effects or to credibly prevail in the maelstrom of war.

Space weapons must be deployed in ways that are useful and meaningful in the practical pursuit of space warfare if they are to be deployed at all. Otherwise their deterrent effect in the mind of the enemy will be minimal. How can the deterrent effects of 'space weapons' be fully anticipated when their effects and potentials as weapons of war are not fully understood? State A could believe in an all-or-nothing, massive opening anti-satellite attack like a Space Pearl Harbor attack (discussed below) by State B, where the opening moves will decide the rest of the war's progress. Meanwhile, leaders of State B know the limitations of any major first strike and therefore downplay the consequences of their movements towards such a strike. Thinking that State A would not find such moves alarming because a first strike could never be so effective, State B has unintentionally escalated the situation. The flaws of perceptions and perfect information over capabilities in deterrence theory should not be replicated in space.[25]

It is far harder to find out someone's intent than it is to estimate actions based on material capabilities, such as worst-case actions based on known weapons and military power.[26] Yet thinking through the worst-case scenario based on known capabilities risks creating a fatalistic pessimism, a spiral of deteriorating relations and possible war even if both sides wish to avoid it.[27] All the same, disaster and miscalculation can occur from an ignorance of actual material capabilities. The so-called Security Dilemma, where states face the dilemmas of interpretation and action regarding others, can be worsened or improved with greater knowledge of capabilities. If, after assessing all possible options of a possible space warfare campaign, the military advice is 'it will be bad but we can manage indefinitely', it can make a world of difference compared to 'we will not manage after a few days'. It also matters whether only one side knows that, or both, as knowing what is known by whom also factors into strategy. Either way, knowledge of the conduct of war if it were to come about is necessary for such net assessments to inform decision-making. Showing a 'sensibility' towards how one's own actions impact the actions and views of the other side, and perhaps avoiding war, requires understanding the likely course of events after the shooting starts.[28]

Compounding calculations over space warfare is that, unlike nuclear war, a state can choose to make a mess of parts of Earth orbit for everyone without directly killing any people (with the exception of the small number of astronauts in LEO at any given time). Attacking machines and robotic infrastructure, however useful, is not the same as directly killing people and may make space warfare more politically acceptable. If destroying machines helps save more of your own people, yet apparently protecting space systems seems to outweigh saving the lives that matter, a leader could face political costs in *not* attacking space systems. Of course, there could be indirect humanitarian consequences as supply chains and certain infrastructural services suffer some shocks, but no weapons will have been fired in anger at people. If sweeping critical infrastructural services are disabled at scale and for a long duration, such as disabling water and energy networks within which disrupting space systems would be a part, the retaliation could be dire. Fouling orbit may be 'worth it' if the political goals of a war on Earth are sufficiently important relative to the effects caused, but such actions risk inviting heavy retaliation on Earth that may include the use of nuclear weapons. Walter McDougall rightly asked 'why is it more important to protect pristine space, where nothing lives, than the crowded Earth?'[29] We may never know the balance of these interests and possibilities on decision-makers at crisis moments. The devil, as always, is in the detail of war plans and in the confidence of leaders in particular courses of (in)action should the moment come.

'Space deterrence' is just a thematic or geographic and technological variant of deterrence in general.[30] The PLA's *Science of Military Strategy* helpfully describes space deterrence as 'deterrence implemented by means of space military forces'.[31] Translated into the terms of deterrence, 'threats of punishment' corresponds to offensive space warfare capabilities to destroy enemy space infrastructure, and 'capabilities for denial' translate into more defensive or passive capabilities that make space infrastructure harder to destroy, harass, or more able to survive attacks, transfer to backup or redundant systems, or be repaired and rebuilt in a hurry. Karl Mueller argued that 'it is not obvious whether space deterrence is a useful construct in the first place, it may not be that separate from other deterrence challenges.'[32] Focusing too narrowly on the concept of 'space deterrence' can mislead analysis to isolate space from Earth. It is as misleading to think in terms of 'air deterrence' or 'sea deterrence'. Deterring a state from taking a particular action in any environment requires more than one method of deterrence by denial or punishment. Rather, modern deterrence needs to account for the role space systems play in building holistic warfare capabilities in every environment, the role satellites and space infrastructure may play in triggering, exacerbating, or resolving crises, as well as winning wars. As part of an overall deterrence posture, spacepower supports deterrence not only on the nuclear level, but also on the 'lower' levels by providing deterrent effects to spacepower-supported conventional forces.

Michael Quinlan pointed out that deterrence by means of nuclear weapons alone was never credible.[33] In the same way, space warfare is just another means and method of supporting conventional and nuclear military campaigns, deterrence, and other interests of the state. Gray emphatically stated that 'if a World War III is waged, it will be waged in space, as everywhere else.'[34] A major orbital conflagration will likely happen alongside and as a consequence of the bloody horrors of a major conflict on Earth, not instead of it, and not necessarily as a fundamental cause of it. The visceral difficulties stemming from the paradoxes, insecurities, psychology, and uncertainties of deterrence and coercion are, in principle, the same in space as on Earth. As Harrison et al. summarise, 'Forrest Morgan, Karl Mueller, Evan Medeiros, Kevin Pollpeter and Roger Cliff argue that "a more reliable strategy for deterring deliberate escalation is one that buttresses threats of punishment with visible capabilities for denial"'.[35] This is not a novel suggestion and could be proposed for any form of deterrence. Punishment and denial can take many forms and a spread of each increases an overall deterrent effect. Punishment could be a tit-for-tat in attacks on satellites, or bombarding sites and targets on Earth may be a more appropriate response. Denial can include hardening satellites to withstand any electronic assault, or in increasing resiliency, redundancy, and replenishment across ever-larger satellite constellations or terrestrial 'work-arounds' and stop-gaps. The USAF's 'disaggregation' proposals to reduce their reliance on low numbers of important satellites in GEO by proliferating a higher number of smaller satellites to provide the same service was floated for many years, though it did not tangibly change U.S. military space architecture.[36]

Regardless of the actual correlation of military forces at any given time, one should not take the first step without considering what might be the last, as in war the result is often not final.[37] Though fighting may stop and start, political intercourse never ceases. The paradoxical nature of strategy, or the conduct of war, means that sometimes because something is expected and believed to be likely, it will simply not be done because of that supposed expectation.[38] Sometimes the unexpected or unlikely course of action will be chosen precisely because it is unexpected, and the art of deterrence morphs into the art of war.

Space warfare is merely the continuation of Terran politics by other means; a shooting war in space does not occur in a political vacuum. Space arms-control is in a similar quandary to deterrence in space. Arms control is a process which reflects prevailing political interests. Arms control cannot prevent warfare when one party is determined to retain the ability to wage warfare in some way. The high point of comprehensive arms-control talks between Moscow and Washington in the 1980s produced a reduction in armaments, going beyond limitation. But the abolition of nuclear weapons was not achieved.[39] Space arms-control is often focused on regulating certain kinds of armaments in a way that satisfies signatories and often does not undermine core war-fighting capabilities. Unless space arms control abolishes all ASAT technologies, space arms-control cannot fully prevent space warfare. If it did, it would be *disarmament* rather than arms *control*, which is a very different notion. It can only hope to regulate specific behaviour or technologies in specific ways without fully removing the ability of states to interfere with enemy space systems.

The Russian and Chinese Draft Treaty on the Prevention of the Placement of Weapons in Outer Space (PPWT) is one such proposal that is often criticised for its lack of provisions on addressing terrestrial ASATs as well as lacking feasible verification mechanisms for such a ban. Whilst the PPWT does prohibit the use of terrestrial ASATs against satellites, it does not contain provisions on preventing their testing or deployment.[40] Space arms-control focusing on specific weapons technologies should not get caught up in rhetoric over preventing 'space warfare' when such proposals like the PPWT are in fact only seeking to prohibit specific kinds of weapons, not all. It is also often interpreted as pertaining to kinetic or hard-kill ASATs when definitions on what kinds of soft-kill ASAT techniques constitute an attack or 'space warfare' remain unaddressed.[41] In part, space arms-control is dogged by the extensive dual-use character of space technologies. In particular, ballistic missile defence systems which are often not on the negotiating table for the U.S. may hinder any ASAT arms control proposal. This is a continuation of old technological dynamics, where the Carter administration proposed arms control on both ASATs and BMD, but the Reagan administration rejected controls on both.[42]

Controlling one without the other may indeed be impossible given the ease with which the orbital interceptions technologies cross over between them.

Dual-use technology is hardly unique to space and has been overcome elsewhere because specific composite devices, processes, behaviour, or vehicles have been defined rather than broad, ubiquitous, and abstract classes of technologies. Indeed, 'weapons' were not a problematic concept when prohibiting the placement of any kinds of weapons on the Moon or any other celestial body in the 1967 OST. Dual-use is a problem but not a roadblock to regulation and agreements. The problems of defining weapons are more political rather than technological, and an act of social construction and political-military bargaining. IRBMs and MRBMs have been defined for treaty purposes despite the problems raised by dual-use technologies, such as in the now-defunct Intermediate Nuclear Forces (INF) Treaty. Through clear definitions on ranges between 500 and 5,000km and accompanied by intrusive verification mechanisms, the INF Treaty succeeded in destroying all such ground missiles (though the bulk of 'strategic' nuclear arsenals remained).[43] Any definition which carries weight in international law or softer interstate conventions requires political will and consent to reach shared understanding.

Similar to Clausewitz's frustrations about 'logical hair-splitting' over definitions of war in his own time,[44] arguing over precise definitions on 'space weapons' and 'space weaponisation' is unnecessary and ultimately pedantic when there are certain realities in the twenty-first century: anti-satellite capabilities are spreading, and Earth orbit could become a hostile zone or a battlefield if a major war broke out between the space powers tomorrow. The perspective of space warfare should add to our understanding of what kinds of arms control may be beneficial, which may be necessary, and which may be harmless tokens to demonstrate goodwill at little cost. Even if space weapons can be clearly defined, it does not necessarily make the situation in space more stable or predictable because what happens in space is driven by wider forces of geopolitics on Earth, for good and ill.

A kinetic ASAT weapons deployment ban, whether from Earth or from space, will not prevent other methods of space warfare or continued development. Banning debris-generating tests above an altitude of approximately 250km may be plausible and verifiable but may simply entrench an advantage to states that have already conducted the necessary tests for their arsenals.[45] Such a move may generate some diplomatic momentum in various space security negotiations, as the United States hopes with Vice President Kamala Harris's declaration of a unilateral direct-ascent ASAT test ban in April 2022. Once war is afoot, however, self-imposed restraints may be ignored if the war's occurrence tramples on other dearly held conventions and treaties. When approaching the issue from the perspective of space warfare, the basing location is not particularly important beyond whether they are better or not than other alternative weapons. Approaching space warfare and attacking space infrastructure from the perspective of war planners highlights that there is nothing special about putting weapons 'in space'; they do not promise an era of certain victory from the ultimate high ground nor of certain doom from above.

Technocratic and liberal institutional approaches based on moral outrage or affecting spiritual or religious transgressions will not exorcise the spectre of space warfare when power politics, perceptions of military insecurity, and strategic logic serve to make space warfare a feasible and potentially useful feature of modern warfare.[46] Space technology is firmly part of the ways in which states wage wars and threaten nuclear annihilation. Eliminating all possibilities of space warfare and attacking space infrastructure requires moving away from military and economic dependence on space infrastructure, seeking a way to overcome the original sin of space technology. However, that may not happen without a revolution in political and technological conditions.

Efforts at diplomacy are not futile. Trying to identify and regulate better behaviour where agreements can be found in specific areas may be better than trying to outlaw classes of technologies that are too useful to do without or too difficult to police.[47] It may be better not to rigidly define space weapons but to agree to restrain from unwanted behaviours such as dazzling satellites, flying too close to another operator's satellite without permission, and agreeing on rights of way between satellites on a collision course. It may be easier, but still not easy, to agree on what counts as 'too close' compared to ubiquitous technologies that can be improvised as weapons. Initial agreements do not need to address everything or be perfect, and may be the stepping-stone to bigger agreements. Beyond the military sphere, there have been limited successes in agreeing to voluntary standards of construction, operations, and behaviour in order to reduce debris generation, such as with the Inter-Agency Space Debris Coordination Committee (IADC) and the Long-Term Space Sustainability Guidelines.[48]

Recent activity at the UN General Assembly appears promising for those wanting to 'prevent' the 'weaponisation of space':

> The Philippines asserted that [any resolution] must not be misconstrued as allowing weapons in space so long as such behaviour is regulated, a sentiment echoed by Costa Rica... 'No first placement of weapons in outer space,' calls on states to make political declarations not to be the first to place weapons in space... New Zealand, which has a national policy against the weaponisation of space... [is] concerned that the resolution provides tacit approval for subsequent weapons in space. Indeed, several states voting in favour of the resolution... emphasised that it must not be used to legitimise the placement of any weapons or arms race in space.[49]

Yet, such statements are not particularly helpful for honest discussions on the matter by implying space is not currently 'weaponised'.

'Space weaponisation' is not some threshold or point of no return in international relations. The proverbial horse of space warfare has bolted through the barn door, and current political conditions do not seem amenable to reverse that situation. On the positive side, the Cold War shows how space weapons can be rolled back, run out of development momentum, or how their deployment is not certain to start a war or usher in a new irreversible phase of great power acrimony. Many Cold War systems were deployed but fell into disrepair or were mothballed like so many other weapon systems on Earth. Some genies can be put back into their bottles if the conditions are right.

Diplomatic posturing on preventing space weaponisation is symptomatic of the wider problems surrounding the perception of space warfare and the original sin of space technology because many states on Earth, not only the major space powers, are exploiting the benefits of spacepower for military advantage and political-economic self-interest in the same way they use the seas and the air.[50] Discussing space in terms of its 'militarisation' and 'weaponisation' distract and hide the practical realities of military spacepower. The argument that space has long been militarised, but not weaponised yet,[51] does not do justice to the reality that several kinds of ASAT systems have been sought and deployed in various ways over the years. It also provides an unwarranted special quality to orbital weapons deployment above terrestrial ASATs. Despite the experiences of the Cold War and the long-term, piecemeal nature of space weapons development, as well as contemporary theoretical critique, many assume that a U.S. deployment of 'space weapons' of some kind will inexorably trigger an 'arms race in space', when it is not necessarily so. Such arguments on predicting an arms race in space wrongly assume that the situation in the past up until now was stable, that responses to U.S. plans will be like-for-like or 'symmetrical', that U.S. allies will become more unilateral, and that space arms-control is the only method to prevent an arms race.[52]

Arms racing is a reflection of prevailing political disputes and interests, both domestically and internationally, military-economic liabilities, and perceptions between the leaders and factions involved. They are not a simple, mechanistic, action-reaction dynamic determined solely by the deployment of weapons that inevitably ends in war. Even when 'armed to the teeth', states can still choose to seek conciliation or some measure of understanding that eases tension. Such was the experience of the British-German naval arms-race which effectively ended in 1912.[53] In 2007, Helen Caldicott and Craig Eisendrath argued against a U.S. policy of placing 'weapons in space' because 'it will weaken rather than enhance our national security. American weapons in space will incite other countries, such as China and Russia, to take countermeasures, including placing their own weapons in space'.[54] It is not necessarily the prospect of a space-based arms race triggered by the U.S. that encourages others to develop counterspace methods, but rather U.S. dependence on those space infrastructures for its massive conventional military advantages.[55] The previous chapter has shown how multiple countries have engaged in ASAT programmes of many kinds for many years, regardless of the level of activity from other major space powers on ASAT development. In the intervening years, the United States cannot be accused of having launched major new crash programmes in ASAT development, though its incrementally improving BMD-related sensor and intercept technologies have provided a maturing KE-ASAT capability. Russia, China, and India have significantly increased their capabilities over the last 20 years, and China and Russia in particular have embarked on a far more elaborate ASAT flight-testing programme.

There is little political will among the major space powers to change the fundamental strategic reality of military and economic dependence on space infrastructure. Instead, seeking to create rules to manage that reality rather than transforming it, or reaching some form of understandings, should be the frame of mind for space arms control, treaty-making, and norm-building. It may well be the first step to something greater that follows, as is the hope with the UN General Assembly effort on a resolution on responsible behaviour in space.

It would be over-simplistic and perhaps ethnocentric to pin the blame for the decades-long investments of ASAT weapons on the piecemeal ASAT development programmes of the United States alone. China's modern ASAT programme has its origins in the mid-1980s, during a peak in Chinese-U.S. relations and cooperation. Arguably, the U.S. has shown considerable restraint in the development of various ASAT capabilities since the end of the Cold War.[56] What has significantly increased in the last 30 years is not only U.S. dependence on space systems for conventional military power and economic infrastructure, but also the dependence of the most powerful military and economic states on Earth on space systems, increasing the quality of Earth orbit as a target-rich environment for war planners from many states targeting many other states, not just the USA.

As both India and China field more orbital infrastructures for economic and military power projection, both Asian space powers as well as Pakistan will have greater incentives to pursue various ASAT options relative to each other, regardless of what the U.S. does or does not choose to do.[57] Chinese military forces are modernising with spacepower support, but an Indian ASAT capability will do little to alter the course of events in a city-busting nuclear war.

Picking apart the PLA's burgeoning space infrastructure is an increasing priority for the Indian military in time for the next potential border clash and anything that might not go nuclear. As both states compete for economic influence across the Indian Ocean, space systems will play a role in modern naval power projection between the PLA Navy (PLAN) and the Indian Navy there too, especially as the PLAN turns towards more surface fleet activities.[58] As Indian forces modernise with space systems, China and Pakistan will be provided with more space targets in a conflict with India. Therefore seeing ASAT development solely through the lens of nuclear war is flawed. They must be seen through the lens of conventional war-fighting and their practical effects should hostilities occur. There are practical, meaningful rationales for developing a range of ASAT capabilities that will impose military and economic costs on the foe, without having to go nuclear or having to directly target people with weapons fire. ASAT capabilities are ultimately meant to neutralise enemy satellites and space systems that are useful to conventional and nuclear military forces, and any space-based platforms that may frustrate such an objective, such as enemy counterspace capabilities. For several states, U.S. spacepower support for its conventional military forces is in part driving their own development of counterspace capabilities. Without satellites or space systems worth hitting, an ASAT capability loses its military purpose and has zero deterrent effect, if any. Therefore 'space warfare' should be recognised as its own area of study and practice that cannot be reduced to deterrence, nuclear strategy, and arms control.

The Banality of the Ultimate High Ground

ASAT weapons may generate some deterrent effects when both sides possess them, but war-fighting rationales are behind ASAT technologies.[59] However, the common refrain of space as the 'ultimate high ground' in warfare is doing damage to strategic thought, education, and public communication. Space as a high ground is an idea with a long history and is intuitive in its meaning. In 1958, Lyndon B. Johnson proclaimed that:

> there is something more important than any ultimate weapon. That is the ultimate position—the position of total control over Earth

> that lies somewhere out in space. That is... the distant future, though not so distant as we may have thought. Whoever gains that ultimate position gains control, total control, over the Earth, for the purposes of tyranny or for the service of freedom.[60]

Two years later, John F. Kennedy stated that the:

> control of Space will be decided in the next decade. If the Soviets control space they can control earth, as in past centuries the nation that controlled the seas dominated the continents. We cannot run second in this vital race.[61]

In his influential work, *Astropolitik*, Everett Dolman produced an astropolitical dictum which states: 'Who controls low-Earth orbit controls near-Earth space. Who controls near-Earth space dominates Terra. Who dominates Terra determines the destiny of humankind.'[62] Such imperial thought still shapes the way many think about outer space today as some zone of final control, space empire, and world domination.[63] Some proponents of space-based missile defence and ASATs have long believed that space could become a way to 'win' major wars on Earth by ensuring no nuclear retaliation could be thrown back by the other side. Indeed, such was the cornerstone of SDI.[64]

Everett Dolman refers to space as the 'ultimate high ground' as it 'offers the side that holds it commanding overviews, fields of fire, and defensive position. In this view, space is the "ultimate high ground" for the terrestrial battlefield' because there is nothing 'higher' than space.[65] Clausewitz remarked that a high ground makes it harder for the attacker to shoot up at the defender on the hill, and the defender has a commanding view and an easier time of firing upon the enemy below, demonstrating 'the feeling of superiority and security in him who stands on a mountain ridge and looks down on his enemy below, and the feeling of weakness and anxiety in him who is below'.[66] There may well be psychological effects to weapons in space, but a dispassionate analysis of space warfare and of the history of high grounds in war shows this can also be little more than a reassuring myth for those holding the high ground. A complacent enemy can provide opportunities all the same.

Despite the need to distinguish space warfare from nuclear strategy, it is important to remember that the major space powers are also nuclear weapon states or are within nuclear alliances. The 'winner' of a major conflagration (if there can be one) between major space powers will risk suffering millions of deaths and glassed cities. 'Wins' and 'losses' may be 'determined by the absence or presence of nuclear clouds over the homeland'.[67] No amount of the command of space, or controlling or denying the use of space, will stop a major nuclear attack. Ballistic missile defences are simply not credible in such a task today or in the near future. Missile defence technologies have never reached such levels of capability; no amount of spacepower or controlling the high ground of Earth orbit can prevent nuclear annihilation. Such concepts are running far ahead of credible technical capabilities. Technologies that determine which cities cease to exist or not require a high level of proven capability to be taken seriously, not fantasy projections. Only one fusion bomb getting through is needed to lay waste to a major city and throttle an economy.

Early in the 1950s the U.S. Army was discussing high-ground concepts for the Moon—a place which 'had' to be 'dominated' to prevent Soviet domination of Earth by placing nuclear missile bases there. Part of such thinking was to demonstrate a nuclear detonation on the lunar surface to show how nowhere was safe from the reach of U.S. military power.[68] As laughable as such a concept may be, such thinking is prevalent in astropolitical commentary on China's lunar exploration missions today, and what robotic missions on the Moon may mean for America's 'dominance' of the 'ultimate high ground' of outer space.[69] In the early twenty-first century, the idea of intercepting enemy munitions in space and holding a supreme orbital defensive position was developed further with John Klein's 'barrier' concept that space may be 'cut off' from the enemy like a naval blockade, leaving the blockader free to enjoy the commanding heights of orbit and denying the use of supporting 'celestial' lines of communication to the enemy.[70] This turned space warfare from merely the task of nuclear warhead interception to attacking satellites in order to deny the use of spacepower for terrestrial conventional military power by the enemy.

USSF General John Raymond said in a 2021 interview that 'space underpins every instrument of national power, our diplomacy, our economy, it is the ultimate high ground for national security.'[71] However, seizing the 'ultimate high ground' of space is as partial and restricted a view of military space strategy as defining the role of seapower in modern warfare through naval victory or dominance alone. Media interviews and public statements should always be treated with a degree of caution because senior military leaders are constantly participating in budgetary politics within military and civilian institutions, leading to stronger, more deterministic language than academic works or doctrinal theories may allow.

The U.S. Space Force's capstone publication argues in more moderate tones that:

> The value of high ground is one of the oldest and most enduring tenets of warfare. Holding the high ground offers an elevated and unobscured field of view over the battlefield, providing early warning of enemy activity and protecting fielded forces from a surprise attack. Furthermore, forces on elevated terrain hold a distinct energy advantage, increasing the efficiency and longevity of military operations. Finally, control of the high ground can serve as an effective obstacle to an opponent's military, diluting combat power by forcing the enemy to dedicate time and resources away from the main effort in order to dislodge an entrenched force. The space domain encompasses all of these attributes, making military spacepower a critical manifestation of the high ground in modern warfare.[72]

As an introductory metaphor, the 'high ground' is not without merit in quickly capturing the role of spacepower in a joint warfare campaign. It is an easy image to understand. The 'high ground' concept is correct enough in a general but sadly banal sense—it is a useful vantage point worth holding or denying to the enemy if possible because it confers some advantages. Like any high ground, those advantages only matter when exploited by competent commanders and appropriate strategies where such advantages are relevant and useful. This is not guaranteed for all possible space powers, terrestrial military forces, and campaigns.

The negative connotations of the metaphor are often overlooked. Clausewitz warned that the high ground does not always provide an unobstructed view nor prevent surprise attacks by the enemy.[73] What if the enemy has gone to great lengths to not appear where the view from the high ground is a good one? What if what is in obvious view from on high are merely feints and misdirections? The limitations of the high ground translate into space as well as we have seen through various deception ploys and countermeasures, yet this is rarely seen in public discourse that venerates the 'ultimate high ground' of space. The most important 'high grounds' also invite determined competition or workarounds by the enemy. Colin Gray rightly criticised the most ardent 1970s and 1980s 'High Frontier' enthusiasts in their belief the U.S. could indefinitely dominate the high ground of Earth orbit without inviting a strong effort by the Soviet Union to undermine or overturn that dominance if it was so important.[74] Gray believed that the danger of increased U.S. reliance on space systems for military power in the 1980s, without pursuing terrestrial backups, may have offered the Soviet Union an Achilles Heel to attack. This predates modern concerns today over a 'Space Pearl Harbor' attack from China, something of a transplant of fears from an enemy conducting a large-scale attack against U.S. space systems at the outset of a conflict as seen in the 2001 Rumsfeld Space Commission.[75] This transplant of a classic Cold War all-out space attack should not be surprising given the neoconservative Cold Warrior senior staff of the George W. Bush-era Pentagon. The penchant for military space literature to mix metaphors and allegories produces several visions that are at odds with each other. Earth orbit would struggle to be the ultimate high ground, an Achilles Heel, and a Pearl Harbor attack all at the same time.

The high ground metaphor can over-emphasise the importance of space warfare in terrestrial war plans. Strategists must be more flexible to anticipate situations where space operations are secondary to what happens on Earth's surface, rather than being the linchpin that 'must' be taken out first or a 'centre of gravity' that must draw in the focus of enemy fire so as to easily topple the rest of the military's ability to fight, as some argue.[76] Contemporary U.S. conventional military power is dependent on spacepower today, not only through material C4ISR infrastructures but also through habitual dependencies and procedures that have developed over the past 30 years. Short of a nuclear war, it is unlikely that a space power such as the United States will suffer a complete breakdown in its spacepower support to terrestrial forces, and therefore the impacts of space warfare on the ground should be interpreted as a matter of degree based on specific services as outlined earlier in this book. From SATCOMs to ISR to PNT, these services provide layered communications and data streams for a multitude of users. Not only does it make the job of any space warfare planner wanting to attack U.S. space systems a taxing prospect, but the sheer scale and diversity of U.S. space systems makes it difficult for any single command or organisation to keep abreast of all relevant and possibly useful space capabilities in a campaign. Different space systems use different radiofrequencies and coded encryptions, different software and operators, some military, some civilian. One electronic warfare or cyber assault will not likely make a significant dent—the amount of effort required for an attacker of U.S. spacepower is very high. As with soft-kill methods, hard-kill attack planners will have to contend with which satellites to target in which orbits, how terrestrial and possibly space-based assets are better placed to attack them and when in their relative orbital phases, and also consider what the environmental consequences of such attacks would be.

This poses a real challenge to conceptions of space infrastructure as a 'centre of gravity', a focal point that if struck will topple the enemy's fighting capability. Like the 'high ground' or 'Space Pearl Harbor' concepts, the centre of gravity concept as applied to space suffers from a quest to find a silver bullet or single key to victory by finding one particular fatal flaw in the enemy. Such a focal point may not be found, or may be well-guarded, or not have the decisive impacts if struck. With this in mind, space warfare clearly becomes a specialism of modern warfare and an aspect of modern strategy in its own right, a conceptual and professional equivalent to land, naval, and aerial warfare.

The Space Pearl Harbor fear is absurd when it is treated as an inevitability, instead of just one possibility of ways states may conduct space warfare and counterspace operations to support terrestrial warfare. Depending on the situation on Earth and in orbit, there may be good reasons to keep major ASAT strikes in reserve for a later, critical moment in the campaign by using a 'counterspace in being' strategy and not going all in with ASAT attacks at the start.[77] Wars can start in any number of ways. If there is a major attack at the outset, it could be in electronic warfare and 'cyberspace', or a major terrestrial missile strike, or an amphibious landing, not necessarily a massive strike against space systems. A war does not need to start with a single, massive blow either. Wars can begin after a gradual escalation of actions, costs, and stakes over time with Special Forces, proxies, militias, and guerrilla warfare that draw in a wider range of organised, more capable military forces over time. ASAT weapons and counterspace operations may be 'nuisance' attacks that gradually escalate over time as political constraints are broken in tandem with increasing stakes in the conflict, leading to desperate attempts by both sides to keep their ASAT forces hidden, secure, and 'in play' until they are needed to strike at a later, more opportune time. Adhering to the Space Pearl Harbor threat too closely as a certainty rather than a possibility among many is sure to blinker strategic analysis just as much as calling space the 'ultimate high ground'.

The word 'ultimate' adds an undeservedly decisive and fatalistic quality to the 'high ground', meaning that it is somehow sufficient or essential for victory in war. 'Ultimate' gives something a finality or insurmountable quality that simply does not accurately reflect space warfare in Earth orbit and its influence on strategy. 'Dominating space' risks becoming an end in itself which is not sufficient for winning a war on Earth. Commanding space, whether through controlling or denying space infrastructure, is one of the many means of a modern military strategy. The sufficiency (dominating space is enough to win) or essentialism (wars cannot be won without dominating space) of 'the ultimate high ground' in achieving victory may be useful in budgetary battles and public relations, particularly in domestic and bureaucratic politics in the United States, but it is not a universal truth of modern strategy for all military and political actors.

Such bad metaphors and ideas travel. For decades, PLA writing has defined outer space as 'the new commanding heights'.[78] Some PLA strategists argue that 'whoever is the strongman of military space will be the ruler of the battlefield; whoever has the advantage of space has the power of the initiative; having "space" support enables victory, lacking "space" ensures defeat.'[79] Prudent war planners should not expect the Chinese leadership or the PLA to give up the fight just because the United States may succeed in neutralising much of China's spacepower in open warfare. Indeed, *The Science of*

Military Strategy states that 'space military systems serve as the commanding heights of the modern military field, and are certain to be key point targets of the opposing sides' attack and defense confrontation', showing recognition that space will be significantly contested in wartime.[80] However, great care is needed when sifting through Chinese doctrinal language from military and academic sources, many of which use open-source U.S. literature as their base for commentary on military space activities. Military officers, technicians, and analysts writing opinion pieces or essays, or even official doctrine, are not official state policy and do not outweigh what the Foreign Ministry or Chinese President thinks or says about arms control and matters of war and peace.[81]

The ultimate high ground is a rather vague metaphor that does not help us get at the heart of spacepower's role and impact in war. Whereas high grounds are often simply held or not, space is more like aerial volumes and waterways where control can be contested or shared to varying degrees by opposing sides at the same time. The 'ultimate' high ground is an intellectual dead-end which encourages seeing space dominance as an all-or-nothing affair, where victory or defeat is inevitable if it is held or lost, where war resembles a simple process of machines and physics, rather than a realm of human creativity, intellect, and emotions. Battling such misconceptions in his own time, Clausewitz was exasperated at 'one-sided systems as a veritable code of laws' and complained that:

> our critical books, instead of being simple, straightforward treatises... are brimful of these technical terms... But they are often something much worse still, being nothing but hollow shells without any kernel. The author himself has no clear perception of what he means and contents himself with vague ideas, which if expressed in plain language would be unsatisfactory even to himself.[82]

He further railed against the overuse of the principle of superior positions or the high ground by military thinkers and writers in his own time:

> They have become the darling themes of learned soldiers, the magic wands of strategical adepts, and neither the emptiness of these fanciful conceits nor their contradictions by experience has

> sufficed to convince authors and readers that they were here drawing water in the leaky vessel of the Danaids... The occupation of such and such a position or tract of country has been looked upon as an exercise of power like a thrust or a blow, and the position or tract of country as in itself a real quantity... This thrust and blow, this object, this quantity, is a *victorious engagement*. That alone really counts, that alone can we reckon with and that we must keep constantly in view... the influence of the ground can only be a subordinate one.[83]

Battle and destroying enemy forces, their capacity to fight, and the will of the enemy by whatever means necessary is what matters, not holding 'good positions'. Holding good positions is a means to the end of strategy—not the strategy or end in itself. For space strategies, Clausewitz's lesson is that holding the 'high ground' of Earth orbit, whilst generally useful and no doubt helpful in some ways, is not the sole deciding factor in winning wars and battles and only has value in how it supports and influences other military operations and the overall military strategy.

Holding good positions over a battlefield is no guarantee of strategic or operational success. The French Army learned that lesson all too well in 1870 when the Prussian Army negated their *positions magnifique* by moving around them and encircling them, such as in the Battle of Sedan.[84] Wars can be so varied as to make what were once thought to be commanding positions irrelevant. The previous 20 years of experience in Afghanistan and Iraq should be an ample, obvious, and brutal lesson on the limits of American power even with a dominant command of space. The Taliban never had any say on the 'ultimate high ground' of Earth orbit yet still brought about the downfall of the U.S.-backed Afghan regime and outlasted U.S. military presence there. In fighting Chinese military forces, their dependence will be different to that of U.S. military forces, meaning that the high ground will be more or less useful depending on the exact situation, therefore simply 'holding the ultimate high ground' is a banal and ultimately a supposedly poetic 'magic wand' for dull strategists. Yet, as any cursory glance at military statements will show, the 'ultimate high ground' of space has become the darling of many strategists, thinkers, politicians, and commentators.

Beyond the banal and mechanistic thinking of the high-ground concept, it is important to recognise that some advantages do accrue to weapons platforms in space, as Dolman highlights. Once deployed in a stable orbit, 'station keeping' or maintaining the desired orbit requires minimal fuel, and moving to other orbits (such as from LEO to GEO and back) is more fuel-efficient once the platform is already in an orbit compared to launching from Earth to that desired orbit. In basic terms, the more you move *with* local gravitational forces and your own inertia, and make them at the correct location and time, the less fuel you will spend, and the longer your vehicle will maintain operational flexibility. The more you move against prevailing gravitational and inertial forces, the more effort of velocity, acceleration/deceleration, delta-V, or fuel and impulse you will need, and the faster the vehicle will run out of fuel. The 'energy advantage' that weapons platforms in higher altitudes or stable gravitational points between two bodies enjoy (e.g. the Lagrange points) does mean that kinetic weapons or projectiles can move elsewhere with relatively low energy expenditure.[85] Coupled with the greater views of and around Earth afforded from higher altitudes, it can make for a feasible orbital weapons platform. In terms of the economy of force it is one possible advantage, and having a position that forces the enemy to plan against you rather than the other way round is generally a good thing, as Sun Tzu stated.[86]

Having a good position is hardly the only consideration in strategy, though. Strategy requires *net* assessments, good political-military command, an appropriate strategy, sufficient resources, and not analytically jumping on isolated advantages or disadvantages to champion a specific kind of weapon or force posture. One of the large brakes on such visions of orbital warfare and utilising Lagrange points is the fact that there is no way of getting around the punishing costs of launching anything into orbit in the first place, and doing anything quickly to meet immediate crisis or military needs requires less efficient fuel burns. Reaching LEO can cost between $2,700 and $20,000 per kilogram, with most providers offering launches at around $7,000–$10,000 per kilogram.[87] Higher orbits are more expensive still. Complex, versatile, and elaborate weapons platforms will weigh tons and many of them will be needed to ensure redundancy and depth of capability. Satellite refuelling is an emerging capability, but as with all else in space, such logistics require lots of mass to be lifted to orbit. Replacing lost satellites at short notice, or 'Operationally Responsive Space' as it was once dubbed in USAF, has been a long-established idea for responding to ASAT attacks,[88] but that too is limited in implementation due to the expense of keeping spare SLVs with satellites ready to launch at a moment's notice. Furthermore, rolling something out at scale and sustaining it is a different thing to proving a concept or building a prototype. Compounding issues is the fact that moving efficiently in orbit or from extremely distant Lagrange points can be slow compared to a less efficient but rapid direct-ascent path from Earth towards LEO, MEO, and GEO. Often, proponents of space-based weapons oversell their technical potential on manoeuvrability and responsiveness to unpredictable operational military needs.[89]

Maximising fuel efficiency and manoeuvring capability in one area requires sacrifices in other areas of capability—such as weapons mass, hardened components, and draining the financial budget from other programmes to pay for more massive systems. Nuclear-electric propulsion may provide more manoeuvrability without seriously shortening a satellite's lifespan, and nuclear-powered satellites may generate enough power for more dangerous lasers. However, there are serious drawbacks to nuclear-powered satellites, such as with the Soviet Union's RORSATs. Putting nuclear generators in orbit will be extremely costly as they are still bulky by satellite standards with the smallest reactors approaching tens of tons producing tens of megawatts.[90] Putting even that modestly-sized reactor into orbit will be extremely expensive. If these are to be weapons or targets of war, adding further sources of electronics-damaging radiation when these systems blow up or are damaged, or not decommissioned properly, will only add to environmental problems in orbit. Nuclear decommissioning is extremely expensive and difficult on Earth, let alone in space.

Orbital energy efficiencies do not provide insurmountable *net* advantages in space-based weapons fire in order to control Earth orbit, not when hurling anything up Earth's gravity well is so expensive and most critical targets in space warfare will be no higher than 40,000km altitude—far away from Lagrange points and other celestial orbital zones such as the Moon (which is ten times further away than GEO). Not until a 'leak-proof' space-based anti-satellite and anti-warhead interceptor system that can withstand nuclear and kinetic barrages from Earth's surface is deployed in orbit will such major advantages come from sitting atop Earth's gravity well for terrestrial wars. If meaningful economic and military activity happens beyond Earth, towards the Moon, and into interplanetary space, then the advantages of orbital basing may increase. But that is not a concern for practical astro-strategists and military space units today. Lunar exploration and science have little bearing on warfare and the global economy for the moment. The most likely future for the Moon is a lunar version of Antarctic research bases, such as McMurdo station, rather than celestial equivalents of oil rigs and service stations.

War is always expensive and logistically taxing. Orbital warfare cannot be done 'on the cheap'. Military weapons programmes and military strategies are not solely determined by what is most efficient or the cheapest. The desire to rely on effectiveness (what reliably 'gets the job done' and 'what works') to generate the practical effects needed at the right time are also important considerations. Trusting in known, proven technologies that can be relied upon to work when needed and can be maintained and replaced easily are important features of any technological system. David Edgerton stresses that maintenance and repair are 'mundane and infuriating, full of uncertainties, they are among the major annoyances surrounding things. The subject is left in the margins, often to marginal groups'.[91] This point should resonate with the typical prestige roles of fighter pilots compared to cargo pilots, logisticians, and aircraft maintenance crew in air service cultures. Relative costs, logistics, and other 'mundane' technological considerations means not doing something in space even though it is possible. In the 'high ground' of space, terrestrial ASAT options for attacking some orbits may be cheaper and more reliable than space-based alternatives. The Soviet FOBS is an example where just because something could be done with orbital weapons does not mean it made sense. ICBMs and SSBNs got the job of threatening the U.S. with nuclear apocalypse effectively. Whilst orbital bombardment is certainly dramatic as a weapons capability and may massage the sci-fi-driven egos of many space enthusiasts, the grim reality of entropy, maintenance, and economic alternatives insists on being recognised by practical, logistically-minded strategists and politicians controlled by budgets.

The same caution is needed on point-to-point rocket logistics. Such rapid flight times between points on Earth would be an interesting development, though hardly free of drawbacks as space-launch vehicles need a large, complex infrastructure for any reusability. This means any rocket that might transport materials or troops to a theatre or combat zone will need a launch pad that can handle complex fuelling systems on-site to ensure a return flight, and any landing craft will be vulnerable to local anti-aircraft systems as any airstrip in a hostile zone would be. Deploying troops via orbital routes is far more cumbersome and logistically taxing than portrayals of the Orbital Drop Shock Troopers in the *Halo* video game franchise may make it appear. Methods of delivering weapons effect or logistical capabilities beyond data from and through space—such as destructive hard-kill weapons fire or heavy materials or troops—will never be a silver bullet to strategic problems. Much like the 'bomber mafia' in the Allies of World War II claiming strategic bombing to be a war-winner, it was in fact a product of rather unique constraints on Allied power projection between 1941 and 1944.[92] Spacepower's effects are wide-ranging but it will not be decisive by itself, regardless of the question of placing weapons in space to bombard Earth or to target other assets in orbit. It is important that discussion of targets, weapons fire, and effects do not lose sight that strategy is about winning wars, not targeting. Even when considering targeting alone, existing military forces on Earth struggle to get enough bandwidth for all data needs in light of the explosion in sensory data and equipment in the twenty-first century, leading perhaps to 'more sensors than sense' in military planning.[93] Regardless of precision-strike methods, controlling the future of the enemy's territory and political life requires methods of control and coercion on land, not simply destruction alone: occupation.

Seemingly obvious advantages can also become liabilities. Efficient routes of transportation can become intellectual blinkers for dull strategists. Warfare is in part the art of surprise, deceit, attacking weak points, and rendering the enemy's strengths irrelevant. If you rely too much on an obvious 'good road' to get to your objective, the enemy may figure out a way to nullify the advantages you may get from it, or make every effort to turn that 'good road' into a 'bad road' by aiming enfilading weapons fire at it or making it a major objective in any offensive. Space-based weapons systems share the same vulnerabilities as the satellite systems they target: physical vulnerabilities in the space environment and also reliance on secure radio communications which can themselves be targeted with electronic warfare. Such is the paradox of strategy where advantages become disadvantages. If the enemy believes their obvious attacks are anticipated, they will act *against* such expectations and not do 'energy efficient' things if they are too predictable. As movement begins, 'so does the fog of war', making pre-war plans rough guides at best, not flawless timetables.[94] The classic work of strategy by Sun Tzu is clear: 'War is such that the supreme consideration is speed. This is to take advantage of what is beyond the reach of the enemy, to go by way of routes where he least expects you, and to attack where he has made no preparations.'[95] Basing principles of space strategies on the mathematics of maximum energy efficiency in orbit is conducting war as the mechanistic 'algebra of action' that Clausewitz roundly condemned.[96] Energy efficiencies are important but are merely one factor among a multitude that strategists must account for.

Holding the high ground is not enough for victory. Intelligent and determined adversaries will find ways to undermine or counteract the advantages derived from any 'high ground', especially if they do not work to exploit the advantages that the high ground gives in other operations and follow-on actions. Just as many military actors have learned to fight under conditions of enemy air superiority, fighting in conditions of enemy space superiority should also feature in all military planning and education. This is prudently established in joint U.S. doctrine on space operations under the moniker of 'degraded space operations'.[97] If the command of space is lost or threatened, you should not give up and seal your own fate through the choice of surrender on Earth when the actual impact of space loss is undetermined, mixed, and may even be reversed. The U.S. would no doubt have quite the crisis on its hands if it lost the use of space in its entirety. But short of nuclear war, it is unlikely to face a complete halt in its use of the 'high ground' of Earth orbit, which shows how the either-or nature of the 'ultimate high ground' image is not very useful, especially when the alternative of commanding space allows for degrees of control, denial, and influence in space, as explained in the next section.

In wars between proxies or clients where third-party forces are supported by and tethered to space systems that are politically 'off limits' to military attack, then acting in the face of enemy spacepower support may be the only option available. The high ground metaphor merely means somewhere where there's an advantage to be gained, just like 'good ground' or a command of the seas or the air provides useful supporting effects and capabilities to winning battles or imposing certain trade-offs on the enemy.[98] In 2022, Ukraine, with little military space infrastructure of its own, managed to frustrate the Russian military which has its own elaborate space infrastructure in the 'ultimate high ground'. As far as is known, Russia has not suffered any major challenges to its command of space in the conflict yet. The true extent of space support provided by the U.S. and its allies to Ukraine remains to be seen.

Military space strategy and space warfare must not be reduced solely to questions of targeting and weapons deployments, as necessary as they are. They are merely tactical and operational considerations, and focusing on them at the expense of the politics and strategy of the entire war risks the tacticisation of strategy, as Michael Handel warned. An obsession in the military on targeting tactics and operations whilst ignoring politics came into much criticism following the failures to build a durable peace in both Afghanistan and Iraq.[99] It is no less a danger to strategic analysis and planning in high-intensity or state-on-state warfare. Reducing space warfare to its technical and targeting elements alone, as the 'ultimate high ground' does, risks the 'operational or military tail wagging the political-strategic dog'.[100] This includes targeting and weapons systems both in space and on Earth—securing or denying the C4ISR benefits from space are not ends in themselves, just as much as destroying satellites in space is not.

Leaders and decision-makers must have some idea of what is actually possible in space warfare, and what its limits are. Space is 'just another' environment in warfare alongside the land, sea, and air, that has tactical, operational and strategic levels that feed into higher levels of war planning and political goals set by governments and cabinets that make use of other levers of power, such as economic tools, diplomacy, and intelligence, what some may term as 'grand strategy' or 'policy'.[101] Roughly speaking, strategy refers to the general art of the 'supreme commander' linking military means to political objectives; linking lower levels of command and activity (tactics and operations) to ultimate political goals of the highest echelons of the state or organisation.[102]

There has to be a strategic value and purpose to such activities to maximise the military's best possible contribution to winning the war, which is ultimately a political activity. Sometimes winning battles and causing destruction is particularly valuable to winning wars, other times they are less so and may make reaching the political objectives of the conflict harder. Tactics are of course important, and consistently losing battles, especially significant ones, increases the odds of strategic defeat. Gray captured this tension when he argued 'it is an error to believe that tactics are more important than strategy… [b]ut, tactical competence is the material of which strategic effect is made.'[103] Tactics is being able to physically fight battles, but strategy must ensure operations string tactical battles together to serve and contribute towards the political goal, otherwise they are pointless.[104] Therefore we cannot consider space warfare without considering the impact of 'commanding space' or 'seizing the high ground of orbit', or the lack of it, on achieving the general political-military strategy in winning the war on Earth. Yet strategy and policy cannot expect things that tactical activity and operational campaigns cannot achieve.[105]

Commanding the Cosmic Coastline

We are in a geocentric strategic and political world. When all of humanity's polities and resources for life are on Earth itself, and not out in space, Earth remains the most important referent point. Any combat operations against or in space must be seen in light of influencing the control or denial of the effects of the support from that orbital flank onto terrestrial activities. Earth orbit is not as much of an ocean, which tends to be distant and isolated from land-based power projection, and only controllable by largely having a presence in that environment itself. Earth orbit is something of a cosmic coastline, a littoral environment, or orbital flank defined by its proximity and intimacy with Earth itself.

In short, the analogy of Earth orbit as a coastal or littoral zone is useful to highlight:[106]

1. The proximity of orbital space to Earth means weapons effects from Earth can project influence over the command of space—who gets to use or deny the use of spacepower. The space infrastructure that is commanded projects direct and indirect influence over military, political, and economic power on Earth in turn. The hostile cosmic coastline of Earth orbit works both ways. Space-based weapons will share the same hostile coastal environment as targets of terrestrial ASAT weapons. The degree of command is not absolute, and warring parties may still have limited use of space even in a losing war.
2. As an adjunct environment or flank, Earth orbit is not a primary theatre in terrestrial wars. It is a secondary theatre that provides supporting, logistical, and infrastructural effects. Controlling or denying outer space—commanding it—is about the exploitation or denial of these effects on Earth where the bulk of the war happens.
3. Modern space powers are terrestrial and may share borders with their foes and with third parties on Earth, like many coastal maritime powers that may share land borders with others. This reinforces the secondary nature of space when terrestrial lines of communication may be more readily available to use or more immediate avenues of attack posed by the enemy.
4. Like other coastal zones, Earth orbit features defined and constrained spaces, converging routes of popular military and civilian traffic, belonging and accessible to a multitude of actors both large and small. Earth orbit and hostile actions against objects there are not restricted to the major powers alone but are open to smaller military powers too. The lines of communication through space do not cross vast expanses like oceans, but rather resemble busy chokepoints and straits.

Like all analogies this one breaks at some point. Earth orbit does not share some other elements of coastal operations and environments. A self-critical engagement with any analogy should help educate strategic minds, something missing in much of the 'high ground' discussion. The weaker points of analogy in the celestial coastline are:

1. Unlike on Earth, in space 'celestial lines of communication' mostly concern the routes along which satellites travel, as well as the pathways of their information and radio communications streams. Heavy materials and logistics cannot be transported via space unlike with sealift and airlift.
2. Unlike coastal logistics, amphibious assault operations have no obvious corollary at this time given the expense and difficulty of accessing Earth orbit and the stresses of atmospheric re-entry.
3. The degree of influence projected from space infrastructure over Earth's surface goes to a far greater extent than coastal seapower, more closely resembling airpower support to land and naval forces rather than coastal fire support to land forces.

These differences are of a minor scale compared to the larger, more appropriate strategic analogy: that Earth orbit is defined by its closeness to Earth and the extremely intimate nature of the connections between activities in orbit and on Earth, whether through supporting infrastructure for satellite constellations, C4ISR from space, or weapons fire from Earth into orbit. The cosmic coastline analogy is more useful for in-depth thought about space warfare and military space strategy because it pushes thought into the specific possibilities and constraints that spacepower presents, far beyond the abstract, banal, generalities of the 'ultimate high ground'.

Like terrestrial coastlines and oceans, military action towards, in, and from space can have economic, psychological, and political impacts and consequences that complement military actions against the enemy. In astroeconomic warfare, where the more dependent economic infrastructure, such as financial services, the energy grid, and maritime cargo transport, rely on space systems to function efficiently, it increases the potential payoffs of attacking those space systems that support it. Yet such attacks could invite significant retaliation. Against less space-dependent economies, the utility of astroeconomic warfare would naturally decrease.[107] Such economic shocks may result in short, sharp economic convulsions, or accumulate over time as supply chains grind to a halt and economic inefficiencies mount.

Terrestrial and orbital ASAT weapons amount to parallels to maritime coastal or air defences, and specialised craft designed for littoral activities. In strategic terms, terrestrially-based ASAT systems resemble coastal guns designed to harass or check the activities of hostile vessels in coastal waters, which are nearby and a flank to land operations. Space-based weapons and satellites are *orbital* spacecraft, not deep-space or interplanetary spacecraft. In that sense, we should see orbital space vehicles, weaponised or not, as vehicles designed for operations in an 'adjunct' or 'flank' environment, rather than the isolated expanses of interplanetary space. Orbital weapons share the same vulnerabilities as the satellites they target because they share the same 'cosmic coastline' which is within reach of weapons fire from Earth, or can be negated by other responses and adaptations on Earth itself.

The proximity of Earth orbit to the primary theatres of Earth provides more opportunities for smaller actors to conduct their own operations. Military forces without significant space-based infrastructure can still pursue significant levels of terrestrial ASAT capabilities, especially in terms of electronic warfare and cyber capabilities. Space Situational Awareness (SSA) networks have been terrestrial for most of the Space Age, and rudimentary space-tracking systems can be devised with a modest budget. New, emerging, or latent nuclear weapon states and space powers such as North Korea and perhaps, eventually, Iran could also engage in nuclear-based ASAT options. Space warfare is not simply the business of the big space powers or the richest states. On the receiving end of the hostile coastline of Earth orbit, small powers can use terrestrial workarounds to orbital problems in the same way armies could attack land targets to support coastal naval operations. If the political consequences are tolerable, why attack a satellite when a ballistic missile salvo at a satellite ground control station can destroy it instead? Earth orbit is populated by a range of state and non-state actors, many of which provide useful services and data for warring parties. Smaller actors can invest in counterspace capabilities such as electronic warfare, computer network operations, as well as Special Forces or sabotage operations against physical space infrastructure or space personnel on Earth behind the front lines. Cuba, Indonesia, and Iran are known to have conducted electronic warfare against satellites. It is worth remembering that not all space powers are separated from their most likely adversaries by two oceans, but rather share terrestrial borders like coastal states with shared land borders. This would place even more pressure for commanding similar orbital regimes and zones from both sides.

The command of space is an important feature of military strategy, like the command of the seas and the air. Many spacepower theorists draw from the precedents of seapower to explain the basic strategic concepts underpinning the exploitation of and resistance to spacepower.[108] In terms of base strategic concepts such as the command of the sea and the command of the air, there is not that much different about the command of space in terms of them referring to controlling and denying the use of transitory mediums or places that machines, materials, and communications travel through between the lands where people live and states reside. Since at least 1958, parts of the U.S. military have considered that space needs to be controlled in the same way they control sea and air in a time of conflict.[109] Engaging in space warfare to impose a command of space is done to influence 'who gets to use Earth orbit to what degree, and how.'[110] The command of space is the founding aspect of any spacepower theory and thinking in terms of military strategy in space as it:

> forms a two-way connection between spacepower and grand strategy, much in the same way that the command of the sea makes the student or practitioner of war think of the role of seapower in a wider war based on the control, denial and exploitation of the sea. Both the control and denial of space can be sought in order to secure or contest the command of space and exploit that command on Earth. Whether through controlling space infrastructure or denying its use, the command of space refers to the influence one can project upon the use and non-use of the medium and theatre of Earth orbit. The command of space must in turn be exploited for spacepower to have strategic effects on Earth. That is, without a degree of the command of space, you cannot shape or influence how you or others can use outer space to meet objectives on Earth... Waging space warfare and commanding space is meaningless without tying it to supreme political objectives on Earth.[111]

The command of space is therefore about commanding the littoral, adjunct, or flank environment of Earth orbit and the tethers that link orbiting machines with infrastructure and users on Earth. Combat with and against these systems relies on or helps determine who commands space. The spread of terrestrially-based ASAT weapons is effectively making the cosmic coastline or Earth orbit a hostile one for satellites in a similar way to how coastal defences could keep enemy naval and maritime traffic in check from the land. This is more accurate and detailed than the common analogy of viewing outer space as the 'ultimate high ground' that, according to its advocates, will ensure victory in future wars, and will therefore necessitate a mass armaments deployment in space to both target other satellites in space about bombard Earth from orbit.

In terms of the degradation of services, attackers of space systems will seek to decrease the efficiency with which the enemy can organise themselves, move about, detect incoming enemy forces or missiles, and generally increase the attrition and exhaustion by making everyday activities more difficult and slower than normal. Successful space warfare can cut off terrestrial forces from the tethers that support them through the cosmic coastline in orbit, making them unaware of events elsewhere and unable to coordinate and adapt as the wider situation develops. This may return some usefulness to numerically large units and reinforcements on Earth, perhaps bringing 'mass' back to conventional forces. Reducing the bandwidth of SATCOMs, making some ISR information unavailable or simply overtaxing them, as well as breaking the integrity and accuracy of navigation signals blunts the 'cutting edge' of modern military forces. Making long-range missiles and aircraft slower to acquire targets, less accurate, and less survivable makes them less efficient by increasing the amount of munitions and vehicles needed to reliably hit targets, especially over very long terrestrial distances. Space warfare operations aimed at reducing one-shot one-kill strike capabilities will place greater strains on logistics and flexibility as more vehicles and missiles will be needed, as well as more time, to ensure the correct targets are hit. This continues classic operations of confusing the enemy, disrupting their supporting infrastructure, and disabling their communications methods. Today, doing so requires extending harassment and sabotage activities against celestial lines of communication in the coastline of Earth orbit.

The actual impact of the command of space on modern warfare is like a contest to impose the dispersing effects of spacepower on the modern battlefield. Space warfare directly impacts the exploitation and denial of this dispersing effect on Earth, and why space weapons may be useful in conventional military campaigns.[112] The orbital infrastructures explored in the first five chapters create a diverse series of effects on the ground from their data and services. For global infrastructure and commerce, they have furthered processes of post-industrial globalisation and information infrastructures. For modern military forces, space infrastructure enhances their ability to physically disperse and downsize whilst increasing their mobility, long-range precision strike capabilities, and improving survivability through improved C4ISR. Concentration does not have to be about massing physical forces themselves, but rather the effects of their precision firepower over larger and larger distances. Though there are limits when cities and territories need occupying or amphibious landing zones are few and far between, for example. In terms of high-intensity mechanised combat, precision warfare imposes a dispersing effect on the enemy because massing heavy forces provides obvious targets for precise munitions, whether through close support aircraft or standoff missiles. Dispersing and presenting less obvious targets is a routine response to enemy fire. In a war between two major space powers, their ground forces will be simultaneously wanting to benefit from their own space infrastructure to support their ability to disperse whilst remaining networked and able to coordinate firepower and manoeuvres over vast distances, whilst trying to deny the ability of their enemy to do the same. Both space powers will be seeking to take advantage of dispersed and isolated enemy forces once their space support tethers are cut and they can no longer see attacks coming, or call for help from friendly units.

Waging space warfare for the command of space by launching various kinds of ASAT attacks on specific systems at the right time and place in the context of the terrestrial campaign resembles coastal operations where support from the sea has influenced land operations, often through logistical support and the confidence, morale, and security such support has afforded. Those that enjoy logistical support from the coastal flank have tended to enjoy greater security, morale, and flexibility, whilst those that lose it have faced demoralisation, entrapment, and paralysis.[113] Without support from the cosmic coastline, dispersed military units lose their ability to network and coordinate effectively with friendly units over large distances, fail to offer support to others or be supported by them, struggle to provide early warning of enemy activities, and fail to effectively bombard obvious enemy units.

Russia, China, India, and the USA will face difficult decisions in terms of when to employ major counterspace offensives, if any. With each military deploying 'networked', dispersed, and long-range strike capabilities and sophisticated C4ISR networks built in part with space infrastructure, they will face tensions between a 'Space Pearl Harbor' first-strike tendency and a 'Counterspace in Being' strategy. It will not necessarily be obvious when major counterspace operations or ASAT attacks of various kinds should be launched. These will depend upon the effectiveness and survivability of the ASAT weapons themselves, of course, but much will also be determined by when they would make the biggest difference in military terms on Earth when the dispersing effects of spacepower needs to be guaranteed or denied for terrestrial military operations. Attacking U.S. space systems after a U.S. Navy task force has arrived at and concentrated in the Taiwan theatre may be preferable to when they are dispersed around the Pacific or USAF assets are in hardened shelters. Rather, attacking U.S. C4ISR systems when the task force is concentrating (in relative terms) in the combat theatre around Taiwan presents a more obvious target for Chinese precision-strike weapons and reduces the warning provided. This assumes that restraint has been seen in space until such a point in a Taiwan conflict, and China has successfully conducted a 'Counterspace in Being' strategy with its more destructive ASAT arsenals. This is not necessarily the best course of action or is certain to happen. Rather, like the Space Pearl Harbor scenario of an all-out ASAT campaign from the start of hostilities, 'Counterspace in Being' is a possibility that may happen if conditions are deemed right for it.

ASAT or space infrastructure attacks may be better-timed depending on the situation on Earth, again reinforcing the secondary and supporting nature of the littoral nature of Earth orbit. However, if both sides fear their space systems or ASAT capabilities may not last long, a 'use it or lose it' situation may hasten counterspace operations in spite of whether or not it is beneficial for terrestrial operations in a specific way.[114] The relationship between Earth and space works in both directions, but the winner on Earth ultimately wins the greatest prize because that is where people live and where all meaningful resources are. Taking spacepower support tethers away from one side, or at least severely disrupting them, may help tip the scales in one's favour on the ground. However, such effects from space must be relevant to an appropriate, larger terrestrial campaign, and consciously linked to enhance terrestrial military capabilities. Terrestrial military capabilities should not necessarily be sacrificed for spacepower investments. A fully functioning space-based ISR capability may only lead to a high-resolution view of your ground forces being torn to pieces by enemy artillery if ground forces cannot withstand superior enemy forces because their armoured units are not good enough, too few in number, have run out of ammunition, their air defences saturated, long-range precision strike assets tasked elsewhere, or the troops have simply not had a decent meal in weeks. Possessing the high ground alone matters little when it merely provides an excellent view of your own impending demise. As the first months of the 2022 Russian invasion of Ukraine has shown, commanding the cosmic coastline is therefore not sufficient in itself; it is not 'ultimate' if you have no way to make that command relevant to battles and more importantly the total war effort on Earth. Strategic depth, mass, numbers, and competently waging mechanised terrestrial warfare still matter in the Space Age. Spacepower adds to that; it does not supersede it.

The remarkable victories over Iraq in 1991 and 2003 are not problem-free cases to project the future of major wars from. The 'First Space War' has two aspects that may not reflect many space warfare scenarios today, though it is often invoked in thinking about space warfare. First, one side did not do much to counter the spacepower of the other. Competent leaders and military forces are planning for the loss of specific space systems, have redundancies or mitigation measures in place, and plan on how to fight against a space-supported enemy along the lines described in Chapter 5. Second, the 1991 Gulf War was an ad-hoc assembly of spacepower to the immediate needs of USCENTCOM, not a system based on existing structures and plans. Subsequent U.S. military operations are more instructive on that front, but even so greater spacepower integration was still not good enough for the Rumsfeld Commission, 10 years after the victory of 1991. Today, space warfare in the cosmic coastline is becoming a more 'normal' part of modern war planning and strategic thought, a far cry from the nuclear strategy and ballistic missile defence approaches that dominate public discussions of spacepower.

Conclusion

Spacepower is diverse in its influence and its impact is ubiquitous alongside the complexity of all political, military, and economic activity on Earth and its use in wartime is a specialism in its own right. Space warfare provides perspectives on matters of war and peace in the Space Age that the approaches of deterrence and arms control do not. Space infrastructures are extensions and providers of many forms of military, political, and economic power on Earth. We can approach the Global Space Age from the perspective of a 'geography' or astrography in the same way that seapower and airpower experts contribute to wider discussions of military strategy and defence policy.

Unlike nuclear weapons, space weapons can provide a range of effects that may be more politically and economically acceptable in certain situations. Space warfare's diverse array of methods and effects means that we may witness it occurring in many forms during crises and extensive hostilities before a nuclear threshold is crossed, if at all. Space warfare is not an all-or-nothing approach as many scenarios of nuclear war lend themselves to. Military forces prepare for a range of conventional military contingencies against nuclear-armed opponents—space is not immune from this as there are many options for space warfare under the conditions of the thermonuclear revolution. However, the risks of entanglement produced by the ubiquity of space systems to nuclear and conventional war fighting does not mean space warfare is a less risky method of war, and may indeed lead to nuclear war when the major space powers are nuclear weapon states or are shadowing under a nuclear umbrella through alliances.

The astrographic environment of Earth orbit resembles that of a coastline due to its littoral, flanking, or adjunct nature. It is near to

Earth's surface, and the projection of power from Earth to space and space to Earth flows easily in both directions. The cosmic coastline of Earth orbit can be a hostile one that makes it easier to destroy your forces, or a place where supporting C4ISR tethers enable your most precise and capable military forces. The dispersing effects of spacepower on military forces on Earth by controlling and denying the use of space infrastructure is the rationale for space warfare. The command of space has no meaning in itself, it must be consciously exploited to make a difference to wars on Earth. This is a superior vision of Earth orbit compared to the 'ultimate high ground' metaphor that pervades much discussion in the United States today. The ultimate high ground invites simplistic and mechanistic thought because it merely highlights a banal truth that space is a place where advantages can be gained.

The mainstay of space weapons today are Earth-to-space weapons; some of these are deployed in significant numbers, but others at a relatively small scale, across soft- and hard-kill options. Even then, these efforts do not compare to the scale of effort the major space powers are pouring into building satellites for terrestrial military modernisation, alongside supporting ground and launch infrastructure, as seen in previous chapters.

Original sin ensures 'business as usual'—the continuation of Terran politics by other means. Once it is understood that space warfare is merely the continuation of terrestrial politics by other means due to the original sin of space technology, it becomes apparent that prohibiting space warfare is not something that can be done without prohibiting the socio-political institution of war itself, and bringing about a revolutionary change in the nature of international anarchy and power politics.

CONCLUSIONS

ANARCHY IN THE GLOBAL SPACE AGE

Teg yw edrych tuag adref

It's good to look towards home

Seventy years since the Global Space Age began, spacepower has earned, in the words of astropolitics scholar Robert Harding, 'a permanent place at the table in matters of international conflict, peace, national and international development, and international law.'[1] Spacepower's time has come yet it is a major aspect of human politics and society that remains largely underappreciated in the study of International Relations. As this book has shown, any number of sub-disciplines in IR and the humanities and social sciences more broadly have much to offer us in helping understand the realities of our Global Space Age. There are vast research agendas waiting for the humanities in space. Space is not just a place for those with scientific or engineering backgrounds when astropolitics increasingly impacts global political systems and technological infrastructures. In the words of Michael Sheehan, studying the international politics of space provides a 'corrective to the idea that space programmes are science-driven bureaucracies somehow aloof from the harsher realities of world politics… There are few, if any, features of contemporary global politics that do not have their echoes in the utilisation of space.'[2] Moving away from an idealistic view of space as a place merely of exploration, curiosity, harmless prestige competition, and international science cooperation is an important step in trying to make things better. The militarised heritage of space technology is its original sin and the military exploitation of space is not going away any time soon unless world politics dramatically changes.

Most of the early space powers' efforts were intimately tied to nuclear and missile projects, and never more than an arm's length from creating a latent indigenous nuclear and missile infrastructure. In economic and social terms, space technologies have moved on from a luxury to a necessity as telecommunications, imagery, and eavesdropping systems found more civil and commercial uses beyond the military, intelligence, and state security institutions.[3] The promise of nuclear missiles committed the original sin of space technology as they provided the foundational technological systems needed to develop the plethora of satellite constellations and space systems that have grown among many of Earth's most capable states since 1957. Spacepower has now matured, extending the tendrils of Earth's political-economy into the cosmic coastline of Earth orbit, manifested in the radio tethers between military and economic power on Earth and orbital infrastructures. This stretches world politics and the supposed Anthropocene into volumes well beyond Earth's surface and atmosphere.[4]

The Space Age has always been 'global' and never was the story of just two superpowers in the second half of the twentieth century. The twenty-first century is likely to see even more space activity as more states and their registered companies and institutions are able to access space in various ways. Spacepower is making its presence felt in more local, national, and regional economies and polities. Whilst some believe that space is being 'democratised' as more states and non-state actors get involved in space and joyrides into suborbital flights for the super wealthy of Earth are developed,[5] such views ring hollow when the benefits of such space technologies continue to accrue to the companies and states that own and control that infrastructure. Private companies still have to play by the rules and the interests of the states that pay for their services or allow them to exist within their borders, therefore the private sector is no salvation whatsoever from the power politics of the original sin of space technology. Spacepower reflects or exacerbates many of the injustices and inequalities of power on Earth today just as it can help address many pressing issues such as arms control verification, missile warning, ecological monitoring, agricultural modernisation, and global warming. The original sin of militarised space is ever-present, and the potential struggle over the command of the cosmic coastline is only becoming more probable as more states choose to exploit Earth orbit for terrestrial combat advantages and economic infrastructure.

Competition and self-serving behaviour in space and the political values put into space technologies reflect the anarchical nature of the international system. Space technologies and orbital infrastructures that are literally global in scope epitomise Geoffrey Herrera's point that 'technology is both systemic and political in character'.[6] It is no coincidence that the spread of multiple orbital infrastructures reflects the systemic political nature of international politics and falls along the fault-lines of global economic blocs. States, societies, and people take their terrestrial baggage with them into orbit. Space technologies are ultimately products of some people and their interests, not all of humankind. Who possesses and controls what technologies, and who influences whom and to what ends, matter. Technology is not separated from human will, values, and influence. Once unleashed upon the world, technical systems help structure human agency (the options that are open to people to meet their objectives) in certain ways. This may strike some readers as obvious, but directly challenges prevailing assumptions about technology in IR scholarship and much of the space community.

The development of space technologies was not some inevitable, natural, or linear process separate from politics and society, to which society must adapt as technologies 'progress' over time.[7] Space technologies and their original sin have played their part in globalising world politics and economics since the Cold War years. Who gets to control those technologies and where funds are allocated to develop newer technologies influences international relations but also takes place within the anarchy of the state-based international order. This concluding chapter briefly analyses the conditions of spacepower today and places the material aspects of spacepower in the context of international anarchy. This continues to shape the Global Space Age and will hamper efforts to overcome the original sin of space technology, even as some eyes turn towards the Moon and Mars for humanity's future in space.

* * *

The 'anarchy' of the international system of states extends to space and in turn space infrastructures feed into the politics and struggles in the international system on Earth. It is why dozens of states have built duplicates of space infrastructure that they have more control and influence over, rather than relying solely on what others have already built and could provide. International anarchy refers to the absence of a credible authority that sits above that of states in the international system. There is no formal world government that can impose its will on states.[8] Only states can check the power of other states and enforce 'international' rules and punish breaches of those rules, which themselves are products of political-economic elites, leaders, and state institutions. This anarchic international system does not necessarily mean chaos. There can be some semblance of 'order' or hierarchy between states based on perceptions of relative power, 'the balance of power' or 'pecking order', and how polities choose to associate with each other.

The absence of an effective, forceful common government between states does not mean there is no international society. Rather, there is an international society and forms of international governance that reflect the power, interests, and values of the states that make up that anarchical international society and enforce its 'norms', agreed values, and rules. Power politics and self-serving behaviour happen within international society.[9] The most powerful states in international society are able to get away with 'bad behaviour', or more selfish, self-aggrandising actions because nobody else other than the most powerful states can stop them and impose significant costs on such behaviour. Powerful states are able to impose significant costs—military, economic, and political—on smaller states that defy them on matters of great importance to them. The 'major' or 'great' powers of the international system today—a small group of powerful states that not only possess significant economic and military capabilities but are socially recognised by others as such[10]—are also recognised as major space powers and possess significant material and technological capabilities.

That does not mean that there is no room for cooperation. The Cold War superpowers found room for extensive cooperation in space science, which needed careful political navigation.[11] Cooperation in space does not happen *in spite* of politics—cooperation is still political. Seeing space science, development, or exploration as a solution to problems caused by international anarchy or politics puts reality the wrong way around. The bulk of space science, organisations, industries, and funding comes *from* states, their military-industrial complexes, and political rationales and interests. Cooperation and agreement is not something apart from politics, yet many utopian dreams in space tend to express this belief. Space science, especially collaborative international space science, *is* politics.

The mid-twentieth century scholar E.H. Carr wrote that since power cannot be expelled from politics—the way humans govern themselves and control each other—the 'planes of utopia and of reality never coincide'.[12] This need not doom any effort at bringing about beneficial change, but visions of utopia in space will run into the same political problems as those debating utopia on Earth. There is no need to accept the fatalism of the Frankfurt School of thought on the totalitarian tendencies of industrial technologies and centralised bureaucracies. The conditions of world politics can change over time, yet any imagined better future in space cannot ignore the power politics needed to bring about such changes. Hans Morgenthau argued that 'a realistic assessment' was needed to bring about changes in the international system if better rules are to be created and enforced on the global level, and perhaps bring about a world state that puts an end to interstate wars and may be able to resolve the apocalyptic risks of nuclear war because states could not protect their people from nuclear annihilation.[13] What John Bew calls 'real' Realpolitik requires understanding the situation as it is in order to achieve transformative goals, including those led by ideology and not brutal power calculations alone.[14]

Practitioners at the meeting points of foreign policy, space policies, and military strategy must anticipate how their actions can be received and interpreted for others, as they will respond accordingly. A naïve utopian ideal about space marginalises the people who will lose out or suffer when spacepower is used in the interests of those who wield it. Any positive transformative strategies must be sensitive to who holds power and what the prevailing socio-economic forces of the day are, lest they be co-opted by the powerful. Ideas and idealism matter for practitioners seeking to change things for the better, but these are not the same as utopianism which is detached from the realities of power. Human expansion into deep space, if it is to occur, will not escape these problems of anarchy and

Table 2: The Distribution of Spacepower between the Major Powers

Satellite Types	**China**	**Europe**	**India**	**Japan**	**Russia**	**USA**	**Total**
Total	*467 (5)*	*611 (29)*	*58 (3)*	*83 (6)*	*164 (2)*	*2,939 (32)*	*3,801*
Government Communications	34	23 (6)	19	4	49	138 (9)	251 (10)
Government Earth Observation/ISR	169 (3)	53 (15)	23 (3)	25 (4)	37	113 (18)	385 (28)
Position, Navigation, Timing	48	26	8	4	27	35	176
Space-Based SSA	0	0	0	0	1	5	6
Commercial Communications	34	127 (3)	1	19	35 (2)	165 (2)	286 (4)
Megaconstellation: Communications	0	287	0	0	0	1,953	1,953
Commercial Earth Observation/ISR	69	31 (2)	0	10 (1)	0	39	118 (1)

Megaconstellation: EO/ISR	0	0	0	0	0	306	306
Technology Development	91 (1)	57 (1)	5	18 (1)	7	147 (2)	272 (4)
Science	22 (1)	17 (3)	2	4	8	36 (1)	81 (2)
Refuelling Satellites	0	0	0	0	0	2	2
Independent Launch to LEO	Yes	Yes	Yes	Yes	Yes	Yes	–
Independent Launch to GEO	Yes	Yes	Yes	Yes	Yes	Yes	–
Kinetic ASAT Weapons Tests	Yes	No	Yes	No	Yes	Yes	–

Notes

Bracketed numbers are jointly operated/owned satellites.

EO/ISR includes SIGINT, IMINT.

Europe includes ESA, EU, and select member states (see Table 3).

Source: Union of Concerned Scientists Satellite Database, July 2021.

the thermonuclear revolution so long as humanity's international system is still made up of sovereign states that possess nuclear weapons. The threat of nuclear 'omnicide' will remain even in a multi-planetary civilisation.[15]

International anarchy is the product of people, their ideas, and their machines.[16] As the distribution of material power is an important aspect of world politics, it is worthwhile looking at the spread of spacepower as states jockey for influence in governing the use of Earth orbit and planning the wars of the future. Material capability and possession is a basis for discussing immaterial or less tangible aspects of power, such as cultural attraction, ideology, norms, and values. A significant limit in the quantitative data presented across tables 2, 3, and 4 is that it does not reveal any qualitative information. Neither the quality of the space systems, nor whether they are actually meeting the needs of their users, can be addressed by metrics. Detailed case-by-case analysis is needed for that. Having lots of satellites does not mean that you are more powerful. Rather, having 'enough' of the correct kind of satellites matters more in terms of enhancing state power and achieving terrestrial goals. Some satellites are more important than others. Table 2 does not capture the highly transnational nature of many satellite and space systems between the U.S., Europe, and Japan, such as through transnational corporations and public investments. The operator by state often does not convey the transnational nature of many, particularly western space companies, which may have HQs, manufacturing plants, and control stations in different states.[17] The bracketed numbers of jointly operated satellites show arguably some of the highest levels of cooperation but the numbers as a whole do not show how reliant upon or integrated some powers are with the space systems of others, such as GPS for military and civilian infrastructure.

Such an approach also does not include the willingness or skill to use such capabilities. There is also the problem of focusing on one form of power that does not have the same importance in every situation, or what David Baldwin calls the 'low fungibility' of power resources and the single-dimension fallacy.[18] Therefore when making net assessments on power in the international system, any single focus such as this one on spacepower should bear in mind that it will be more or less useful in different circumstances. Spacepower is not immune from the political-strategic reality that no one form of power is more important than the others at all times.[19] Not everything can be quantified but what can be quantified is still useful to know as a starting point for further detailed qualitative investigation. Material capabilities generate effects in three ways: first, they influence the possibilities and probabilities of outcomes; second, material capabilities constrain and enable different activities, and enable the art of the possible in political life; and third, geography and natural resources constrain and enable further actions.[20] Yet actual power in disputes and in a relationship do not rely on what one possesses alone, of course.[21]

Within any power relationship or identity is a material capability to do things in, from, or in spite of, outer space and the machines there. To do things or influence the world or people around it, an actor needs material capabilities to underwrite them. Technology can shape and structure the processes and abilities of humans to interact with the social and natural world.[22] In the other direction, however, whether a technology is a success or failure, good or bad, is still a subjective human decision.[23] Mapping out who has what kind of space technologies and infrastructure is useful in mapping political influence and how political choices are being created through technological structures.

The 'balance of spacepower' in material, tangible terms today is not unipolar—the United States is not the only space power of significance or not without credible competitors. Washington cannot reliably impose its will on the behaviour of the other major space powers and dictate the terms of their development. It is not a bipolar order either—China and the United States will be unlikely to be able to carve up Earth into 'space power alliances' between them. As can be seen in Table 2, the United States retains a significant quantitative lead in space based on the state the satellite operator is located within. This lead is now particularly pronounced with the emergence of the first megaconstellations in recent years, such as Starlink and Planet. The OneWeb constellation distorts the UK's 'holdings' in space relative to its European peers, as seen in Table 3. Discounting megaconstellations, the United States still maintains a numerical lead with China in a safe second place, with Russia now trailing third. Russia used to be second in this regard until the mid-

2010s, an expression of Russia's general trajectory of serious, long-term relative economic decline.[24] There is a physical scale in space systems with the U.S. and China that others have not matched, yet. Indeed, China's wider investment and infrastructure programmes, dubbed 'Belt and Road Initiative' or variations thereof, are reliant on space infrastructures too. With a massive space infrastructure others will struggle to match bar the United States, it will increase Chinese influence in macroeconomic and global infrastructural matters. Though there may be doubts over such a programme's future after Xi Jinping's rule because it is intimately tied to his own efforts to consolidate power.[25] If 'Europe' was a single state it would be second only to the United States, pushing China into third. On space industrial regulations, space governance, and specific critical space infrastructures, the EU already acts as a major space power and coordinator of European states. We should not simply assume that the assets of the other states are insignificant, as important as U.S. and Chinese infrastructural powers are. What these 'major space powers' have demonstrated with their investments across the entire gambit of crucial space technologies is that they have a firm base for future investment and growth if they wish to develop further along these lines, particularly for Europe, Japan, and India.

In the presented tables, 'Government' refers to any state- or publicly-owned or operated satellite, whether military, civil, or for intelligence agencies. Military-civil distinctions in communications and EO/ISR satellites are blurred and difficult to maintain in practice, as argued earlier. The blurred boundaries between military, government, and civilian space systems also makes it difficult to find out exactly how much any government is spending on space, particularly when some aspects of military space spending is kept secret. Much space-relevant spending may not come under a single space budget. Additionally, commercial space revenues are not necessarily generating private sector revenues, rather from public spending on their services. For example, the Iridium telecoms system is classed as 'military' by the UCS database, as the military is a significant client and the Pentagon was a saving investor in Iridium's creation in the 1990s. Economic and military power are both 'integral parts of political power; and in the long run one is helpless without the other'.[26] Commercial satellite holdings show a snapshot of the health of commercial space activities, with India in particular not faring as well as the others here. Despite U.S. attempts to stall Japanese SATCOMs developments in the 1990s, Japan now has a mix of commercial and government holdings in space systems. Whether operated or owned by a private company or not, if a satellite system provides crucial services to military forces, the government, and intelligence services then it is highly likely to be targeted by the enemy. Just because something is private does not mean it is immune from war—as is the case on Earth. Intelsat and Inmarsat reflect some of the earliest commercial activities in space, and in the 1990s the wave of deregulation in Western economies had reached outer space with the privatisation of those companies.[27] However, despite their commercial nature, they still act as telecoms infrastructure for many government and military branches and receive a good deal of revenue from public spending. A private global space economy, therefore, is not quite here yet.

The 66-satellite Iridium mobile satellite phone constellation in LEO was completed shortly after the turn of the millennium. The project came close to complete failure and bankruptcy, with its proponents and leaders becoming known to some as the 'Iridiots' leading a $6bn white elephant space project.[28] It was saved from bankruptcy through an eleventh-hour deal for services that involved the Pentagon. The Pentagon saw the value of Iridium for 'national defense' matters, showing the porous boundaries between the military and civilian dimensions of spacepower.[29] Iridium's success possibly serves as a template for modern megaconstellations. At the time of writing, the Pentagon is OneWeb's single largest customer and the UK Government maintains a stake in the company following its £400m bailout contribution in 2020.[30] The U.S. Army has signed a research cooperation agreement to explore the utility of Starlink as a broadband relay network for Army systems.[31] SpaceX and NASA's so-called 'commercial crew' programme is ultimately paid for by the taxpayers of the International Space Station member states. 'Commercial space' may be the wrong word to use when 'big customers' are still states and government agencies spending public funds on them.

Technology development and science satellites serve as rough indicators of innovation and wider investments in space, as well as a wider pool of skills and knowledge production in space science and engineering through the work of universities and government laboratories. The interrelationship between the military and scientific communities in space can also be seen in satellite control rooms, particularly in the post-Cold War world in Europe and the United States where their 'elements were re-infused with the expertise, norms and technologies of defense through an emerging sector that bundled and expanded military and civilian portfolios.'[32] Ground-based infrastructure is an element of spacepower that is not detailed in these tables, but is less useful in a quantitative analysis. Sufficient ground control and SSA capabilities must go hand in hand with increasing constellation sizes.

The U.S., China, Russia, India, Japan, and Europe represent significant but unequal poles of spacepower. The basis for grouping these six 'powers' together on a higher rung is based on the spread of capabilities in the most important technological areas for independent action in space and in future spacepower development: launch to the most useful orbits, secure SATCOMs, PNT systems, and EO/ISR industries, as well as LEO and GEO launch capabilities. This does not mean the overall power of each state is the same on Earth as it is in space. Spacepower is only part of the overall power of a state. Some arguments could be made that the USA and China (and perhaps Russia) are in their own class, but such a view overly downplays the relative importance and agency of others in contemporary astropolitics, and also downgrades the importance of globalised high technology supply chains.

SATCOMs provide essential celestial lines of communication that are crucial for C4ISR, both nuclear and non-nuclear military forces and intelligence capabilities. PNT systems, and GNSS in particular mark out four entities that have poured in significant resources to developing their own versions of what is arguably one of the most important critical infrastructures of the post-Cold War world. Japan and India have made significant investments in their own regional PNT systems or GNSS enhancers, and could develop further space-based PNT systems in future if they wish, perhaps even a full GNSS—although that is no small undertaking. EO/ISR, including SIGINT and IMINT capabilities, are crucial intelligence collection platforms from space, as well as crucial observation systems for civilian uses and economic planning.

Launch systems are essential, with GEO access assuring a state that it can deploy a satellite into some of the most prized volumes of orbital space around Earth for critical SATCOMs or SIGINT satellites. Roger Launius said as much in no uncertain terms: 'Access. No single word better describes the primary concern of everyone interested in the exploration and development of space. Every participant in space activities... needs affordable, reliable, frequent, and flexible access to space.'[33] Up until the mastery of reusable first-stage rockets by SpaceX, all rockets for satellite launches were expendable, single-use disposables. In the years to come Europe and China may follow suit. Whilst the global launch market can help governments make returns on their investments or help cover the costs in their SLVs, they are such an important capability that any independently-minded spacepower must maintain their own options even if it is not commercially attractive on a global level. In that way it is an infrastructure investment for state power. Russia may be going down this route now as its global competitiveness for commercial launches is in severe decline.

Each of these capabilities are important to mark out the most capable space powers, and six of them listed here have at least a significant and perhaps permanent foothold in the most important, critical services and capabilities spacepower has to offer for military, political, economic, and intelligence needs. A high level of dependence on others in these areas marks out a serious limit on the freedom of action a state has in a crisis, or in open warfare, and also in setting out rules for the global governance of space. As seen in the tables below, other states do not quite meet these capabilities in breadth or depth, especially European states when treated in isolation. European states are interdependent and integrated in space. Collective capabilities via ESA, EU, and to a lesser extent NATO enhance their individual capabilities into a more coherent whole, raising 'Europe' as a whole to a similar level as the other five in Table 1 in terms of breadth and depth of capability in the most essential technological areas. The EU Commissioner for the Internal Market claims that 'Europe is already a great space power',[34] which has some merit but the meaning of 'Europe' is complex and its political incoherence holds Europe back as a spacepower relative to the other states in the top ranks in orbit.

These six 'major space powers' could be interpreted as the hierarchs or gatekeepers of the modern 'space club', looking at material capability. Between them, the six are powerful in terms of possessing the lion's share of global spacepower. The 'space club', according to Deganit Paikowsky, shapes global preferences and policies regarding space, the rules and dynamics of cooperation and competition within that club, and projects power in a number of global governance and international rule-making issues.[35] More broadly, these six are also many of the most powerful on Earth, and also have nuclear weapons or are allied to nuclear powers. According to Hedley Bull this means they can regard themselves, and be regarded by others, as 'having special rights and duties'.[36] Such a dynamic is clearly at play in deciding what counts as 'responsible behaviour' in outer space, as is currently being deliberated at the UN. Any agreement will lack teeth and credibility if one of the six does not agree. The EU Code of Conduct on voluntary rules of the road for satellite operations failed due to a lack of the EU's persuasive powers in 2015, with India, China, and Russia being among the most active opponents of such a code despite U.S. support of the EU's proposals.[37] Smaller powers have limited options to challenge such an order if all six agree amongst themselves and get to enjoy special rights conferred by the 'rules of the game' that they have agreed on.[38]

States may struggle to trust that a single launch provider, a single GNSS provider, a single C4ISR provider, will never use that infrastructural power for leverage on them. Even among allies, a high level of dependency has drawbacks and does not always sit comfortably, as seen between the U.S., Western Europe, and Japan. Europe and Japan were balancing their own independent development with dependencies and cooperation. Those who control infrastructure set the terms of use and gain power and influence through technology they have chosen to build. Every chapter in the book has been showing the tangible manifestations of international anarchy in space and space technologies. The Global Space Age has been shaped by the desires of states to seek security and survival, the ability to threaten harm upon others, and to compete economically with the new industries of space, information technology, and advanced materials. Some can choose cooperation, but as we have seen even between allies economic competition and political disagreements can be particularly acute with space technology.

Might does not always make right—the great powers can instead be interpreted as 'the great irresponsibles' as they might not conduct themselves with the behaviour members of the wider international system approve of, and suffer minimal consequences. However, great powers need to seek some degree of perceived legitimacy and consent, or the management of the system falls apart with excessive and numerous 'irresponsible' acts.[39] Kinetic, destructive ASAT-testing by Russia, the United States, China, and India fit into what some may describe as irresponsible acts. However, it is important to recognise that the U.S. and China are far beyond any of the other space powers in terms of net capabilities, and this is reflected in Table 2. As Zhang Yongjin argues, China, 'the second among the great irresponsibles... has significantly increased its bargaining power in the negotiations for constructing the institutional and normative architecture of global governance'.[40] If Beijing and Washington could come to an equitable arrangement between them, their consensus may be hard to ignore in determining the future in space, possibly ushering in a new bipolar order.

As the states of Earth try to set out new rules, norms, or codes of conduct for outer space, they will do so under the conditions of international anarchy and the thermonuclear revolution. Any international space laws that emerge, whether 'hard laws' such as treaties, or 'soft laws' such as conventions or norms of behaviour, need a critical mass of powerful states to agree on the terms of such laws or rules if they are to be effective. Many states and institutions will seek to ensure that these rules do not make their pursuit of spacepower and terrestrial interests more difficult at least, or actually further their goals at best. In this basic sense, 'celestial geography mirrors the power relations of terrestrial politics.'[41] The way the major powers have set out to use space perpetuates that, and they will have the biggest say due to their ability to 'carry on regardless' or develop and sustain alternative models to others without undermining their own security. It may well be that truly equitable global governance needs a world state built on a supranational community.[42] In this first quarter of the twenty-first century, the 'major space powers' are those that began their journeys into the Global

Table 3: The Distribution of Spacepower between Select European Actors

Satellite Types	**Finland**	**France**	**Germany**	**Italy**	**Luxembourg**	**Netherlands**	**Norway**	**Spain**	**Switzerland**	**UK**	**ESA/ EU**	**Total**
Total	*15*	*16 (13)*	*44 (2)*	*14 (7)*	*40*	*15 (1)*	*8*	*22 (2)*	*13*	*334 (4)*	*90 (1)*	*611 (29)*
Government Communications	0	2 (2)	2	3 (2)	1	1	2	5	0	6 (1)	1 (1)	23 (6)
Government Earth Observation/ISR	0	6 (8)	10 (2)	7 (3)	0	1	3	6 (2)	0	2	18	53 (15)
Position, Navigation, Timing	0	0	0	0	0	0	0	0	0	0	26	26
Space-Based SSA	0	0	0	0	0	0	0	0	0	0	0	0
Commercial Communications	0	0	1	0	31	7 (1)	3	9	10	35 (2)	32	127 (3)
Megaconstellation: Communications	0	0	0	0	0	0	0	0	0	287	0	287
Commercial Earth Observation/ISR	13	5 (2)	0	1	8	0	0	0	0	4	0	31 (2)
Megaconstellation: EO/ISR	0	0	0	0	0	0	0	0	0	0	0	0
Technology Development	2	3 (1)	28	2	0	6	0	2	2	8	4	57 (1)
Science	0	1	3	1 (2)	0	0	0	0	1	2 (1)	9	17 (3)

Refuelling Satellites	0	0	0	0	0	0	0	0	0	0	0	0
Independent Launch to LEO	No	No	No	No	No	No	No	No	No	No	Yes	–
Independent Launch to GEO	No	No	No	No	No	No	No	No	No	No	Yes	–
ESA Membership	Yes	Yes	Yes	Yes	Yes	Yes	Yes	Yes	Yes	Yes	N/A	–
EU Membership	Yes	Yes	Yes	Yes	Yes	Yes	No	Yes	No	No	N/A	–
NATO Membership	No	Yes	Yes	Yes	Yes	Yes	Yes	Yes	No	Yes	N/A	–

Source: Union of Concerned Scientists Satellite Database, July 2021.

Space Age during the Cold War. They are already attempting to set rules and prevent unwanted rules from being set. The PPWT, the failed EU Code of Conduct, and the ongoing Resolution on Responsible Behaviours in Outer Space at the UN General Assembly are examples of efforts at regulating space activities reflecting terrestrial political divides today.

Europe is a peculiar entity with regard to spacepower. Neither 'Europe', ESA, nor the EU are states. Yet collectively European states and those two institutions build, control, and develop essential space technologies to a high standard and at considerable scale, second only to the USA if 'Europe' was a singular state. The lack of political coherence means that European spacepower cannot be interpreted like the state spacepower of the others. John Krige refers to a 'European Space System', a 'complex of institutions, artifacts, national and international networks, production facilities and commercial activities.'[43] In Europe, military and intelligence space capabilities are often created and controlled on a state basis, and sometimes on a bilateral or multilateral basis between states, with additional data-sharing agreements, as is often the case between France, Germany, and Italy—Europe's three biggest state spenders in space. ESA and the EU collectively provide platforms for their members to conduct affordable activities and investments in joint programmes that none could fund individually, and ESA and the EU do not share the same state members. Through the EU and ESA, 'European' spacepower is being manifested in major infrastructural space projects such as the Galileo GNSS and the Arianespace launch capability, which have ubiquitous military and civilian applications. Several European militaries retain significant investments in military space infrastructure relative to their terrestrial size.

It is difficult to describe the current distribution of spacepower as unipolar or bipolar because the smaller space powers, such as in Tables 3 and 4, have plenty of opportunities for 'forum shopping', and many choices in military, industrial and scientific partners or economic infrastructure providers because each of the six entities in Table 2 have mastered the necessary, essential technologies for themselves to establish a significant presence in the cosmic coastline of Earth orbit. Each of the six entities can compete with each other to attract other states and companies on Earth for launch services, satellite manufacturing, components supply, as well as data and services. Smaller states and space powers are able to focus on specific areas of investment, such as Israel's focus on LEO launches, SATCOMs, and ISR for specific needs, or Taiwan's cooperation with the United States in EO satellites, Argentina's significant commercial EO holdings, and Canada's commercial SATCOMs and high-technology sensors industry. Table 4 shows a variety of investments by other countries in terms of satellite holdings, but beyond these many of these states take part in the wider global space industry through component manufacture, and stimulate a global space economy through supply and purchasing data and services to other states and the open market. Satellite holdings should be seen as something of a high point for investments in certain areas, or, as in the case of Luxembourg and other commercially attractive states, a more permissive regulatory regime.

Multipolar perhaps better describes contemporary political realities in the current Space Age, rather than 'democratising', as more actors enter the world of spacepower. Some use 'democratisation' in space as the diffusion of information, technology, and finance with spacepower.[44] However, given the number of space powers that are authoritarian or increasingly so, democratisation is a loaded term, and a misnomer when multipolar is an accepted term referring to a greater distribution of power among more states, rather than making claims on regime types, whether domestically or internationally. Democracy also implies that members of that polity have methods and systems of representation in the governance of that system. In the global governance of space today, that is not the case, casting further doubts on a trend towards democratisation. The UN and other international institutions that govern space and its finite resources, including those in Earth orbit, such as the International Telecommunications Union (ITU), are shaped by the most powerful political and economic forces generated by the major space powers. They are not inherently fair or politically equitable democratic international institutions, even before considering the levels of democratic representation specific states possess, meaning many governments may not be fully representative of the diverse views of their citizens.

More states are becoming more influential in space. In particular, South Korea, Argentina, and Brazil could become more significant

Table 4: Distribution of Spacepower between Selected States

Satellite Types	Argentina	Australia	Brazil	Canada	Indonesia	Iran	Israel	Mexico	Nigeria	North Korea	Saudi Arabia	South Korea	Taiwan	Turkey	UAE	Vietnam
Total	*28 (4)*	*14*	*13*	*53 (1)*	*8*	*1*	*18 (1)*	*6 (1)*	*3*	*0*	*12*	*18*	*1 (11)*	*9*	*12*	*2*
Government Communications	0	1	0	0	0	0	3	2	0	0	1	2	0	0	3	2
Government Earth Observation/ISR	2	0	4	4 (1)	3	1	8 (1)	3	2	0	4	8	1 (11)	4	6	2
Position, Navigation, Timing	0	0	0	0	0	0	0	0	0	0	0	0	0	0	0	0
Space-Based SSA	0	0	0	2	0	0	0	0	0	0	0	0	0	0	0	0
Commercial Communications	5 (4)	7	9	35	5	0	0	1	1	0	4	3	0	4	3	0
Megaconstellation: Communications	0	0	0	0	0	0	0	0	0	0	0	0	0	0	0	0
Commercial Earth Observation/ISR	19	0	0	1	0	0	0	0	0	0	0	0	0	0	0	0
Megaconstellation: EO/ISR	0	0	0	0	0	0	0	0	0	0	0	0	0	0	0	0

Technology Development	2	6	0	4	0	0	6	1 (1)	0	0	3	2	0	1	0	0
Science	0	0	0	7	0	0	1	0	0	0	0	3	0	0	0	0
Refuelling Satellites	0	0	0	0	0	0	0	0	0	0	0	0	0	0	0	0
Independent Launch to LEO	No	No	No	No	No	Yes	Yes	No	No	Yes	No	Yes	No	No	No	No
Independent Launch to GEO	No	No	No	No	No	No	No	No	No	No	No	No	No	No	No	No

Notes

Bracketed numbers are jointly operated/owned satellites.

EO/ISR includes SIGINT, IMINT.

Source: Union of Concerned Scientists Satellite Database, July 2021.

space powers if government and industry decide to invest in those areas. The Finnish commercial space sector has established itself as a reliable source of radar imagery and industry. Israel and South Korea already possess a LEO-capable SLV, and like Iran and North Korea could launch particular kinds of satellites from their own soil if they decide to spend the resources to do so. These are examples of space powers with significant levels of autonomy in certain areas, but will be held back in terms of scale due to their absolute demographic and economic size, in particular Israel. Yet they are significant partners for other space powers.

Africa is poised for further growth in spacepower. Nigeria, Algeria, and South Africa are already operating their own satellites having purchased their manufacture, and Rwanda recently set up its own space agency. Egypt is becoming host to a new African Space Agency under the auspices of the African Union. Indonesia is an example of a very large, populous state with great potential, but at present a relatively small direct presence in space. Developing and developed states, large and small, have many opportunities to develop spacepower according to their needs, and the top six space powers will be eager to sell their services or develop 'partnerships' as they wish to extend their global influence and commercial competitiveness through spacepower, whilst checking the interests of the established major powers of the Global Space Age. The six major powers will need to make any order in space work well enough to satisfy the interests of these smaller or emerging space powers, lest they entice more organised technopolitical and industrial opposition to their domination.

* * *

The French political theorist Raymond Aron asked, 'short of a revolution in the heart of man and the nature of states, by what miracle could interplanetary space be preserved from military use?'[45] That revolution cannot come about until the prevailing political conditions that should be changed are identified and targeted for lasting change. Those political conditions are shaped by the anarchy of the international system where states form the most effective and powerful types of governance and cannot rely on any authority above themselves to provide for their security and material needs. The Global Space Age has not absolved itself of the original sin of space technology. Humanity has failed to escape the political-military rationales of states and the military-industrial complexes that gave rise to and sustain spacepower. The Space Age has not overcome the forces of international anarchy. Space is just another environment that states and people use for war, development, and prestige. In that context, the spectre of space warfare is entirely 'normal' and unremarkable. Yet it does not have to be this way for ever. Politics is the art of the possible. The first step towards trying to build something better is to recognise things as they are, which means tempering visions of space either as a realm exclusively for science as an inherent 'good' and cooperative action, or a place of certain utopia in the future. Scholars in the field of IR have plenty to offer in the exploration and development of outer space, and are needed if any political revolution for the better is to happen. I hope this book has shown that international anarchy under the conditions of the thermonuclear revolution is the key idea, problem, and feature of astropolitics that we have to tackle for a better future, and not domination.

Carl Sagan's *Pale Blue Dot* is one of the better-known pleas for common humanity and global solidarity from a cosmic perspective. Beyond the orbit of Uranus and hurtling to the edge of the Solar System, the Voyager 1 spacecraft turned around to look at Earth. From that distance, it only looked like a pale blue dot. Sagan instructed his readers to:

> look again at that dot. That's here. That's home. That's us. On it everyone you love, everyone you know, everyone you ever heard of, every human being who ever was, lived out their lives... Think of the rivers of blood spilled by all those generals and emperors so that, in glory and triumph, they could become the momentary masters of a fraction of a dot. Think of the endless cruelties visited by the inhabitants of one corner of this pixel on the scarcely distinguishable inhabitants of some other corner, how frequent their misunderstandings, how eager they are to kill one another, how fervent their hatreds... Our posturings, our imagined self-importance, the delusions that we have some privileged position in the Universe, are challenged by this point of pale light. Our planet is a

> lonely speck in the great enveloping cosmic dark... There is perhaps no better demonstration of the folly of human conceits than this distant image of our tiny world. To me, it underscores our responsibility to deal more kindly with one another, and to preserve and cherish the pale blue dot, the only home we've ever known.[46]

The sentiment of common humanity and ecological stewardship is hard to disagree with. Yet such an anti-politics condescension towards political divisions has problems. What common humanity and ecological stewardship mean in practice require political discussion, engagement, and education. Whilst we should not be chauvinist and see ourselves as having an objectively important place in the universe, our home and how we live on it matters most of all to us. Observing that we are small and insignificant in the universe and that we are not really that different in the grand scheme of the cosmos belittles efforts to challenge, prevent, and reduce political violence, harm, and prejudice visited on Earth's inhabitants towards each other, and efforts to prevent nuclear Armageddon itself. Sagan's views produce contempt for ourselves as people, but value for our machines and works instead; perhaps it should be the other way around, McDougall argues.[47] The control of resources and the terms of life are supreme *political* issues that everyone should have an interest in because space technologies and who controls them play an important part in determining political issues for everyone on Earth, no matter the insignificance of humanity on the cosmic scale.

The 'overview effect', a cosmopolitan, solidarist, and environmentally aware perspective of all humanity brought about by space travel for astronauts suffers from the same problem.[48] This was epitomised by the billionaire Jeff Bezos who claimed, after his own short 'hop' above the atmosphere, that 'when you look at the planet, there are no borders... It's one planet, and we share it and it's fragile.'[49] Such profound comments are politically banal and naïve. It provides an illusion of global solidarity through observations on interconnectedness by claiming state boundaries are 'invisible', 'artificial', and therefore unimportant. It glosses over the fact that wars rage on Earth and massive economic inequalities continue to accrue along borders and political dividing lines that directly shape the terms of life on the various fractions of our pale blue dot.[50] It portrays global solidarity like an easy 'no-brainer' rather than an extremely difficult political, economic, and social project that requires people to study politics, power, and the problems of human governance and injustice that benefit some inhabitants of the pale blue dot over others. Space technology is portrayed as a road to absolution when it has entrenched inequalities on Earth and made the most powerful militaries more efficient at killing people and breaking things. This disconnect is as uncomfortable as it is ubiquitous.

The technopolitics of Earth—the politics surrounding technological systems that are designed by some people for specific reasons—matter. Some terms of life and ideologies are superior to others in terms of diversity, plurality, political representation, equity, and self-determination within a tolerant community. Preventing a totalitarian regime from conquering a fraction of the pale blue dot and purging its inhabitants matters. Controlling the global means of production in food, energy, and material goods produces political and economic power on Earth for certain people and certain organisations are supremely important political matters and are impacted by space technologies and how we decide to govern Earth orbit in the decades to come. Borders, territory, and political control matter, regardless of how high one's altitude or wealth when cogitating on Earth, humanity, and the cosmos.

Herbert Marcuse argued that 'authentic self-determination' requires 'effective social control over the production and distribution of the necessities'.[51] Spacepower provides much of what is deemed a necessity in the production and consumption of resources today, but its control will always be in a few centres or 'poles' of political-economic power, lending itself to influence over others as they come to depend on essential infrastructure that is beyond their means to challenge it as 'progress' and increasing material comforts from space services become expected and customary. Even in democratic societies, there are un-democratic, technocratic forces at work that are passed off as a 'token of technical progress' that undermine democratic control.[52] Yet more involved democratic populations need to be educated and informed about the complex systems which impact them. This tension is easily identified but seldom solved.

With an increasingly 'multipolar' or multi-actor orbital environment in the twenty-first century, we should approach international astropolitics as a global negotiation over the building, control, and management of multiple large technological infrastructures, which at worst threatens to repeat existing technopolitics of power on the 'haves' and 'have nots' on Earth, delivering no absolution from the original sin of space technology. Spacepower is being created and shaped from several centres of world power. The Global Space Age is just another outgrowth of the international system and the multitude of actors within it. Spacepower and other global infrastructures exist in a geocentric global system fraught with unpredictable and chaotic results, and the uneven patterns of adoption, adaptation, imitation, or resistance to technopolitical change and power. The superpowers tried and failed to control the early spread of space technology, sharing, or cooperating with others only in specific and clearly defined areas, and often with strings attached.

The competitive and militarised original sin of space has consolidated various locations of Earth orbit 'as sites for critical infrastructure. If it ever was, space is no longer the locale of optimistic projection, a backdrop for a grander narrative of the human condition or an aspirational destination. Instead of an end, it has truly become a means'.[53] We can still recognise the good in these systems and enjoy the triumphs of space exploration, but we must not absolve space technologies from the original sin that they at best cannot move away from and at worst exacerbate. We must not ignore or forget the suffering caused by or committed in the service of such activities. The tethers of modernity in orbit and the technological infrastructures projecting and enhancing military and economic power on Earth will not remain in the cosmic coastline of Earth orbit if similar economic and military interests are developed on the Moon and beyond at any significant scale.

International organisations and institutions short of a world state are no easy solution either, since they tend to reflect the interests of their most powerful members, powerful coalitions within them, and any bargains struck as they vie for influence and concessions over governance and investments. Space and international organisations or forums regulating space activities are no different. The ITU, founded in 1865 and now containing over 190 member states, is already apportioning orbital resources—GEO slots of satellites and the parts of the radiofrequency spectrum that they can use. Whilst the ITU is seen as a custodian of an international good and concern, it is still the scene of power-political bargaining between states, as Guilhelm Penent writes:

> when requests for new allocations or expansion of old ones are received, consideration is given in priority to the impact on previously established users under the so-called 'first-come, first-served' practice… thus theoretically enabling the technologically advanced countries to gain a permanent hold over the resources. … The developing countries began asking, much to the alarm of the ITU [personnel], for equity to be introduced into deliberations on equal [footing] with the 'rational use' of telecommunications and encouragement of efficiency after they had come to realize that decisions made behind the technical veil of the organization were not necessarily in their own best interests. Some member states might appear today more than willing to remind the ITU that it does not operate in a political vacuum.[54]

Whereas having the ITU may be preferable to no international body at all, since it can provide some frequency slots for all states regardless of their state of development, the political and economic forces shaping ITU's governance of limited resources in the GEO belt does not augur well for a truly equitable use of lunar resources, if that ever becomes feasible. The Moon Treaty of 1979 attempted to bring in an equitable regime for lunar resource allocation, but failed to get many states to sign onto it and is today considered a defunct treaty. Anarchy and self-interest have already established themselves in the orbital regime as it apportions limited resources and does not augur well for the rest of the Solar System if significant economic activity happens beyond orbit. If we are to avoid the problems of world politics in space, we need to overcome the problems of international anarchy. Utopian dreamers of the human future in space will make no progress if they do not engage with the practice and study of international politics.

The original sin of space technology stands ready to move beyond Earth's cosmic coastline if we choose to let Earth's anarchic system extend its political-economic tendrils even further into cislunar space. Those arguing for Martian 'colonisation' or 'settlement' do so on the grounds that the resources are 'there' to sustain technological life, whilst it will become some ill-defined engine of technological 'progress' and the scene for the 'free development of a new society' according to libertarian ideals.[55] The Starlink terms of service agreement which speaks of a 'Free Mars' from any Earth government is not only legally invalid (as UN law applies on Mars) but leaving the details of governance to be sorted out 'in good faith at the time of the Martian settlement'[56] betrays a naïve and ignorant approach to astropolitics and the crucial issues of how we govern ourselves and make life worth living for everyone. Martian libertarian utopians, such as Robert Zubrin and Richard Wagner, long for a new colonial frontier because here on Earth 'the cops are too close'.[57] Human history on the frontiers was brutal and witnessed extreme power inequalities, criminality, and hierarchical, imperial, and authoritarian political structures. The colonial, homesteading frontier was a product of the imperial state, not a fantastical libertarian escape from it. In the nineteenth century, new transportation technologies ushered in centralised, dystopian, and often genocidal political and military regimes for 'frontier', enslaved, and indigenous peoples. For those arguing that Martian habitation is necessary to avoid extinction events on Earth, the negative consequences are simply ignored, as well as the ability to use nuclear weapons to destroy habitats beyond Earth. Rarely are the questions over potential negative consequences of such technological developments asked. Daniel Deudney, one of the few to do so systematically, argues that politically ignorant visions of a human space-faring civilisation are not just a series of bad proposals, but exhibit a fanaticism that disregards the potential of technological totalitarianism and the horrors they present.[58]

The most pressing questions about a human future in space is not how it can be done and how soon, but *why, and for whose benefit?* Will the further opening of space benefit the few or the many? Without critical political thought and knowledge that can challenge and check the powerful, we will only perpetuate the original sin of space technologies as it has already happened in Earth orbit, leading to a political hell in heaven for humans as it is here. Our inability so far to escape the original sin of space technology seems likely to bring about C.S. Lewis's fears of 'fallen man' spreading to the cosmos.[59] That future is not certain, and to avoid it a greater political understanding and awareness of our actions there are needed. At worst, Deudney's fears will materialise where a 'major movement into the heavens promises a descent into the depths of hell.'[60]

The Global Space Age was not the result of some inevitable force of technological progress, rational scientific quests, or satisfying innocent curiosity. Rather, it was the result of successive choices made by people in the chaotic universe of politics navigating conflict, competition, and self-interest. Space is not a special place separated from terrestrial politics. What we do in space reflects what we do on Earth, and what happens in orbit is done to further specific interests on Earth in turn. Spacepower is still the job of centralised authorities marshalling massive amounts of resources to meet the needs of war, development, and prestige. Governments and corporations have only just managed to build a functioning orbital infrastructure merely in Earth's cosmic coastline. Any effort to build an extra-terrestrial political-economy will be an even greater undertaking with ever-greater stakes, and the worlds of science and engineering will benefit from whatever funds those states and large private investors pour into doing so.

As the First World War drew to a close, Georges Clemenceau famously remarked that war is too serious to be left to soldiers. Humanity's use of outer space is too serious to be left to scientists. With increasing awareness and knowledge of the militarised and political nature of space activities within an anarchical and thermonuclear international system, we can be more optimistic that feasible pathways for more equitable forms of space development and security can be discussed and imagined. Whatever form political changes may take, they cannot come about without studying astropolitics as it is today and accepting the original sin of space technology.

NOTES

INTRODUCTION

1. BBC, 'Space Force'.
2. North Atlantic Treaty Organization, 'NATO's Approach to Space'.
3. Park, 'We go together'.
4. French Ministry of the Armed Forces, 'Space Defence Strategy', p. 4 (English edition).
5. Greene, 'RAAF planning for new military space command'.
6. Broad, 'How Space Became the Next "Great Power" Contest'.
7. For a discussion of hyperbolic claims on Chinese 'dominance' of outer space based on its space exploration activities, see: Hunter, 'China and the US'; Day, 'Red Moon Revisited'.
8. Sheehan, *The International Politics*, p. 2.
9. Epstein, *Torpedo*, p. 229.
10. Billings, 'Q&A'.
11. Baldwin, *Power and International Relations*, p. 29.
12. United Nations Office of Outer Space Affairs, The Outer Space Treaty, Article I.
13. Sheehan, *The International Politics*, pp. 1–2.
14. Peoples and Stevens, 'At the Outer Limits', p. 299.
15. Sheehan, *The International Politics*, p. 2.
16. McFarland, *Cambridge Dictionary of Christian Theology*, p. 473.
17. McDougall, ...*The Heavens*, p. 44.
18. Peoples, 'Haunted Dreams', pp. 95–6.
19. Deudney, *Dark Skies*, p. 146.
20. For example: MacDonald, *The Long Space Age*, pp. 49–52.
21. For example: Lee and Singer, 'China's Space Program Is More Military Than You Might Think'.
22. Herrera, 'Technology and International Systems', p. 573.
23. For example: Moltz, *The Politics of Space Security*; Johnson-Freese, *Space as a Strategic Asset*; Dolman, *Astropolitik*; Preston, *Plowshares and Power*; Burrows, *This New Ocean*; Launius, *Apollo's Legacy*; Stares, *Space and National Security*; Gray, *American Military Space Policy*; Launius and Jenkins, *To Reach the High Frontier*; Norris, *Spies in the Sky*; Brown,

Spacepower Integration; Klein, *Space Warfare*; Lambakis, *On the Edge of Earth*; Logsdon, *John F. Kennedy*; Temple, *Shades of Gray*; Easton and Frazier, *GPS Declassified*; David, *Spies and Shuttles*; Hays, *Space and Security*; Johnson-Freese, *Space Warfare*; Hays et al., *Spacepower for a New Millennium*; Richelson, *America's Space Sentinels*; MacDonald, *The Long Space Age*.

24. For example: McDougall, …*The Heavens and the Earth*; Sheng-Chih, *Transatlantic Space Politics*; Andrews and Siddiqi, *Into the Cosmos*; Siddiqi, *Challenge to Apollo*; Hörber and Stephenson, *European Space Policy*; Harvey, *The Rebirth of the Russian Space Program*; Harvey, *European-Russian Space Cooperation*; Neufeld, *Spaceflight*.
25. Notable exceptions: Sheehan, *The International Politics*; Gorman, *Dr Space Junk*; Paikowsky, *The Power of the Space Club*; Hardy, *Space Policy*.
26. Powel, 'Blinkered Learning, Blinkered Theory', p. 977.
27. Enloe, *Bananas, Beaches, and Bases*, p. 35.
28. Herrera, 'Technology and International Systems', pp. 582, 585.
29. Baldwin, *Power and International Relations*, p. 85.
30. Bowen, *War in Space*, p. 22.
31. Baldwin, *Power and International Relations*, p. 82.
32. Neufeld, 'Cold War—But No War—in Space', p. 63.
33. Gökalp, 'On the Analysis', pp. 58–63.
34. Edgerton, *The Shock*, p. ix.
35. For example: Krige et al., *NASA in the World*; Harvey et al., *Emerging Space Powers*; Singh, *The Indian Space Programme*; Gopalaswamy, *Final Frontier*; Harvey, *China in Space*; Handberg and Zhen, *Chinese Space Policy*; Redfield, *Space in the Tropics*; Neufeld, *Spaceflight*; Bormann and Sheehan, *Securing Outer Space*; Sadeh, *Space Strategy in the 21st Century*; Coletta and Pilch, *Space and Defense Policy*.
36. McCarthy, 'Introduction', p. 1.
37. For example, McCurdy, *Space and the American Imagination*; Geppert, *Imagining Outer Space*; James and Siddiqi, *Into the Cosmos*; Kilgore, *Astrofuturism*; Poole, *Earthrise*.
38. McCarthy, 'Technology and "the International"', p. 476.

1. THE DAWN OF THE GLOBAL SPACE AGE

1. Kennedy, Speech at Rice University.
2. On women and crewed spaceflight, see: Weitekamp, *Right Stuff, Wrong Sex*; Nolen, *Promised the Moon*; Kevles, *Almost Heaven*; Shayler and Moule, *Women in Space*.
3. Marcuse, *One-Dimensional Man*, p. 252.
4. Paikowsky, *The Power of the Space Club*, p. 229.

5. McDougall, ...*The Heavens*, p. 5.
6. For example: Zubrin and Wagner, *The Case for Mars*.
7. Neufeld, 'Cold War—But No War—in Space', p. 45.
8. Orsekes, 'Introduction', in *Science and Technology*, p. 2.
9. Ibid. p. 6.
10. Bateman, 'Technological Wonder', p. 329.
11. Moltz, *The Politics of Space Security*, p. 4.
12. On the Global Cold War approach, see: Westad, *The Cold War*, pp. 1–17.
13. For example: Starling et al., 'The Future of Security in Space'.
14. On a deconstruction of the 'frontier' analogy to outer space, see: Trevino, *The Cosmos is Not Finished*.
15. Siddiqi, *Challenge to Apollo*, p. 653.
16. Wolfe, *Competing with the Soviets*, p. 92.
17. Burrows, *This New Ocean*, pp. 44–8.
18. Neufeld, *Spaceflight*, pp. 17–18.
19. Siddiqi, *Challenge to Apollo*, pp. 109–10.
20. Neufeld, *Spaceflight*, p. 25; See also: Crim, *Our Germans*; Neufeld, *Von Braun*.
21. Neufeld, *Spaceflight*, pp. 29–30.
22. Peoples, 'Haunted Dreams', pp. 97–9.
23. Kilgore, *Astrofuturism*, pp. 54–5.
24. Craig, 'Solving the nuclear dilemma', p. 349.
25. Ibid. pp. 349–50.
26. Brodie, 'Implication for Military Policy', in *The Absolute Weapon*, p. 62.
27. Brodie, *Strategy in the Missile Age*, p. 269.
28. Craig and Logevall, *America's Cold War*, p. 56.
29. MacKenzie, *Inventing Accuracy*, p. 107.
30. Craig and Logevall, *America's Cold War*, p. 168.
31. Siddiqi, *Challenge to Apollo*, pp. 59–62.
32. Richelson, *America's Space Sentinels*, p. 3; Siddiqi, *Challenge to Apollo*, p. 107.
33. Siddiqi, *Challenge to Apollo*, p. 109.
34. Ibid. p. 110.
35. Ibid. p. 119.
36. Ibid. pp. 59–62, 87, 351.
37. Gainor, 'The Nuclear Roots of the Space Race', p. 84.
38. Jervis, *The Meaning*, pp. 23–9; Futter, *The Politics of Nuclear Weapons*, pp. 87–91.
39. Deudney, *Bounding Power*, p. 245.
40. RAND Corporation, 'Preliminary Design of an Experimental World-Circling Spaceship'.
41. Spinardi, *From Polaris to Trident*, p. 17.

42. Spinardi, *From Polaris to Trident*, pp. 15–18.
43. MacKenzie, *Inventing Accuracy*, pp. 162–3.
44. Gainor, 'The Nuclear Roots of the Space Race', p. 82.
45. Ibid. p. 75–7.
46. Ibid. p. 86.
47. McDougall, ...*The Heavens*, p. 187.
48. Neufeld, 'Cold War—But No War—In Space', p. 47.
49. Launius, 'National Security, Space', p. 5.
50. Bateman, 'Technological Wonder', p. 329.
51. Wolfe, *Competing with the Soviets*, p. 89.
52. McDougall, ...*The Heavens*, p. 65.
53. McQuaid, 'Sputnik Reconsidered', pp. 372, 375–6.
54. Massey and Robins, *History of British Space Science*, p. 39.
55. Zubok, *A Failed Empire*, p. 131.
56. Brzesinski, *Red Moon Rising*, pp. 42–4; Siddiqi, *Challenge to Apollo*, p. 143.
57. Siddiqi, *Challenge to Apollo*, p. 351.
58. Ibid.
59. Logsdon, *John F. Kennedy and the Race to the Moon*, pp. 206–7.
60. Craig and Logevall, *America's Cold War*, p. 174.
61. Temple, *Shades of Gray*, pp. 96–7.
62. Ibid. p. 97.
63. Craig and Logevall, *America's Cold War*, p. 175.
64. Launius, 'Interpreting the Moon Landings', pp. 226–7, citing: Logsdon, *The Decision to Go to the Moon*.
65. McQuaid, 'Sputnik Reconsidered', p. 389.
66. Logsdon, *John F. Kennedy and the Race to the Moon*, pp. 33, 71–3, 156.
67. Launius, *Reaching for the Moon*, p. 142.
68. Logsdon, *John F. Kennedy and the Race to the Moon*, pp. 182, 204.
69. Siddiqi, *Challenge to Apollo*, pp. 84, 102, 106, 118, 120–1, 128, 143–4, 396; Podvig, *Russian Strategic Nuclear Forces*, pp. 120–2.
70. Jenks, *The Cosmonaut*, p. 92.
71. Ibid. p. 95.
72. Siddiqi, *Challenge to Apollo*, p. 117–18, 122–4.
73. Jenks, *The Cosmonaut*, pp. 95–7.
74. Siddiqi, *Challenge to Apollo*, p. 151.
75. Zubok, *A Failed Empire*, p. 131.
76. Ibid. p. 131.
77. Podvig, *Russian Strategic Nuclear Forces*, p. 122.
78. Zubok, *A Failed Empire*, p. 377; Scott, *The Cuban Missile Crisis*, pp. 28–31.
79. Siddiqi, *Challenge to Apollo*, p. 305.
80. Jenks, *The Cosmonaut*, p. 99.

81. Siddiqi, *Challenge to Apollo*, p. 313.
82. Ibid. pp. 395–396, 436–8.
83. Ibid. pp. 151, 210–11, 237–8, 240.
84. Ibid. pp. 786–7.
85. Ibid. pp. 516–17.
86. Neufeld, *Spaceflight*, pp. 58–9.
87. Launius, *NASA: A History*, pp. 29–32.
88. Temple, *Shades of Gray*, p. 218.
89. Logsdon, *John F. Kennedy*, p. 40.
90. Ibid. p. 83.
91. Levine, *The Space and Missile Race*, pp. 102–3.
92. Temple, *Shades of Gray*, p. 220.
93. Launius, *NASA*, p. 34.
94. Ibid. pp. 34–5.
95. Temple, *Shades of Gray*, p. 219.
96. Ibid. p. 232.
97. Launius, *NASA*, p. 35.
98. Tyson, *Accessory to War*, p. 370.
99. Gavroglu and Renn, 'Positioning the History of Science', p. 3.
100. Burrows, *This New Ocean*, p. 229–31; Hays, *Struggling Towards Space Doctrine*, p. 181.
101. Immerman, *The Hidden Hand*, p. 61.
102. Hays, *Struggling Towards*, p. 181.
103. Sheehan, *The International Politics*, p. 92.
104. Dickey, 'The Rise and Fall'; Hays, *Struggling Towards Space Doctrine*, pp. 19–23, 63, 175–6, 186; The Outer Space Treaty, 1967, Article III.
105. Temple, *Shades of Gray*, pp. 102, 150.
106. Ibid. p. 199.
107. Neufeld, 'Cold War—But No War—In Space', p. 52.
108. Temple, *Shades of Gray*, pp. 308–9.
109. Sheehan, 'West European Integration', p. 109.
110. Ibid. p. 109.
111. Hörber, 'Introduction', in *European Space Policy*, p. 3.
112. Temple, *Shades of Gray*, p. 248.
113. Dickey, 'The Rise and Fall'.
114. Temple, *Shades of Gray*, p. 209.
115. Ibid. p. 220.
116. Ibid. p. 299.
117. Richelson, *Soldiers, Spies and the Moon*.
118. Launius, *Reaching for the Moon*, p. 20.
119. US Air Force, *AU-18 Military Space Primer*, pp. 8–9.
120. Wicken, 'Space science and technology', pp. 207–8, 210.

121. McDougall, ...*The Heavens*, pp. 110–11.
122. Launius, *Reaching for the Moon*, p. 21.
123. Temple, *Shades of Gray*, pp. 138–9.
124. McDougall, ...*The Heavens*, pp. 180–1, 275.
125. Hays, *Space and Security*, p. 29.
126. Temple, *Shades of Gray*, pp. 480–1.
127. Eleazer, 'Battle of the Titans'.
128. Temple, *Shades of Gray*, p. 488.
129. Ibid. p. 488.
130. Ibid. p. 508.
131. Launius, *Reaching for the Moon*, p. 209.
132. Griffin, 'The spaces between us', p. 74.
133. Immerwahr, 'Twilight of Empire', p. 129.
134. Ziarnick, *Developing National Power in Space*, pp. 19–22.
135. Nelson, State of NASA Address, 02/06/2021.
136. McCurdy, *Space and the American Imagination*, p. 155.
137. Ibid. pp. 157–63.
138. Trevino, *The Cosmos Is Not Finished*, pp. 120–6, 174.
139. Siddiqi, 'Transcending Gravity'.
140. Brown, 'Violence', in *The Oxford History of the American West*, pp. 396–8.
141. Kilgore, *Astrofuturism*, p. 2.
142. Ibid. pp. 51–62, 78–9.
143. Gorman, 'La Terre et l'Espace', p. 155.
144. Brodie, *Strategy in the Missile Age*, p. 269.
145. Oldenziel, 'Islands' in *Entangled Geographies*, p. 21.
146. Hecht, 'Introduction' in *Entangled Geographies*, p. 11.
147. Oldenziel, 'Islands', p. 29.
148. Richelson, *America's Space Sentinels*, p. 51.
149. Ibid. p. 54.
150. Gorman, *Dr Space Junk*, p. 97.
151. Ibid. p. 234–5.
152. Maile, 'Resurgent Refusals', pp. 57–69.
153. Kennedy, Speech at Rice University, Houston, Texas, 12 September 1962.
154. Northrop, *Veiled Empire*, pp. 10, 344.
155. Siddiqi, *The Challenge to Apollo*, pp. 133–6.
156. Northrop, *Veiled Empire*, p. 14.
157. Cameron, *The Hungry Steppe*, pp. 143–5.
158. Lieven, 'Russian Empires', in *Sovereignty After Empire*, p. 33.
159. Gorman, *Dr Space Junk*, p. 241.
160. Deudney, *Dark Skies*, esp. pp. 56–7.

161. Launius, *Reaching for the Moon*, p. 209.
162. McDougall, ...*The Heavens*, pp. 5–6.
163. Ibid. pp. 7–8.
164. Peoples, 'Haunted Dreams', pp. 106–7.
165. Neufeld, *Spaceflight*, p. 63.
166. Krige, 'Embedding the National in the Global', in *Science and Technology*, p. 228.

2. BEYOND BIPOLARITY

1. Neufeld, 'The Nazi aerospace exodus', p. 49.
2. Stroikos, *China, India in Space*, p. 256.
3. Krige, 'Embedding the National in the Global', pp. 244–5.
4. Gainor, 'The Nuclear Roots of the Space Race', p. 78.
5. For example: Hecht, *Being Nuclear*; Hecht, 'On the Fallacies'.
6. Krige et al., *NASA in the World*, p. 24.
7. Baylis and Stoddart, 'The British Nuclear Experience', p. 333.
8. Hennessey, *The Secret State*, pp. 50–1; Baylis, *Wales and the Bomb*, pp. 23–4.
9. Baylis, *Wales and the Bomb*, p. 24.
10. Krige et al., *NASA in the World*, p. 25.
11. Hecht, 'On the Fallacies', p. 77.
12. Paikowsky, *The Power of the Space Club*, p. 13.
13. Baylis, *Wales and the Bomb*, p. 27.
14. See: Jones, *The Official History, Volume I*; Jones, *The Official History, Volume II*.
15. Jones, *The Official History, Vol. I*, pp. 198–9.
16. Bronk, 'Britain's "Independent" V-Bomber Force', p. 976.
17. Harvey, *Europe's Space Programme*, p. 50.
18. Moore, 'Bad Strategy and Bomber Dreams', p. 146.
19. Ibid. pp. 147, 149, 158.
20. Hill, *A Vertical Empire*, p. 103.
21. Moore, 'Bad Strategy and Bomber Dreams', p. 160.
22. Stoddart, *Losing an Empire*, p. 51.
23. Jones, *The Official History, Vol. I*, p. 210.
24. Ibid. p. 209.
25. Pagedas, 'The afterlife of Blue Streak', pp. 19–21.
26. Hill, *Vertical Empire*, pp. 110–12; Moore, 'Bad Strategy and Bomber Dreams', p. 153; Massey and Robins, *History of British Space Science*, p. 21.
27. Harvey, *Europe's Space Programme*, p. 90.
28. Moss, 'There Are Many Other Things', 51:4, pp. 457–8.

29. Billaud and Journe, 'The Real Story', p. 356.
30. Pelopidas and Philippe, 'Unfit for purpose', pp. 1–2, 9–10, 13–15.
31. Hill, *Vertical Empire*, pp. 125–7, 135.
32. Ibid. p. 135.
33. Ibid. p. 138.
34. Neufeld, 'Cold War—But No War—In Space', p. 49.
35. Harvey, *Europe's Space Programme*, p. 25.
36. Krige et al., *NASA in the World*, p. 27.
37. Sheehan, 'West European Integration', p. 110.
38. Harvey, *Europe's Space Programme*, p. 163.
39. Krige et al., *NASA in the World*, pp. 51–2.
40. Gainor, 'The Nuclear Roots of the Space Race', p. 79; Wang, *Transatlantic Space Politics*, pp. 48–9; Harvey, *Europe's Space Programme*, p. 45; Pagedas, 'The afterlife of Blue Streak', pp. 3–4.
41. Billaud and Journe, 'The Real Story', pp. 367–8.
42. Harvey, *Europe's Space Programme*, pp. 96–102.
43. Wang, *Transatlantic Space Politics*, p. 50.
44. Harvey, *Europe's Space Programme*, pp. 52–6.
45. McLean, *Western European Military Space Policy*, p. 70.
46. Hill, *Vertical Empire*, p. 230.
47. Sheehan, 'West European Integration', p. 93.
48. Harvey, *Europe's Space Programme*, p. 121.
49. Hörber, 'Post-war space policy in Europe', pp. 17–19; Oikonomou, 'The EMWF', p. 164.
50. Sheehan, 'West European Integration', p. 95.
51. Krige, 'Embedding the National in the Global', p. 233.
52. Sheehan, 'West European Integration', pp. 105–8.
53. Krige et al., *NASA in the World*, pp. 58–60.
54. Sheehan, 'West European Integration', p. 111.
55. Wang, *Transatlantic Space Politics*, p. 52.
56. Krige et al., *NASA in the World*, pp. 60–1.
57. Ibid. pp. 52–7.
58. Hörber, 'Post-war Space Policy in Europe', p. 19–20.
59. U.S. Central Intelligence Agency, 'The Ariane Space Launch Vehicle'.
60. Krige et al., *NASA in the World*, pp. 16, 65–6, 95.
61. Harvey, *Europe's Space Programme*, pp. 29–30.
62. Ibid. p. 58.
63. Gorman, 'La Terre et l'Espace', p. 164.
64. Harvey, *Europe's Space Programme*, p. 48.
65. Gorman, 'La Terre et l'Espace', p. 159.
66. Redfield, *Space and the Tropics*, p. 128–37.
67. Siddiqi, 'Dispersed sites', p. 176.

68. Ibid. pp. 180, 187–8.
69. Ibid. pp. 189–91.
70. Gorman, 'La Terre et l'Espace', p. 165.
71. Oikonomou, 'The EMWF', pp. 170–1.
72. Harvey et al., *Emerging Space Powers*, pp. 2–5, 8–10.
73. Krige et al., *NASA in the World*, pp. 186–7.
74. Harvey et al., *Emerging Space Powers*, p. 12.
75. Krige et al., *NASA in the World*, pp. 188, 190.
76. Harvey et al., *Emerging Space Powers*, pp. 14–15.
77. Krige et al., *NASA in the World*, pp. 187–9.
78. Ibid. p. 191.
79. Harvey et al., *Emerging Space Powers*, pp. 21–3.
80. Krige et al., *NASA in the World*, pp. 191, 197–8, 206.
81. Gainor, 'The Nuclear Roots of the Space Race', p. 80.
82. Harvey et al., *Emerging Space Powers*, pp. 24–5.
83. Ibid. p. 31.
84. Ibid. pp. 68–75.
85. Krige et al., *NASA in the World*, p. 210.
86. Harvey et al., *Emerging Space Powers*, p. 10; Krige et al., *NASA in the World*, p. 197.
87. Krige et al., *NASA in the World*, p. 195.
88. Pekkanen and Kallender-Umezu, *In Defense of Japan*, p. 137.
89. Ibid. p. 1.
90. Ibid. pp. 2, 4–5.
91. Krige et al., *NASA in the World*, p. 186.
92. Ibid. pp. 197–210.
93. Sheehan, *The International Politics of Space*, p. 158.
94. Handberg and Zhen, *Chinese Space Policy*, pp. 11–13.
95. Harvey, *China in Space*, p. xi.
96. Hunter, 'The forgotten first iteration', p. 158.
97. Kulacki and Lewis, *A Place for One's Mat*, p. 5.
98. Harvey, *China in Space*, pp. 42–3.
99. Ibid. p. 44.
100. Kulacki and Lewis, *A Place for One's Mat*, p. 30.
101. Faligot, *Chinese Spies*, pp. 49–50.
102. Kulacki and Lewis, *A Place for One's Mat*, p. 7.
103. Zuoyue, 'The Cold War', p. 352.
104. Faligot, *Chinese Spies*, p. 299.
105. Kulacki and Lewis, *A Place for One's Mat*, pp. 6–7, 9.
106. Ibid. pp. 9–10, 46.
107. Ibid. p. 11.
108. Hunter, 'The forgotten first iteration', pp. 159–60; Alexis-Martin, 'The nuclear imperialism-necropolitics nexus', pp. 152–3.

109. For example, see: Sheetz and Yun, '"Adjust your location quickly"'.
110. Kulacki and Lewis, *A Place for One's Mat*, p. 12.
111. Ibid. pp. 13–14.
112. Harvey, *China in Space*, p. 49.
113. Kulacki and Lewis, *A Place for One's Mat*, pp. 11–14.
114. Harvey, *China in Space*, p. 49.
115. Hunter, 'The forgotten first iteration', p. 160.
116. Gainor, 'The Nuclear Roots of the Space Race', p. 80.
117. Handberg and Zhen, *Chinese Space Policy*, p. 11.
118. Ibid. pp. 13, 63–5.
119. Ibid. pp. 53, 82–3.
120. Schmalzer, 'Self-reliant Science', pp. 78–9.
121. Harvey, *China in Space*, p. 63.
122. Sheehan, *The International Politics of Space*, p. 162.
123. Harvey, *China in Space*, p. 172.
124. Ibid. pp. 502–3.
125. Kulacki and Lewis, *A Place for One's Mat*, p. 29.
126. Ibid. p. 30.
127. Harvey, *China in Space*, p. 508.
128. Kulacki and Lewis, *A Place for One's Mat*, p. 32.
129. Sheehan, *The International Politics of Space*, p. 147.
130. Lele, *ISRO*, pp. xiv-xv.
131. Singh, *The Indian Space Programme*, pp. 50–1.
132. Lele, *ISRO*, p. 2.
133. Ibid. pp. 23–4.
134. Maher, 'Grounding the Space Race', p. 141.
135. Singh, *The Indian Space Programme*, p. 24.
136. Ibid. pp. 117, 123–5.
137. Singh, *The Indian Space Programme*, p. 193.
138. Lele, *ISRO*, p. 4.
139. Krige et al., *NASA in the World*, p. 220.
140. Ibid. pp. 221–2.
141. Singh, *The Indian Space Programme*, pp. 144–7.
142. Sheehan, *The International Politics of Space*, p. 146.
143. Lele, *ISRO*, pp. 6–7, 25–6.
144. Singh, *The Indian Space Programme*, p. 194.
145. Lele, *ISRO*, p. 8.
146. Sheehan, *The International Politics of Space*, p. 142.
147. Gopalaswamy, *Final Frontier*, p. 4.
148. Sheehan, *The International Politics of Space*, p. 142.
149. Ibid. pp. 142–4, 151.
150. Gopalaswamy, *Final Frontier*, pp. 5–6.

151. Krige et al., *NASA in the World*, pp. 227–32.
152. Westad, *The Cold War*, p. 423.
153. Ibid. pp. 439–44.
154. Lele, *ISRO*, p. 28.
155. Singh, *The Indian Space Programme*, pp. 226–7.
156. Lele, *ISRO*, p. 32.
157. Indian Space Research Organisation, 'Polar Satellite Launch Vehicle'; Singh, *The Indian Space Programme*, p. 230.
158. Singh, *The Indian Space Programme*, p. 253.
159. Lele, *ISRO*, pp. 36–7.
160. Singh, *The Indian Space Programme*, p. 255.
161. Eisenhower, *Partners in Space*, pp. 40–9.
162. Lele, *ISRO*, p. 38.
163. Singh, *The Indian Space Programme*, pp. 255–6.
164. Gopalaswamy, *Final Frontier*, pp. 23–4.
165. Singh, *The Indian Space Programme*, p. 251.
166. Ibid. p. 256.
167. Sheehan, *The International Politics of Space*, p. 155.
168. Gopalaswamy, *Final Frontier*, p. 8; Singh, *The Indian Space Programme*, p. 396.
169. Gopalaswamy, *Final Frontier*, pp. 33–4.
170. Singh, *The Indian Space Programme*, p. 257.
171. Cohen and Dasgupta, *Arming Without Aiming*, p. 98.
172. Narang, *Nuclear Strategy*, pp. 114–15.
173. Cohen and Dasgupta, *Arming Without Aiming*, p. 99–101.
174. Narang, *Nuclear Strategy*, p. 122.
175. Kristensen and Korda, 'Indian Nuclear Forces 2020', pp. 220–2.
176. Gopalaswamy, *Final Frontier*, pp. 30–3; Cohen and Dasgupta, *Arming Without Aiming*, p. 26.
177. Lele, *ISRO*, p. 29.
178. Singh, *The Indian Space Programme*, pp. 127–9.
179. Ibid. p. 167.
180. Singh, *The Indian Space Programme*, pp. 257–61; Cohen and Dasgupta, *Arming Without Aiming*, pp. 166–7.
181. Gopalaswamy, *Final Frontier*, pp. 37–8.
182. Ibid. p. 42.
183. Singh, *The Indian Space Programme*, p. 131.
184. Ibid. p. 169.
185. Hecht, 'Rupture-Talk', pp. 691–4.
186. Paikowsky, *The Power of the Space Club*, pp. 8, 13.
187. Dolman, *Astropolitik*, pp. 144–65.
188. McDougall, *...The Heavens*, p. 345.

3. APPLIED WITCHCRAFT AND TECHNICAL WIZARDRY

1. Siddiqi, *Challenge to Apollo*, p. 859.
2. Deudney, *Dark Skies*, p. 318.
3. Edgerton, *The Shock of the Old*, esp. p. 209.
4. Stout and Warner, 'Intelligence is as intelligence does', p. 518.
5. Gill and Phythian, *Intelligence in an Insecure World*, p. 18.
6. Gaddis, 'The Long Peace', pp. 123–5.
7. Phythian, *Understanding the Intelligence Cycle*, p. 2.
8. Norris, *Spies in the Sky*, pp. 68–9.
9. Burrows, *This New Ocean*, pp. 229–30.
10. Ibid. p. 234.
11. Richelson, *Spying on the Bomb*, p. 128.
12. Temple, *Shades of Gray*, pp. 132, 143, 186, 200, 315–97.
13. Richelson, *The Wizards of Langley*, pp. 102–3, 128–30.
14. Burrows, *This New Ocean*, p. 234.
15. Zak, 'Russia's Military Spacecraft'; Zak 'Yantar-1KFT'.
16. Gainor, 'The Nuclear Roots', p. 73.
17. Hays, *Struggling Towards*, p. 236.
18. Norris, *Spies in the Sky*, pp. 26–7.
19. Day, 'Applied witchcraft'.
20. Norris, *Spies in the Sky*, p. 96.
21. Bateman, 'Technological Wonder', p. 334.
22. Zak, 'The Tselina'.
23. Zak, 'Russia's Military Spacecraft'.
24. Day, 'Above the clouds'.
25. Day, 'Shipkillers'.
26. Ibid.
27. Hays, *Space and Security*, p. 28.
28. Ibid. p. 12.
29. Bateman, 'Technological Wonder', p. 331.
30. Norris, *Spies in the Sky*, p. 58.
31. Hays, *Struggling Towards*, p. 249.
32. Richelson, *The Wizards of Langley*, pp. 198–200.
33. Ibid. pp. 205–8.
34. Immerman, *The Hidden Hand*, pp. 114–15.
35. Richelson, *The Wizards of Langley*, p. 202.
36. Day, 'Intersections… (part 2)'.
37. Day, 'Intersections… (part 1)'.
38. Norris, *Spies in the Sky*, p. 78.
39. Richelson, *Spying on the Bomb*, pp. 277–316.
40. Preston, *Plowshares and Power*, p. 26.

41. Hunter, 'The Forgotten', p. 162.
42. Black, *The Global Interior*, pp. 184–6, 197, 211–12.
43. Norris, *Spies in the Sky*, p. 107.
44. Richelson, *Spying on the Bomb*, pp. 232–3, 428–35.
45. Ibid. p. 426.
46. Federici, *From the Sea to the Stars*, pp. 100–2.
47. Norris, *Spies in the Sky*, p. 82.
48. Hays, *Struggling Towards*, p. 244.
49. Harvey, *China in Space*, p. 273.
50. Ibid. p. 267.
51. Siddiqi, 'Staring at the Sea', pp. 397, 399, 404–7, 412.
52. Friedman, *Seapower and Space*, pp. 126, 129–33.
53. Day, 'Shipkillers'.
54. Siddiqi, 'Staring at the Sea', p. 412.
55. Hendrickx, 'The status of Russia's signals intelligence satellites'.
56. Union of Concerned Scientists, Satellite Database.
57. Day, 'War at Sea'.
58. Neufeld, *Spaceflight*, p. 64.
59. Harvey, *Europe's Space Programme*, pp. 57, 71–2.
60. McLean, *Western European Military Space Policy*, pp. 96–7.
61. Preston, *Plowshares and Power*, pp. 23–5.
62. Harvey, *China in Space*, pp. 55–7.
63. Ibid. pp. 279–85.
64. Ibid. pp. 323–4, 328–38.
65. Harvey, *China in Space*, p. 303.
66. Gopalaswamy, *Final Frontier*, p. 14.
67. Singh, *The Indian Space Programme*, pp. 273, 282.
68. Ibid. pp. 396–7.
69. Pekkanen and Kallender-Umezu, *In Defense of Japan*, pp. 131–43.
70. Ibid. p. 149.
71. Harvey, *Europe's Space Programme*, pp. 238–9.
72. Reliable data on contemporary satellite systems by state and capability can be found at Gunter's Space Page, https://space.skyrocket.de/index.html
73. ESA, 'SAR-Lupe'.
74. Werner, 'Spacety shares first images'.
75. On 1973 Yom Kippur War and intelligence failure, see: Betts, 'Analysis, War, and Decision'; Betts, *Surprise Attack*.
76. Paikowsky, *The Power of the Space Club*, pp. 163–71.
77. Urban, *UK Eyes Alpha*, p. 57.
78. Day, 'War at Sea'.
79. Finlan, 'British Special Forces', p. 82.

80. Urban, *UK Eyes Alpha*, pp. 60–1.
81. Bateman, 'US-UK Military and Intelligence Cooperation in Space'.
82. Urban, *UK Eyes Alpha*, pp. 62–5.
83. Ferris, *Behind the Enigma*, p. 322.
84. As paraphrased by Urban, *UK Eyes Alpha*, p. 65.
85. McLean, *Western European Military Space Policy*, pp. 99–100.
86. US National Geospatial Intelligence Agency, 'National Imagery and Mapping Agency'.
87. Neufeld, 'Cold War—But No War—In Space', p. 57; Podvig, *Russian Strategic Nuclear Forces*, pp. 428–9; Podvig, 'History and the Current Status', p. 35.
88. Richelson, *America's Space Sentinels*, p. 109.
89. Ibid. pp. 82, 108.
90. Norris, *Spies in the Sky*, p. 97.
91. Hays, *Struggling Towards*, p. 243.
92. Richelson, *America's Space Sentinels*, p. 161.
93. Hendrickx, 'EKS'.
94. Ibid.
95. Richelson, *America's Space Sentinels*, pp. 137–56.
96. Tracy and Wright, 'Modelling the Performance of Hypersonic Boost-Glide Missiles,' pp. 149, 158.
97. Richelson, *America's Space Sentinels*, pp. 243–5.
98. Ibid. pp. 269–70.
99. Missile Defense Agency, 'PTSS Fact Sheet'.
100. Karako and Dahlgren, 'Complex Air Defense', p. 20.
101. Erwin, 'GAO steps up criticism'.
102. Edgerton, *The Shock*, p. 206.
103. Acton, 'Escalation through Entanglement', pp. 84–9.
104. Globalsecurity.org, 'Chinese Ballistic Missile Early Warning'; The Union of Concerned Scientists' Satellite Database also cites potential Early Warning functions for TJSW-2 and TJSW-5.
105. Ramana, Rajaraman, and Zia, 'Nuclear Early Warning in South Asia', pp. 282–3.
106. Siddiqi, *Challenge to Apollo*, p. 856.
107. Siddiqi, 'Soviet Spacepower', p. 148.
108. Deudney, *Dark Skies*, pp. 315–19.
109. Sun Tzu, *The Art of Warfare*, p. 74.
110. Clausewitz, *On War*, pp. 276–7.
111. Sun Tzu, *The Art of Warfare*, p. 74.
112. Clausewitz, *On War*, p. 300.
113. Ibid. pp. 319–20.
114. Gaddis, 'The Long Peace', p. 123.

115. Friedman, *Seapower and Space*, p. 10.
116. Moltz, *The Politics of Space Security*, pp. 44–7.
117. Hays, *Struggling Towards*, pp. 243–4.
118. Deudney, *Bounding Power*, p. 256.
119. Neufeld, 'Cold War—But No War—In Space', p. 46.
120. Norris, *Spies in the Sky*, p. 111.
121. Deudney, *Dark Skies*, p. 329.

4. TETHERS OF MODERNITY

1. Herrera, 'Technology and International Systems', p. 582.
2. Cameron Hunter takes credit for explaining this insight to me.
3. Bowen, *War in Space*, pp. 193–225.
4. Johnson-Freese, *Space as a Strategic Asset*, p. 83.
5. USAF, *AU-18*, p. 10.
6. Krige et al., *NASA in the World*, pp. 51–2.
7. Lele, *ISRO*, p. 6.
8. Singh, *The Indian Space Programme*, pp. 155–7.
9. Ibid. pp. 155–9; Krige et al., *NASA in the World*, p. 239–41.
10. Singh, *The Indian Space Programme*, p. 165; Krige et al., *NASA in the World*, pp. 236, 244–5.
11. Siddiqi, 'Whose India?', p. 467.
12. Krige et al., *NASA in the World*, pp. 236, 245.
13. Ibid. p. 246.
14. Siddiqi, 'Whose India?', p. 453.
15. Ibid. p. 455.
16. Krige et al., *NASA in the World*, p. 246.
17. Siddiqi, 'Whose India?', p. 454.
18. Singh, *The Indian Space Programme*, pp. 283–4.
19. Ibid. p. 398.
20. Harvey, *The Rebirth*, p. 121.
21. Downing, 'The Intersputnik System', pp. 465–8.
22. Ibid. pp. 468–9.
23. Ibid. pp. 469–70.
24. Zak, 'Russian Communications Satellites'.
25. Siddiqi, 'Staring at the Sea', p. 409.
26. Hays, *Struggling Towards*, pp. 244–5.
27. US Space Force, 'AEHF Factsheet'.
28. Graff, 'We're the only plane in the sky'.
29. Krige et al., *NASA in the World*, p. 54.
30. Ibid. pp. 55–6.
31. Ibid. pp. 56–63, 191, 198.

32. Harvey, *Europe's Space Programme*, pp. 113, 160, 182–4.
33. Ibid. pp. 100–1; Finlan, 'British Special Forces', p. 86.
34. Friedman, *Seapower and Space*, p. 258.
35. Krige et al., *NASA in the World*, p. 210.
36. Ball and Tanter, *The Tools of Owatatsumi*, pp. 95–6.
37. Pekkanen and Kallender-Umezu, *In Defense of Japan*, pp. 150–7.
38. Preston, *Plowshares and Power*, p. 143.
39. Ibid. p. 123.
40. Bernard, *The Foreign Missile*.
41. Burrows, *This New Ocean*, p. 225; Norris, *Spies in the Sky*, pp. 94–5.
42. Richelson, *The Wizards of Langley*, pp. 216–18.
43. Day, 'Stealing secrets'.
44. Eves, *Space Traffic Control*, p. 28.
45. Ibid. pp. 29, 81.
46. Weeden and Samson, 'Global Counterspace Capabilities', pp. 1–2—1–7.
47. Ibid. p. 2–2.
48. White, *Into the Black*, pp. 365–6, 383–4, 390–3.
49. Siddiqi, 'Shaping the World', pp. 41–55.
50. Eves, *Space Traffic Control*, p. 33.
51. Singh, *The Indian Space Programme*, p. 184.
52. Weeden, 'Current and Future Trends', p. 36.
53. Weeden and Samson, 'Global Counterspace Report', pp. 1–25—1–27; 2–34—2–36.
54. Lal et al., 'Global Trends', pp. 32–3; Susumu, 'SSA Capabilities and Policies in Japan'; *The Economic Times*, 'ISRO sets up dedicated control centre for Space Situational Awareness'.
55. Peldszus, 'EU Space Surveillance'.
56. Hendrickx, 'EKS'.
57. Weeden and Samson, 'Global Counterspace Report', pp. 2–34.
58. Lal et al., 'Global Trends', pp. 27–9.
59. Davis, 'The Story of the Kettering Radio Group'.
60. Bateman, 'Technological Wonder', p. 339.
61. Ibid. p. 339.
62. Johnson-Freese, *Space Warfare*, p. 12.
63. Eves, *Space Traffic Control*, p. 84.
64. Jonathan's Space Report, 10 May 2021.
65. Day, 'Burning Frost'.
66. Weeden and Samson, 'Global Counterspace Capabilities', pp. 3–6–3–7.
67. Day, 'The NRO and the Space Shuttle'.
68. Kelso, Keynote address.

69. MacKenzie, *Inventing Accuracy*, p. 144.
70. Hays, *Struggling Towards*, pp. 246–7.
71. Ceruzzi, *GPS*, pp. 37–45.
72. MacKenzie, *Inventing Accuracy*, p. 144.
73. Ceruzzi, *GPS*, pp. 77–82.
74. Hays, *Struggling Towards*, p. 247.
75. MacKenzie, *Inventing Accuracy*, pp. 3, 10.
76. Herrera, 'Technology and International Systems', p. 572.
77. MacKenzie, *Inventing Accuracy*, p. 1.
78. Ibid. pp. 128, 149, 196–204, 284.
79. Easton and Frazier, *GPS Declassified*, p. 55.
80. Ibid. pp. 61, 95.
81. Pace et al., *The Global Positioning System*, p. 45.
82. Ceruzzi, *GPS*, pp. 84–6.
83. Friedman, *Seapower and Space*, p. 280.
84. Herrera, 'Technology and International Systems', p. 577.
85. MacKenzie, *Inventing Accuracy*, pp. 170–1.
86. Easton and Frazier, *GPS Declassified*, p. 2.
87. Ceruzzi, *GPS*, pp. 123–33.
88. Easton and Frazier, *GPS Declassified*, p. 178.
89. Twigg, 'Russia's space program', pp. 70–2.
90. Harvey, *The Rebirth*, pp. 7–13.
91. Ceruzzi, *GPS*, p. 159.
92. Easton and Frazier, *GPS Declassified*, p. 180.
93. Bowen, *War in Space*, pp. 193–267.
94. Yonggang, 'Latecomer's Strategy', p. 1; European Global Navigation Satellite Systems Agency, 'GSA GNSS Market Report', p. 12.
95. Yonggang, 'Latecomer's Strategy', p. 7.
96. Kania and Costella, 'Seizing the commanding heights', p. 243.
97. Goff, 'India: A High-Tech Partner'.
98. Murthia et al., 'New developments in Indian space policies', p. 337; Mathieu, 'Assessing Russia's space cooperation', pp. 358–9.
99. Mathieu, 'Assesing Russia's space cooperation', p. 358.
100. Goff, 'Successful Launch of Japan's Second Michibiki'.
101. Suzuki, 'The contest for leadership in East Asia', pp. 102–3.
102. European Global Navigation Satellite Systems Agency, 'GSA GNSS Market Report', p. 6.
103. European Commission, 'Space Strategy for Europe', p. 8.
104. Ceruzzi, *GPS*, pp. 162–3.
105. Ibid. pp. 163–5.
106. Sheng-Chih, *Transatlantic Space Politics*, pp. 152, 160.
107. Ceruzzi, *GPS*, p. 166.

108. Sheng-Chih, *Transatlantic Space Politics*, p. 10.
109. Beidleman, 'GPS vs Galileo', pp. 117–18.
110. Sheng-Chih, *Transatlantic Space Politics*, pp. 122–3.
111. Bolton, 'Neo-realism and the Galileo and GPS negotiations', pp. 199–201.
112. Sheng-Chih, *Transatlantic Space Politics*, p. 128.
113. Sheng-Chih, *Transatlantic Space Politics*, pp. 123–8; Carstens, 'EU and US reach agreement'; Ballantyne, 'Neo-realism and the Galileo', p. 203.
114. The White House, 'Memorandum on Space Policy Directive 7'.
115. Bolton, 'Neo-realism and the Galileo and GPS negotiations', p. 186.
116. Sheng-Chih, *Transatlantic Space Politics*.
117. Ibid. p. 121.
118. Bolton, 'Neo-realism'.
119. Tardy, 'Does European defence really matter?', p. 5.
120. Honig, 'The Tyranny of Doctrine and Modern Strategy', p. 266.
121. Hinsley, *Power and the Pursuit of Peace*, pp. 177–8, 299.
122. Ceruzzi, *GPS*, p. 159.
123. Ibid. pp. 170–2.
124. Hays, *Struggling Towards*, p. 247.

5. SPACEPOWER AT WAR

1. Friedman, *Seapower and Space*, p. 8.
2. Day, 'Shipkillers'.
3. U.S. Navy, 'From the Sea to the Stars', pp. 120–1.
4. U.S. Defense Intelligence Agency, Memorandum: 'Intelligence as a force multiplier'.
5. Walker et al., *Seize the High Ground*, p. 13.
6. Ibid. p. 78.
7. Ford and Rosenberg, 'The Naval Intelligence Underpinnings', pp. 398, 402.
8. Friedman, *Seapower and Space*, p. 305.
9. Day, 'Shipkillers'.
10. Norris, *Spies in the Sky*, p. 188.
11. Richelson, *The Wizards of Langley*, p. 247.
12. Mann, 'Desert Storm', pp. 10–11.
13. Spires, *Beyond Horizons*, pp. 205–245.
14. Lupton, *On Space Warfare*.
15. Woodward, 'CIA Aiding Iraq'.
16. Gallegos, 'After the Gulf War', pp. 63–4.
17. Easton and Frazier, *GPS Declassified*, p. 126.

18. Shimko, *The Iraq Wars*, p. 81.
19. Ibid. p. 70.
20. Zubok, *Collapse*, p. 196.
21. Shimko, *The Iraq Wars*, pp. 97–8.
22. Mann, 'Desert Storm', p. 10.
23. USAF, *AU-18*, p. 183.
24. Friedman, *Seapower and Space*, p. 10.
25. Ibid. pp. 272–3.
26. Watling, 'More Sensors Than Sense', p. 90.
27. Campen, 'Introduction', p. xi.
28. Wentz, 'Communications support', p. 21.
29. McKinley, 'When the Enemy Has Our Eyes', p. 316.
30. Ibid. p. 318.
31. Long and Green, 'Stalking the Secure Second Strike', pp. 56–7.
32. Ibid. p. 58.
33. Friedman, *Seapower and Space*, p. 315.
34. Easton and Frazier, *GPS Declassified*, p. 118; Shimko, *The Iraq Wars*, p. 63.
35. Campen, 'Introduction', p. xi.
36. Shimko, *The Iraq Wars*, p. 75.
37. Ibid. pp. 64–5, 69, 75.
38. Finlan, *The Gulf War 1991*, p. 81.
39. Lee, 'Counterspace Operations', p. 269.
40. Richelson, *The Wizards of Langley*, p. 249.
41. McKinley, 'When the Enemy', p. 303.
42. Zubok, *Collapse*, p. 196.
43. McKinley, 'When the Enemy', p. 319.
44. Shimko, *The Iraq Wars*, p. 93.
45. Ibid. p. 94.
46. Lambakis, 'Space Control', p. 418.
47. Lonsdale, *The Nature of War*, p. 3.
48. Sterling, 'Soviet Reactions'.
49. Krepinevich, 'The Military-Technical Revolution', pp. 7–8.
50. Marshall, Memorandum of 23 August 1993, p. 1.
51. Gray, *Strategy and History*, p. 82.
52. Lonsdale, *The Nature of War*, pp. 49–95.
53. Ibid. p. 1.
54. Shimko, *The Iraq Wars*, p. 110.
55. Ibid. pp. 133–4.
56. Luft, 'The OODA Loop'.
57. Betz, 'The More You Know'.
58. Gray, *Airpower for Strategic Effect*, pp. 206–7.
59. Ibid. p. 358. Emphasis Handel's.

60. Echevarria, *Clausewitz and Contemporary War*, p. 133.
61. Handel, *Masters of War*, pp. 354–6.
62. Gray, *Modern Strategy*, p. 6.
63. Marshall, Memorandum of 23 August 1993, p. 5.
64. Campen, 'Introduction', p. xi.
65. Bowen, *War in Space*, pp. 105–54.
66. Gray and Sheldon, 'Space Power', p. 32.
67. Ibid. pp. 32–5.
68. Kofman et al., *Russian Military Strategy*, pp. i, 55–6.
69. Podvig, 'Russian space systems', pp. 37–41.
70. Renz, *Russia's Military Revival*, p. 63.
71. Kofman et al., *Russian Military Strategy*, p. 59.
72. Bruusgaard, 'Russian nuclear strategy', p. 26.
73. Sukhankin, 'From "Bridge of Cooperation"', pp. 17–18.
74. Ibid. pp. 28–30, 34.
75. Bruusgaard, 'Russian nuclear strategy', p. 22.
76. Renz, *Russia's Military Revival*, pp. 59–61, 78, 172.
77. Ibid. p. 161.
78. Ibid. p. 182–9.
79. Sukhankin, 'From "Bridge of Cooperation"', p. 32.
80. Withington, 'Russian EW'; Watling and Reynolds, 'Operation Z', p. 2.
81. Zak, 'Spooky world of military satellites'.
82. Thomas, 'Russian Lessons Learned in Syria', pp. 7, 10, 12, 17–18.
83. Renz, *Russia's Military Revival*, p. 196.
84. Ibid. pp. 63, 80.
85. Ibid. pp. 67–83.
86. Engvall, *Russia's Military R&D Infrastructure*, p. 43.
87. Renz, *Russia's Military Revival*, p. 83.
88. Grau and Bartles, *The Russian Way of War*, p. 240.
89. Russian MoD, 'Chief of the General Staff'.
90. Grau and Bartles, *The Russian Way of War*, pp. 263–5.
91. Renz, *Russia's Military Revival*, p. 190.
92. Lieven, *Chechnya*, p. 269.
93. Xiaoming, *Deng Xiaoping's Long War*, pp. 1–6.
94. Bowen, *War in Space*, pp. 27–8.
95. Xiaobing, *The Dragon in the Jungle*, p. 265.
96. Wuthnow, 'A Brave New World', p. 188.
97. Kania and Costello, 'Seizing the commanding heights', p. 222.
98. Xiaobing, *The Dragon in the Jungle*, p. 266.
99. Sheehan, 'Counterspace Operations', 12:2.
100. For example, see: Pollpeter, 'Space, The New Domain'; Cheng, 'China's Military Role in Space'; Heginbotham et al., *The U.S.-China*

Military Scorecard; Hagt and Durnin, 'Space, China's Tactical Frontier'; Wortzel, *The Dragon Extends Its Reach*; Cheng, Testimony; Academy of Military Science Military Strategy Department, *The Science of Military Strategy*, pp. 225–37.

101. Academy of Military Science Military Strategy Department, *The Science of Military Strategy*, p. 228.
102. Ibid.
103. Blasko, 'The PLA Army', p. 176.
104. Centre for Strategic and International Studies, 'Missiles of China'.
105. Kania and Costello, 'Seizing the commanding heights', p. 243.
106. Ibid. p. 260.
107. Wuthnow, 'A Brave New World', p. 185.
108. Blasko, 'The PLA Army', pp. 149–54.
109. Wuthnow, 'A Brave New World', p. 172.
110. Kania and Costello, 'Seizing the commanding heights', p. 219.
111. Blasko, 'The PLA Army', p. 151; Kania and Costello, 'Seizing the commanding heights', p. 219.
112. Blasko, 'The PLA Army', p. 160.
113. Ibid. p. 171.
114. Wuthnow, 'A Brave New World', p. 170.
115. Logan, 'Are they reading Schelling', p. 5.
116. Acton, 'Escalation through Entanglement', pp. 57–8, 76, 81–9.
117. Lewis and Xue, 'Making China's nuclear war plan', p. 46.
118. Logan, 'Are they reading Schelling', p. 24.
119. Ibid. p. 27.
120. Woolf, 'Conventional Prompt Global Strike', pp. 3–4.
121. U.S. Defense Intelligence Agency, *North Korea Military Power*, p. 24.
122. CSIS Missile Threat Project, 'Missiles of South Korea'.
123. Lee and Lee, 'South Korea Seeks'.
124. CSIS Missile Threat Project, 'Missiles of Taiwan'.
125. See: Bowen, *War in Space*, pp. 228–67; Easton, *The Chinese Invasion Threat*.
126. Nuclear Threat Initiative, 'The CNS India'.
127. CSIS Missile Threat Project, 'Missiles of Iran'.
128. Byman and Waxman, 'Kosovo and the Great Air Power Debate'.
129. See: Pape, *Bombing to Win*; Clodfelter, *The Limits of Air Power*; Warden, 'Success in Modern War'; Watts, 'Ignoring reality'; Heuser, *The Evolution of Strategy*, pp. 313–50.
130. Shimko, *The Iraq Wars*, pp. 137, 140.
131. Ibid. p. 122.
132. Finlan, *Contemporary Military Strategy*, pp. 86–8.
133. Shimko, *The Iraq Wars*, pp. 153–5.

134. Finlan, *Contemporary Military Strategy*, pp. 134–5.
135. Ibid. p. 136.
136. Brun, 'While You're Busy Making Other Plans', pp. 558–9.
137. Assemblée National, 'Compte rendu Commission'.
138. Bowen, *War in Space*, pp. 215–23.
139. Bateman, 'Technological Wonder', pp. 343–4.
140. Ibid. p. 346.
141. Watling, 'More Sensors Than Sense', p. 98.

6. ARSENALS FOR SPACE WARFARE

1. Jones, *Security, Strategy, and Critical Theory*, pp. 131–2.
2. Sevastopulo, 'China conducted two'.
3. Clausewitz, *On War*, pp. 279–81.
4. Duvall and Havercroft, 'Taking sovereignty'.
5. Weeden and Samson, 'Global Counterspace Capabilities'; Harrison et al., 'Defense Against the Dark Arts'.
6. Hays, *Struggling Towards*, p. 254.
7. Dickey, 'The Rise and Fall'.
8. Gray, *American Military Space Policy*, p. 105.
9. Bateman, 'Spooks and satellites'.
10. Friedman, *Seapower and Space*, p. 10.
11. Bowen, *War in Space*, pp. 228–63; Gray, *American Military Space Policy*, p. 9.
12. Gray, *Another Bloody Century*, p. 307.
13. For example: Lee, *War in Space*; Chalfont, *Star Wars*; Stares, *Space and National Security*; Long, et al., *Weapons in Space*; Bulkeley, et al., *Space Weapons*; Ra'anan and Pfaltzgraff, Jr. *International Security Dimensions of Space*; Payne, *Laser Weapons*; Karas, *The New High Ground*.
14. Bateman, 'Mutually assured surveillance', p. 120.
15. Bateman, 'Intelligence and alliance politics', p. 946; Ellis, *Reds in Space*.
16. For example: Dolman, *Astropolitik*; Johnson-Freese, *Space as a Strategic Asset*; *Astropolitics: The International Journal of Space Politics and Policy*, 2003, Vol. 1 Issue 1; Bormann and Sheehan, *Securing Outer Space*; Coletta and Pilch, *Space and Defense Policy*; Klein, *Space Warfare*; Logsdon and Adams, *Space Weapons*; Krepon et al., 'China's Military Space Strategy'; Duvall and Havercroft, 'Taking sovereignty'; Burris, 'Astroimpolitic'.
17. Gray, *Another Bloody Century*, p. 310.
18. Gray, *American Military Space Policy*, p. 48.
19. Gray, *Another Bloody Century*, pp. 310–11; Moltz, *The Politics of Space*, pp. 42–3; Lambakis, *On the Edge*, p. 135; O'Hanlon, *Neither Star Wars*, p. 109; Bowen, 'Cascading Crises'.

20. Booth and Wheeler, *The Security Dilemma*, pp. 1, 4–5.
21. Karas, *The New High Ground*, pp. 177–9.
22. Stares, *Space and National Security*, p. 120.
23. Gray, *American Military Space Policy*, p. 46.
24. Ibid. p. 7.
25. Bowen, 'Cascading Crises', pp. 49–50.
26. Clausewitz, *On War*, pp. 271–2.
27. Baram, *Technology in Warfare*, p. 4.
28. Air Force Doctrine Document 3–14.1, *Counterspace Operations*, 28/07/2011, pp. 4–5, 34.
29. Weeden and Samson, 'Global Counterspace Capabilities', pp. 1–18.
30. Baram, *Technology in Warfare*, p. 537.
31. McDermott, 'Russia's Electronic Warfare', p. iv.
32. Futter, '"Cyber" semantics'.
33. Rid, 'Cyber War Will Not Take Place', pp. 16–22.
34. Baram, *Technology in Warfare*, p. 35.
35. Hecht, *Beam Weapons*, p. 251.
36. Preston et al., *Space Weapons*, pp. 25–6.
37. Hecht, *Beam Weapons*, p. 243.
38. Ibid. pp. 243–4.
39. Gray, *American Military Space Policy*, p. 47.
40. Temple, *Shades of Gray*, pp. 105–11.
41. Ibid. p. 362.
42. Ibid. pp. 382, 454.
43. Siddiqi, 'Soviet Spacepower', p. 148.
44. National Security Council Memorandum 345, reproduced in: Hays, *Space and Security*, pp. 181–2.
45. Bateman, 'Mutually Assured Surveillance', pp. 126–32.
46. Siddiqi, 'Staring at the Sea', p. 407.
47. U.S. Navy, *From the Sea to the Stars*, p. 104.
48. Pfrang and Weeden, 'US Direct Ascent'; Chun, *Shooting Down a "Star"*.
49. Weeden and Samson, 'Global Counterspace Capabilities', pp. 3–15.
50. Stares, *Space and National Security*, p. 181.
51. Osborn, 'Missile Defense Agency Optimistic'.
52. Podvig, 'Did Star Wars', p. 7.
53. Pfrang and Weeden, 'Russian Co-orbital'; Podvig, 'Did Star Wars', p. 9.
54. Hendrickx, 'Burevestnik'.
55. Weeden and Samson, 'Global Counterspace Capabilities', pp. 2–19.
56. Podvig, 'Did Star Wars', p. 19.
57. Pfrang and Weeden, 'Russian Direct Ascent'.

58. Weeden and Samson, 'Global Counterspace Capabilities', pp. 2–16–2–17.
59. Ibid. pp. 2–18.
60. Ibid. pp. 1–11–1–12.
61. Stokes and Cheng, 'China's evolving space capabilities', p. 9.
62. Lewis and Kulacki, 'A Different View'.
63. Ibid.
64. Weeden and Samson, 'Global Counterspace Capabilities', pp. 1–13—1–14, 1–17.
65. Ibid. pp. 5–3–5–4.
66. Sankaran, 'Limits of the Chinese Antisatellite Threat', pp. 20–6.
67. Weeden and Samson, 'Global Counterspace Capabilities', pp. 1–17.
68. Morgan, *Deterrence and First Strike*, p. 2.
69. Ibid. p. 6.
70. Wawro, *Warfare and Society in Europe*, pp. 136–9, 145.
71. Weeden and Samson, *Global Counterspace Capabilities*, pp. 1–20, 2–23–2–25.
72. Ibid. pp. 3–17–3–20.
73. Ibid. p. 2–28.
74. Ibid. pp. 6–5–6–6.
75. Vizard, 'Safeguarding GPS'.
76. Shaw, 'MarRINav Final Report', pp. 5–6.
77. Bowen, 'UK Spacepower and the Integrated Review'.
78. Weeden and Samson, 'Global Counterspace Capabilities', pp. 1–24—1–25, 2–29–2–34.
79. Hecht, *Beam Weapons*, p. 258–62, 267–92.
80. Hays, *Space and Security*, p. 74.
81. Weeden and Samson, 'Global Counterspace Capabilities', pp. 3–21—3–23.
82. Ibid. pp. 2–29—2–31.
83. Ibid. p. 1–22.
84. Hecht, *Beam Weapons*, p. 258.
85. Ibid. pp. 251–4.
86. Mowthorpe, 'The Soviet/Russian approach', p. 27–8.
87. Siddiqi, *Challenge to Apollo*, pp. 235, 394.
88. Pfrang and Weeden, 'Russian Co-orbital; Zak, 'IS'.
89. Weeden and Samson, 'Global Counterspace Capabilities', pp. 2–2, 2–4.
90. Ibid. pp. 2–4–2–12.
91. Clark, 'US, China, Russia'.
92. Roberts, 'Space-based Missile Defense'.
93. Preston et al., *Space Weapons, Earth Wars*, pp. 43, 104–5.
94. Ibid. p. 26.

95. Baram, *Technology in Warfare*, p. 549.
96. Preston et al., *Space Weapons, Earth Wars*, pp. 26, 118.
97. Weeden and Samson, 'Global Counterspace Capabilities', pp. 3–23.
98. NASA, 'About the Space Station solar arrays'.
99. Hecht, *Beam Weapons*, p. 258.
100. Preston et al., *Space Weapons, Earth Wars*, pp. 115, 121.
101. Bateman, 'Intelligence and alliance politics', p. 942.
102. Hays, *Space and Security*, pp. 70–1, 76.
103. Weeden and Samson, *Global Counterspace Capabilities*, pp. 2–28.
104. Richelson, *America's Space Sentinels*, p. 48.
105. Siddiqi, 'The Soviet Fractional Orbital Bombardment System', p. 28.
106. Bowen and Hunter, 'Chinese Fractional Orbital Bombardment'.
107. Lewis and Hua, 'China's Ballistic Missile Programs', pp. 17–19.
108. Siddiqi, 'Soviet Spacepower', p. 148; Day, 'Nuking Moscow'.
109. Temple, *Shades of Gray*, p. 455.
110. Houchin, *US Hypersonic Research and Development*, pp. 1–2.
111. Day, 'Behind the Blue', p. 91.
112. Houchin, *US Hypersonic Research and Development*, pp. 176–8.
113. Siddiqi, 'Soviet Spacepower', p. 147.
114. Zak, 'Salyut-3'.
115. Preston et al., *Space Weapons, Earth Wars*, pp. 40–3.
116. Ibid. p. 45.
117. Ibid. pp. 111–14.
118. Temple, *Shades of Gray*, p. 461.
119. Preston et al., *Space Weapons, Earth Wars*, p. 55.
120. On gunboat diplomacy see: Cable, *Gunboat Diplomacy*.
121. Kania and Costello, 'Seizing the commanding heights', p. 245.
122. Luttwak, *Strategy*, pp. 34–6.

7. WAR ON THE COSMIC COASTLINE

1. Gray, *Modern Strategy*, p. 17.
2. For example, see: Lupton, *Space Warfare*, pp. 51–63; Johnson-Freese, *Space Warfare*, pp. 48–9, 180; Caldicott and Eisendrath, *War in Heaven*, pp. 117–25.
3. Gray and Sheldon, 'Space Power', pp. 23–4.
4. Caldicott and Eisendrath, *War in Heaven*, p. 4.
5. Kaplan, *The Wizards of Armageddon*.
6. Kahn, *Thinking About*, pp. 58–9.
7. Ibid. pp. 54–7.
8. Clausewitz, *On War*, p. 289.
9. Quinlan, *Thinking About Nuclear Weapons*, p. 20.

10. Snyder, 'Deterrence and power', p. 163.
11. Schelling, *Arms and Influence*, p. 2.
12. Ibid. p. 3.
13. Bowen, *War in Space*, pp. 134–9.
14. Quinlan, *Thinking About Nuclear Weapons*, p. 22.
15. Clausewitz, *On War*, p. 279.
16. Sun Tzu, *The Art of Warfare*, p. 75.
17. Cohn, 'Sex and Death', pp. 690–2; Jervis, *The Meaning*, p. 175.
18. Clausewitz, *On War*, pp. 270–2.
19. Jervis, *The Meaning*, pp. 20–1; Clausewitz, *On War*, p. 282.
20. On non-linearity, chaos, and war, see: Beyerchen, 'Clausewitz, Nonlinearity'.
21. Clausewitz, *On War*, pp. 267–74.
22. Ibid. p. 266.
23. Ibid. pp. 298–9, 320–1.
24. Ibid. p. 292.
25. On the limits perception and knowledge in leadership calculations, see: Jervis, *Perception and Misperception*.
26. Booth and Wheeler, *The Security Dilemma*, pp. 58–9.
27. Ibid. pp. 59–61.
28. Ibid. pp 4–5, 7.
29. Lambakis, *On the Edge of Earth*, p. 259.
30. Harrison et al., *Escalation and Deterrence*, pp. 20–1.
31. Academy of Military Science Military Strategy Department, *The Science of Military Strategy*, p. 229.
32. Mueller, 'The Absolute Weapon', p. 53.
33. Quinlan, *Thinking About Nuclear Weapons*, p. 22.
34. Gray, *American Military Space Policy*, p. 10.
35. Harrison et al., *Escalation and Deterrence*, p. 23.
36. Johnson-Freese, *Space Warfare*, p. 178.
37. Clausewitz, *On War*, pp. 270, 907–8.
38. Sun Tzu, *The Art of Warfare*, pp. 74, 85, 89.
39. Freedman, *The Evolution*, pp. 397–400.
40. Jinyuan Su, 'The Legal Challenge', p. 187.
41. Bowen, 'Space Oddities', pp. 271–6.
42. Bateman, 'Mutually assured surveillance'.
43. Freedman, *The Evolution*, pp. 397–401, 420.
44. Clausewitz, *On War*, p. 268.
45. Liemer and Chyba, 'A Verifiable Limited'.
46. Religious connotations of 'violating heaven' with space weapons, as well as religious terminology when characterising outer space are frequent. For example: Deudney, *Dark Skies*, esp. pp. 6, 13–14, 110; Caldicott

and Eisendrath, *War in Heaven*; Sheehan, 'Profaning the Path'; Johnson-Freese, *Heavenly Ambitions*; McDougall, ...*The Heavens*; Jack Manno, *Arming the Heavens*; Poole, 'The Myth of Progress'; Sheehan, *The International Politics*, p. 183.

47. Krepon, 'What is a Space Weapon...'.
48. UNOOSA, 'Long-term Sustainability of Outer Space Activities'.
49. West, 'Outer space', p. 28.
50. Bowen, 'Space Oddities', pp. 275–6.
51. Neufeld, 'Cold War—But No War—In Space', pp. 45–6; Sheehan, *The International Politics*, p. 94.
52. Lopez, 'Predicting an Arms Race', pp. 53–60.
53. Epkenbans, 'Was a peaceful outcome thinkable?' p. 126; Stevenson, *Armaments and the Coming of War*, pp. 417–18.
54. Caldicott and Eisendrath, *War in Heaven*, p. 83.
55. Lambakis, *On the Edge of Earth*, p. 266.
56. Ibid.; Moltz, *The Politics of Space Security*, p. 351.
57. On India-China-Pakistan tensions as they apply to space, see: Lele, 'Space Security Dilemma'; Chandrashekar, 'Space, War, and Deterrence'; Ahmed, 'Emerging Trends'; Khan and Khan, 'Space Security Trilemma'.
58. See: You, 'The Indian Ocean'.
59. Stares, *Space and National Security*, p. 141.
60. Taken from: Hersch and Steer, 'Introduction', p. 5.
61. Taken from: Kay, 'John F. Kennedy' p. 574.
62. Dolman, *Astropolitik*, p. 8.
63. On the imperial aspects of spacepower and space weapons, see: Duvall and Havercroft, 'Taking sovereignty', pp. 755–7, 763–70; Bowen, *War in Space*, pp. 19–50: Deudney, *Dark Skies*.
64. Lupton, *On Space Warfare*, p. 101; Gray, *American Military Space Policy*, pp. 11–12.
65. Dolman, *Astropolitik*, pp. 151–2.
66. Clausewitz, *On War*, p. 613.
67. Lupton, *On Space Warfare*, pp. 101–2.
68. Levine, *The Space and Missile Race*, p. 104.
69. For example see: Goswami, 'Explaining China's'; Goswami, 'Why the Chang'e-4'; Flewelling, 'Securing Cislunar Space'; Listner, 'A U.S. return to the Moon'.
70. Klein *Space Warfare*, pp. 91–106.
71. Erwin, 'Raymond's progress report'.
72. USSF, Space Capstone Publication, p. 16.
73. Clausewitz, *On War*, p. 613.
74. Gray, *American Military Space Policy*, p. 13.

75. U.S. House of Representatives Committee on Armed Services, 'Report of the Commission', pp. 8, 13–15.
76. On the centre of gravity in space, see: Smith, *Ten Propositions*, pp. 64–9; Bowen, *War in Space*, pp. 76–86, 138.
77. On an in-depth discussion of Space Pearl Harbor and a Counterspace in Being strategy, see: Bowen, *War in Space*, pp. 228–63.
78. Godwin, 'The PLA Faces', p. 70; Pollpeter, 'Space, The New Domain', p. 711.
79. Kania and Costello, 'Seizing the commanding heights', p. 225.
80. Academy of Military Science Military Strategy Department, *The Science of Military Strategy*, p. 229.
81. Lewis and Kulacki, 'A Different View'.
82. Clausewitz, *On War*, p. 381.
83. Clausewitz, *On War*, pp. 614–15. Emphasis Clausewitz's.
84. Howard, *The Franco-Prussian War*, pp. 204–23.
85. Dolman, *Astropolitik*, pp. 71–5.
86. Sun Tzu, *The Art of Warfare*, p. 89.
87. U.S. Government Accountability Office, 'Surplus Missile Motors', p. 30.
88. Johnson-Freese, *Space Warfare*, p. 179.
89. Mueller, 'Totem and Taboo', p. 35.
90. World Nuclear Association, 'Small Nuclear Power Reactors'.
91. Edgerton, *Shock of the Old*, p. 77.
92. Gray, *Airpower for Strategic Effect*, pp. 136–9.
93. Watling, 'More Sensors Than Sense', p. 87–98.
94. Luttwak, *Strategy*, pp. 7, 105–6.
95. Sun Tzu, *The Art of Warfare*, p. 113.
96. Clausewitz, *On War*, p. 266.
97. U.S. Joint Chiefs of Staff, Joint Publication 3–14, pp. IV-1, A-1, A-6.
98. On the maritime roots of the command of space, see: Bowen, 'From the sea to outer space'.
99. Shimko, *The Iraq Wars*, p. 162.
100. Handel, *Masters of War*, p. 355.
101. On the history and use of the concept of grand strategy, see: Morgan-Owen, 'History and the Perils', pp. 352–3; Milevski, *The Evolution*.
102. Freedman, *Strategy*, p. 74; see also Heuser, *Evolution of Strategy*, pp. 3–28.
103. Gray, *Modern Strategy*, p. 22.
104. Clausewitz, *On War*, p. 934.
105. Ibid. pp. 935–9.
106. On the coastline analogy to Earth orbit in full, see: Bowen, *War in Space*, pp. 105–54.

107. Bowen, *War in Space*, pp. 124–30.
108. For example, see: Dolman, *Astropolitik*; Smith, *Ten Propositions*; Gray, *Another Bloody Century*, p. 306; Klein, *Space Warfare*; Lutes et al., *Toward a Theory*; Bowen, *War in Space*.
109. Temple, *Shades of Gray*, p. 209.
110. Bowen, *War in Space*, p. 56.
111. Ibid. pp. 57–8.
112. On the dispersing effects of spacepower, see: Bowen, *War in Space*, pp. 193–267.
113. On logistics and paralysis on Earth through spacepower, see: Bowen *War in Space*, pp. 139–48.
114. On first-strike instability in space, see: Morgan, *First Strike*.

CONCLUSION: ANARCHY IN THE GLOBAL SPACE AGE

1. Harding, *Space Policy*, p. 1.
2. Sheehan, *The International Politics*, pp. 1–2.
3. Burrows, *This New Ocean*, pp. 610–11.
4. Peoples and Stevens, 'At the Outer Limits', p. 312.
5. Kim, 'The "Democratization of Space"'; Thorbecke, 'What to know about'.
6. Herrera, 'Technology and International Systems', p. 562.
7. McCarthy, 'Introduction', pp. 5, 8, 11.
8. Deudney, *Dark Skies*, p. 40.
9. Bull, *The Anarchical Society*, pp. 46–51.
10. Linklater and Suganami, *The English School*, p. 44.
11. Krige et al., *NASA in the World*, pp. 134–49.
12. Carr, *Twenty Years' Crisis*, p. 100.
13. Scheuerman, *Morgenthau*, pp. 11–12; Craig, *Glimmer of a New Leviathan*, pp. 93–116.
14. Bew, *Realpolitik*, pp. 300–3.
15. Deudney, *Dark Skies*, p. 151.
16. McCarthy, 'Introduction', p. 13.
17. Brooks and Wohlforth, 'Power, Globalization', pp. 5–53; Moltz, 'The Changing Dynamics', pp. 66–94.
18. Baldwin, 'Interdependence and power', pp. 500–1; Baldwin, *Power and International Relations*, pp. 70–7, 162; Nye, *The Future of Power*, p. 5.
19. Baldwin, *Power and International Relations*, pp. 52–3.
20. Wendt, *Social Theory*, pp. 96, 110–12.
21. Baldwin, *Power and International Relations*, p. 77.
22. McCarthy, 'Technology and the International', p. 476.
23. McCarthy, 'Introduction', p. 11.

24. Vidal, 'Russia's Space Policy'.
25. Chong and Pham, 'Critical Perspectives', p. 13.
26. Carr, *The Twenty Years' Crisis*, pp. 108, 132.
27. Neufeld, *Spaceflight*, p. 128.
28. Bloom, *Eccentric Orbits*, p. x.
29. Ibid. pp. 423–5.
30. Jewet, 'OneWeb Partners'.
31. Erwin, 'U.S. Army'.
32. Peldszus, 'Architectures of Command', p. 287.
33. Launius, 'Introduction', p. 1.
34. Lory, 'Europe is already'.
35. Paikowsky, *The Power of the Space Club*, pp. 5–7.
36. Bull, 'The great irresponsibles?' p. 437.
37. Johnson-Freese, *Space Warfare*, p. 84.
38. Bull, *The Anarchical Society*, pp. 221, 228.
39. Ibid. p. 229.
40. Zhang, 'China and liberal hierarchies', p. 811.
41. Gorman, *Dr Space Junk*, p. 226.
42. Scheuerman, *Morgenthau*, p. 126.
43. Sheehan, 'West European Integration', p. 94.
44. Johnson-Freese, *Space Warfare*, p. 166.
45. Aron, *Peace and War*, p. 664.
46. Sagan, *Pale Blue Dot*, pp. 6–7.
47. McDougall, *...The Heavens*, pp. 456–7.
48. White, *The Overview Effect*.
49. Chow, 'Jeff Bezos'.
50. Gorman, *Dr Space Junk*, p. 127.
51. Marcuse, *One-Dimensional Man*, pp. 255–6.
52. Ibid. p. 1; see also McDougall, *...The Heavens*, pp. 3–19.
53. Peldszus, 'Architectures of Command', pp. 305–6.
54. Penent, 'Introduction', p. 13.
55. Zubrin and Wagner, *The Case for Mars*, pp. xxxiii–xxxiv, 1, 325.
56. Salmeri, 'Op-ed | No, Mars'.
57. Zubrin and Wagner, *The Case for Mars*, p. 325.
58. Deudney, *Dark Skies*, pp. 37, 46, 369.
59. Ibid. pp. 6, 46.
60. Ibid. p. 355.

BIBLIOGRAPHY

BOOKS

Andrews, James T. and Asif Siddiqi, ed. *Into the Cosmos: Space Exploration and Soviet Culture* (Pittsburgh University Press, 2011).

Aron, Raymond, *Peace and War: A Theory of International Relations*, Richard Howard and Annette Baker Fox, trans. (Weidenfeld & Nicolson, 1966).

Baldwin, David A. *Power and International Relations: A Conceptual Approach* (Princeton University Press, 2016).

Ball, Desmond, and Richard Tanter, *The Tools of Owatatsumi: Japan's Ocean Surveillance and Coastal Defence Capabilities* (Australian National University Press, 2015).

Baram, Abdul Karim, *Technology in Warfare: The Electronic Dimension* (Emirates Center for Strategic Studies and Research, 2008).

Baylis, John, *Wales and the Bomb* (University of Wales Press, 2019).

Betts, Richard K. *Surprise Attack: Lessons for Defense Planning* (Brookings, 1982).

Bew, John, *Realpolitik* (Oxford University Press, 2016).

Black, Megan, *The Global Interior: Mineral Frontiers and American Power* (Harvard University Press, 2018).

Bloom, John, *Eccentric Orbits: The Iridium Story* (Grove Press, 2016).

Booth, Ken, and Nicholas J. Wheeler, *The Security Dilemma: Fear, Cooperation, and Trust in World Politics* (Palgrave, 2008).

Bormann, Natalie, and Michael Sheehan, ed. *Securing Outer Space* (Routledge, 2009).

Bowen, Bleddyn E. *War in Space: Strategy, Spacepower, Geopolitics* (Edinburgh University Press, 2020).

Brodie, Bernard, *Strategy in the Missile Age* (Princeton University Press, 1965).

Brown, Kendall K., ed. *Spacepower Integration: Perspectives from Space Weapons Officers* (Air University Press, 2006).

Brzesinski, Matthew, *Red Moon Rising* (Bloomsbury, 2007).

Bulkeley, Rip, Graham Spinardi, and Christopher Meredith, ed. *Space Weapons: Deterrence or Delusion?* (Polity Press, 1986).

Bull, Hedley, *The Anarchical Society* (Macmillan, 1977).

Burrows, William E. *This New Ocean: The Story of the First Space Age* (Modern Library, 1999).

Cable, James, *Gunboat Diplomacy, 1919–1991* (Palgrave, 1994).

Caldicott, Helen, and Craig Eisendrath, *War in Heaven: The Arms Race in Outer Space* (W.W. Norton, 2007).

Cameron, Sarah, *The Hungry Steppe: Famine, Violence, and the Making of Soviet Kazakhstan* (Cornell University Press, 2018).

Carr, Edward Hallett, *The Twenty Years' Crisis, 1919–1939* (Macmillan, 1974).

Ceruzzi, Paul, E. *GPS* (MIT Press, 2018).

Chalfont, Alun, *Star Wars: Suicide or Survival* (Weidenfeld & Nicolson, 1985).

Chun, Clayton K.S. *Shooting Down a "Star": Program 437, the US Nuclear ASAT System, and Present-Day Copycat Killers*, Cadre Paper No. 6, April 2000 (Air University Press, 2000).

Clausewitz, Carl von, *On War*, O.J. Matthijs Jolles, trans. In *The Book of War* (Modern Library, 2000).

Clodfelter, Mark, *The Limits of Air Power* (Free Press, 1989).

Cohen, Stephen, and Sunil Dasgupta, *Arming without Aiming: India's Military Modernization* (Brookings, 2012).

Coletta, Damon, and Frances T. Pilch, *Space and Defense Policy* (Routledge, 2009).

Craig, Campbell, *Glimmer of a New Leviathan: Total War in the Realism of Niebuhr, Morgenthau, and Waltz* (Columbia University Press, 2003).

Craig, Campbell, and Fredrik Logevall, *America's Cold War: The Politics of Insecurity* (Harvard University Press, 2009).

Crim, Brian E. *Our Germans: Project Paperclip and the National Security State* (Johns Hopkins University Press, 2018).

David, James E. *Spies and Shuttles: NASA's Secret Relationship with the DoD and CIA* (Smithsonian, 2015).

Deudney, Daniel, *Bounding Power: Republican Security Theory from the Polis to the Global Village* (Princeton University Press, 2007).

———, *Dark Skies: Space Expansionism, Planetary Geopolitics, and the Ends of Humanity* (Oxford University Press, 2020).

Dolman, Everett C. *Astropolitik: Classical Geopolitics in the Space Age* (Frank Cass, 2002).

Easton, Ian, *The Chinese Invasion Threat: Taiwan's Defense and American Strategy in Asia* (Project 2049 Institute, 2017).

Easton, Richard D. and Eric F. Frazier, *GPS Declassified: From Smart Bombs to Smartphones* (Potomac Books, 2013).

Echevarria II, Antulio J., *Clausewitz and Contemporary War* (Oxford University Press, 2007).

Edgerton, David, *The Shock of the Old: Technology and Global History Since 1900* (Profile Books, 2008).

Eisenhower, Susan, *Partners in Space: US-Russian Cooperation After the Cold War* (Eisenhower Institute, 2004).

Enloe, Cynthia, *Bananas, Beaches and Bases: Making Feminist Sense of International Politics* (University of California Press, 2014).

Epstein, Katherine C. *Torpedo: Inventing the Military-Industrial Complex in the United States and Great Britain* (Harvard University Press, 2014).

Eves, Stuart, *Space Traffic Control* (AAIA, 2017).

Faligot, Roger, *Chinese Spies: From Chairman Mao to Xi Jinping*, Natasha Lehrer, trans. (Hurst, 2015).

Ferris, Robert, *Behind the Enigma: The Authorised History of GCHQ* (Bloomsbury, 2020).

Finlan, Alastair, *Contemporary Military Strategy and the Global War on Terror* (Bloomsbury, 2014).

———, *The Gulf War 1991* (Osprey, 2003).

Freedman, Lawrence, *The Evolution of Nuclear Strategy* (Palgrave, 2003).

———, *Strategy: A History* (Oxford University Press, 2013).

Friedman, Norman, *Seapower and Space: From the Dawn of the Missile Age to Net-Centric Warfare* (Chatham Publishing, 2000).

Futter, Andrew, *The Politics of Nuclear Weapons* (Palgrave, 2020).

Geppert, Alexander C.T., ed. *Imagining Outer Space: European Astroculture in the Twentieth Century* (Palgrave, 2018).

Gill, Peter and Mark Phythian, *Intelligence in an Insecure World* (Polity, 2018).

Gopalaswamy, Bharath, *Final Frontier: India and Space Security* (Westland, 2019).

Gorman, Alice, *Dr Space Junk Vs the Universe: Archaeology and the Future* (MIT Press, 2019).

Gray, Colin S. *Airpower for Strategic Effect* (Air University Press, 2012).

———, *American Military Space Policy: Information Systems, Weapon Systems and Arms Control* (Abt Books, 1982).

———, *Another Bloody Century* (Weidenfeld & Nicolson, 2005).

———, *Modern Strategy* (Oxford University Press, 1999).

———, *Strategy and History: Essays on Theory and Practice* (Routledge, 2006).

Handberg, Roger and Zhen Li, *Chinese Space Policy: A Study in Domestic and International Politics* (Routledge, 2007).

Handel, Michael, *Masters of War: Classical Strategic Thought* (Routledge, 2001).

Harding, Robert C. *Space Policy in Developing Countries: The Search for Security and Development on the Final Frontier* (Routledge, 2013).

Harvey, Brian, *China in Space: The Great Leap Forward* (Springer, 2019).

———, *Europe's Space Programme: To Ariane and Beyond* (Springer, 2003).

———, *European-Russian Space Cooperation: From de Gaulle to ExoMars* (Springer, 2021).

———, *The Rebirth of the Russian Space Program: 50 Years After Sputnik, New Frontiers* (Springer, 2007).

Harvey, Brian, Henk H.F. Smid, and Theo Pirard, *Emerging Space Powers: The New Space Programs of Asia, the Middle East, and South America* (Springer, 2010).

Hays, Peter L., *Space and Security* (ABC-Clio, 2011).

Hays, Peter L., James M. Smith, Alan R. Van Tassel, and Guy Walsh, *Spacepower for a New Millennium: Space and U.S. National Security* (Institute for National Security Studies, 2000).

Hecht, Gabrielle, *Being Nuclear: Africans and the Global Uranium Trade* (MIT Press, 2012).

Hecht, Jeff, *Beam Weapons: The Next Arms Race* (Plenum Press, 1984).

Hennessy, Peter, *The Secret State: Preparing for the Worst, 1945–2010* (Penguin, 2010).

Heuser, Beatrice, *The Evolution of Strategy: Thinking War from Antiquity to the Present* (Cambridge University Press, 2010).

Hill, C.N. *Vertical Empire: The History of the UK Rocket and Space Programme, 1950–1971* (Imperial College Press, 2001).

Hinsley, F.H. *Power and the Pursuit of Peace* (Cambridge University Press, 1967).

Hörber, Thomas and Paul Stephenson, ed. *European Space Policy: European Integration and the Final Frontier* (Routledge, 2016).

Houchin, Roy F. *US Hypersonic Research and Development: The Rise and Fall of Dyna-Soar, 1944–1963* (Routledge, 2006).

Howard, Michael, *The Franco-Prussian War* (Granada, 1979).

Immerman, Richard H., *The Hidden Hand: A Brief History of the CIA* (Wiley, 2014).

Jenks, Andrew L., *The Cosmonaut Who Couldn't Stop Smiling* (Northern Illinois University Press, 2014).

Jervis, Robert, *The Meaning of the Nuclear Revolution* (Cornell University, 1989).

———, *Perception and Misperception in International Politics* (Princeton University Press, 2017).

Johnson-Freese, Joan, *Heavenly Ambitions: America's Quest to Dominate Space* (Pennsylvania University Press, 2009).

———, *Space as a Strategic Asset* (Columbia University Press, 2007).

———, *Space Warfare in the 21st Century: Arming the Heavens* (Routledge, 2017).

Jones, Matthew, *The Official History of the UK Strategic Nuclear Deterrent: Volume I* (Routledge, 2019).

———, *The Official History of the UK Strategic Nuclear Deterrent: Volume II* (Routledge, 2019).

Jones, Richard Wyn, *Security, Strategy, and Critical Theory* (Lynne Rienner, 1999).

Kahn, Herman, *Thinking About the Unthinkable in the 1980s* (Simon and Schuster, 1984).

Kaplan, Fred, *The Wizards of Armageddon* (Stanford University Press, 1991).

Karas, Thomas, *The New High Ground: Strategies and Weapons of Space-Age War* (Simon and Schuster, 1983).

Kevles, Bettyann, *Almost Heaven: The Story of Women in Space* (MIT Press, 2006).

Kilgore, De Witt Douglas, *Astrofuturism: Science, Race, and Visions of Utopia in Space* (University of Pennsylvania Press, 2003).

Klein, John. J. *Space Warfare: Strategy, Principles and Policy* (Routledge, 2006).

Krige, John, Angelina Long and Ashok Maharaj, *NASA in the World: Fifty Years of International Collaboration in Space* (Palgrave, 2013).

Lambakis, Steven, *On the Edge of Earth: The Future of American Space Power* (Kentucky University Press, 2001).

Lambeth, Benjamin S., *Mastering the Ultimate High Ground* (RAND, 2003).

Launius, Roger D., *Apollo's Legacy: Perspectives on the Moon Landings* (Smithsonian, 2019).

———, *NASA: A History of the U.S. Civil Space Program* (Krieger, 1994).

———, *Reaching for the Moon: A Short History of the Space Race* (Yale, 2019).

Launius, Roger D. and Dennis R. Jenkins, *To Reach the High Frontier: A History of U.S. Launch Vehicles* (Kentucky University Press, 2002).

Lee, Christopher, *War in Space* (Hamish Hamilton, 1986).

Lele, Ajey, *ISRO* (Rupa Publications, 2021).

Levine, Alan J., *The Space and Missile Race* (Praeger, 1994).

Lieven, Anatol, *Chechnya: Tombstone of Russian Power* (Yale University Press, 1998).

Linklater, Andrew, and Hidemi Suganami, *The English School of International Relations* (Cambridge University Press, 2006).

Logsdon, John M. *The Decision to Go to the Moon: Project Apollo and the National Interest* (MIT Press, 1970).

———, *John F. Kennedy and the Race to the Moon* (Palgrave, 2010).

Logsdon, John M. and Gordon Adams, ed. *Space Weapons: Are They Needed?* (Washington D.C.: Space Policy Institute, 2003).

Long, Franklin A., Donald Hafner and Jeffrey Boutwell, ed. *Weapons in Space* (W.W. Norton, 1986).

Lonsdale, David J., *The Nature of War in the Information Age: Clausewitzian Future* (Routledge, 2004).

Lupton, David E., *On Space Warfare: A Space Power Doctrine* (Air University Press, 1988).

Lutes Charles D. et al., ed. *Toward a Theory of Spacepower: Selected Essays* (National Defense University Press, 2011).

Luttwak, Edward N., *Strategy: The Logic of War and Peace* (Harvard University Press, 1987).

MacDonald, Alexander, *The Long Space Age: The Economic Origins of Space Exploration from Colonial America to the Cold War* (Yale, 2017).

MacKenzie, Donald, *Inventing Accuracy: A Historical Sociology of Nuclear Missile Guidance* (MIT Press, 1993).

Manno, Jack, *Arming the Heavens: The Hidden Military Agenda for Space, 1945–1995* (Dodd, Mead and Company, 1984).

Marcuse, Herbert, *One-Dimensional Man* (Routledge, 1991).

Massey, Harrie, and M.O. Robins, *History of British Space Science* (Cambridge University Press, 1986).

McCurdy, Howard E., *Space and the American Imagination* (Johns Hopkins University Press, 2011).

McDougall, Walter, *...The Heavens and the Earth: A Political History of the Space Age* (Johns Hopkins University Press, 1985).

McFarland Ian A., et al., *The Cambridge Dictionary of Christian Theology* (Cambridge University Press, 2011).

McLean, Alasdair, *Western European Military Space Policy* (Dartmouth, 1992).

Mearsheimer, John J., *The Tragedy of Great Power Politics* (W.W. Norton, 2001).

Milevski, Lukas, *The Evolution of Modern Grand Strategic Thought* (Oxford University Press, 2016).

Moltz, James Clay, *The Politics of Space Security: Strategic Restraint and the Pursuit of National Interests* (Stanford University Press, 2011).

Narang, Vipin, *Nuclear Strategy in the Modern Era: Regional Powers and International Conflict* (Princeton University Press, 2014).

Neufeld, Michael J., *Spaceflight: A Concise History* (MIT Press, 2018).

———, *Von Braun: Dreamer of Space, Engineer of War* (Vintage, 2008).

Nolen, Stephanie, *Promised the Moon: The Untold Story of the First Women in the Space Race* (Thunder's Mouth, 2004).

Norris, Pat, *Spies in the Sky: Surveillance Satellites in War and Peace* (Springer, 2008).

Northrop, Douglas, *Veiled Empire: Gender and Power in Stalinist Central Asia* (Ithaca, 2004).

Nye, Joseph S., *The Future of Power* (Public Affairs, 2011).

O'Hanlon, Michael, *Neither Star Wars nor Sanctuary: Constraining the Military Uses of Space* (Brookings, 2004).

Pace, Scott et al., *The Global Positioning System: Assessing National Policies* (RAND, 1995).

Paikowsky, Deganit, *The Power of the Space Club* (Cambridge University Press, 2017).

Pape, Robert A., *Bombing to Win: Air Power and Coercion in War* (Cornell University Press, 1996).

Payne Keith B., ed. *Laser Weapons in Space: Policy and Doctrine* (Westview Press, 1983).

Pekkanen, Saadia M. and Paul Kallender-Umezu, *In Defense of Japan: From the Market to the Military in Space Policy* (Stanford University Press, 2010).

Phythian, Mark, *Understanding the Intelligence Cycle* (Routledge, 2013).

Podvig, Pavel, ed. *Russian Strategic Nuclear Forces* (MIT Press, 2001).

Poole, Robert, *Earthrise: How Man First Saw the Earth* (Yale University Press, 2008).

Preston, Bob, *Plowshares and Power: The Military Uses of Civil Space* (National Defense University Press, 1994).

Preston, Bob, Dana J. Johnson, Sean J.A. Edwards, Michael Miller, and Calvin Shipbaugh, *Space Weapons, Earth Wars* (RAND, 2002).

Quinlan, Michael, *Thinking About Nuclear Weapons* (Oxford University Press, 2009).

Ra'anan, Uri, and Robert L. Pfaltzgraff, Jr., ed. *International Security Dimensions of Space* (Archon Books, 1984).

Redfield, Peter, *Space in the Tropics: From Convicts to Rockets in French Guiana* (California University Press, 2000).

Renz, Bettina, *Russia's Military Revival* (Polity, 2018).

Richelson, Jeffrey T., *America's Space Sentinels: The History of the DSP and SBIRS Satellite Systems* (Kansas University Press, 2012).

———, *Spying on the Bomb: American Nuclear Intelligence from Nazi Germany to Iran and North Korea* (W.W. Norton, 2007).

———, *The Wizards of Langley: Inside the CIA's Directorate of Science and Technology* (Westview Press, 2002).

Sadeh, Eligar, ed. *Space Strategy in the 21st Century: Theory and Policy* (Routledge, 2013).

Sagan, Carl, *Pale Blue Dot: A Vision of the Human Future in Space* (Ballantine Books, 1994).

Schelling, Thomas C., *Arms and Influence* (Yale, 2008).

Scheuerman, William E., *Morgenthau* (Polity, 2009).

Scott, Len, *The Cuban Missile Crisis and the Threat of Nuclear War* (Continuum, 2007).

Shayler, David J., and Ian Moule, *Women in Space: Following Valentina* (Springer, 2005).

Sheehan, Michael, *The International Politics of Space* (Routledge, 2007).

Sheng-Chih Wang, *Transatlantic Space Politics: Competition and Cooperation Above the Clouds* (Routledge, 2013).

Shimko, Keith L., *The Iraq Wars and America's Military Revolution* (Cambridge University Press, 2012).

Siddiqi, Asif, *Challenge to Apollo: The Soviet Union and the Space Race, 1945–1974* (National Aeronautics and Space Administration, 2000).

Singh, Gurbir, *The Indian Space Programme: India's Incredible Journey from the Third World to the First* (AstrotalkUK, 2017).

Smith, Michael V., *Ten Propositions Regarding Spacepower* (Air University Press 2002).

Spinardi, Graham, *From Polaris to Trident: The Development of U.S. Fleet Ballistic Missile Technology* (Cambridge University Press, 2008).

Spires, David N., *Beyond Horizons: A Half Century of Air Force Space Leadership* (Air University Press, 1998).

Stares, Paul B., *Space and National Security* (Brookings, 1987).

Stevenson, David, *Armaments and the Coming of War: Europe 1904–1914* (Oxford University Press, 1996).

Stoddart, Kristan, *Losing an Empire and Finding a Role: Britain, the USA, NATO and Nuclear Weapons, 1964–70* (Palgrave, 2012).

Sun Tzu, *The Art of Warfare*, Roger T. Ames, trans. in *The Book of War* (Modern Library, 2000).

Temple, L. Parker, *Shades of Gray: National Security and the Evolution of Space Reconnaissance* (American Institute of Aeronautics and Astronautics, 2005).

Tyson, Neil deGrasse, and Avis Lang, *Accessory to War: The Unspoken Alliance Between Astrophysics and the Military* (W.W. Norton, 2018).

Urban, Mark, *UK Eyes Alpha* (Faber and Faber, 1996).

Walker, James, Lewis Bernstein, and Sharon Lang, *Seize the High Ground: The U.S. Army in Space and Missile Defense* (U.S. Army Space and Missile Defense Command, 2003).

Wawro, Geoffrey, *Warfare and Society in Europe, 1792–1914* (Routledge, 2000).

Weitekamp, Margaret, *America's First Women in Space Program* (Johns Hopkins University Press, 2004).

Wendt, Alexander, *Social Theory of International Politics* (Cambridge University Press, 1999).

Westad, Odd Arne, *The Cold War: A World History* (Penguin, 2018).

White, Frank, *The Overview Effect: Space Exploration and Human Evolution* (AIAA, 1998).

White, Rowland, *Into the Black* (Penguin, 2016).

Wolfe, Audra J., *Competing with the Soviets: Science, Technology, and the State in Cold War America* (Johns Hopkins University Press, 2013).

Wortzel, Larry M., *The Dragon Extends Its Reach: Chinese Military Power Goes Global* (Potomac Books, 2013).

Ziarnick, Brent, *Developing National Power in Space: A Theoretical Model* (McFarland, 2015).

Zubok, Vladislav M., *Collapse: The Fall of the Soviet Union* (Yale University Press, 2021).

———, *A Failed Empire: The Soviet Union in the Cold War from Stalin to Gorbachev* (University of North Carolina Press, 2007).

Zubrin, Robert and Richard Wagner, *The Case for Mars* (Free Press, 2021).

BOOK CHAPTERS

Baldwin, David A. 'Power Analysis and World Politics: New Trends versus Old Tendencies' in David A. Baldwin, ed. *Theories of International Relations* (Ashgate, 2008).

Bolton, Iain Ross Ballantyne, 'Neo-realism and the Galileo and GPS negotiations', in Natalie Bormann and Michael Sheehan, ed. *Securing Outer Space* (Routledge, 2009).

Bowen, Bleddyn E., 'Space Oddities: Law, War, and the Proliferation of Spacepower', in James Gow and Ernst Djixhoorn, ed. *Routledge Handbook of War, Law, and Technology* (Routledge, 2019).

Brodie, Bernard, 'Implication for Military Policy', in Bernard Brodie, ed. *The Absolute Weapon*, (New Haven, CT, 1946).

Brown, Richard Maxwell, 'Violence', in Clyde A. Milner II, Carol A. O'Connor, Martha A. Sandweiss, ed. *The Oxford History of the American West* (Oxford University Press, 1994).

Campen, Alan D. 'Introduction', in Alan D. Campen, ed. *The First Information War* (AFCEA, 1992).

Chong, Alan and Quang Minh Pham, 'Critical Perspectives from Outside China on the Belt and Road Initiative: An Introduction', in Alan Chong and Quang Minh Pham, ed. *Critical Reflections on China's Belt and Road Initiative* (Palgrave, 2020).

Day, Dwayne, 'Behind the Blue: The Unknown U.S. Air Force Manned Space Program', in Paul G. Gillespie and Grant T. Weller, ed. *Harnessing the Heavens: National Defense Through Space* (Imprint Publications, 2008).

Epkenbans, Michael, 'Was a peaceful outcome thinkable? The naval race before 1914', in Holger Afflerbach and David Stevenson, ed. *An Improbable War? The Outbreak of World War I and European Political Culture Before 1914* (Berghan, 2007).

Gainor, Christopher, 'The Nuclear Roots of the Space Race', in Alexander C.T. Geppert, Daniel Brandau, and Tilmann Siebeneichner, ed. *Militarizing Outer Space: Astroculture, Dystopia, and the Cold War* (Palgrave, 2021).

Gallegos, Frank, 'After the Gulf War: Balancing Space Power's Development', in Bruce M. DeBlois, ed. *Beyond the Paths of Heaven: The Emergence of Space Power Thought* (Air University Press, 1999).

Gavroglu, Kostas, and Jürgen Renn, 'Positioning the History of Science', in Kostas Gavroglu and Jürgen Renn, ed. *Positioning the History of Science* (Springer, 2007).

Godwin, Paul H.B. 'The PLA Faces the Twenty-First Century: Reflections on Technology, Doctrine, Strategy, and Operations', in James R. Lilley and David Shambaugh, ed. *China's Military Faces the Future* (East Gate, 1999).

Griffin, Penny, 'The spaces between us: The gendered politics of outer space', in Natalie Bormann and Michael Sheehan, ed. *Securing Outer Space* (Routledge, 2009).

Hecht, Gabrielle, 'Introduction' in Gabrielle Hecht, ed. *Entangled Geographies: Empire and Technopolitics in the Global Cold War* (MIT Press, 2011).

———, 'On the Fallacies of Cold War Nostalgia: Capitalism, Colonialism, and South African Nuclear Geographies' in Gabrielle Hecht, ed. *Entangled Geographies: Empire and Technopolitics in the Global Cold War* (MIT Press, 2011).

Hersch, Matthew, and Cassandra Steer, 'Introduction', in Cassandra Steer and Matthew Hersch, ed. *War and Peace in Outer Space*, (Oxford University Press, 2021).

Hörber, Thomas, 'Introduction', in Thomas Hörber and Paul Stephenson, ed. *European Space Policy* (Routledge, 2016).

———, 'Post-war space policy in Europe', in Thomas Hörber and Paul Stephenson, ed. *European Space Policy: European Integration and the Final Frontier* (Routledge, 2016).

Jinyuan Su, 'The Legal Challenge of Arms Control in Space', in Cassandra Steer and Matthew Hersch, ed. *War and Peace in Outer Space*, (Oxford University Press, 2021).

Krepon, Michael, Eric Hagt, Shen Dingli, Bao Shixiu, Michael Pillsbury, and Ashley Tellis, 'China's Military Space Strategy: An Exchange', *Survival*, 2008, 50:1.

Krige, John, 'Embedding the National in the Global: US-French Relationships in Space Science and Rocketry in the 1960s', in Orsekes and Krige, *Science and Technology in the Global Cold War* (MIT Press, 2014).

Launius, Roger D. 'Introduction: Episodes in the Evolution of Launch Vehicle Technology', in Roger D. Launius and Dennis R. Jenkins, ed. *To Reach the High Frontier: A History of U.S. Launch Vehicles* (Kentucky University Press, 2002).

———, 'National Security, Space, and the Course of Recent U.S. History', in Paul G. Gillespie and Grant T. Weller, ed. *Harnessing the Heavens: National Defense Through Space* (Imprint Publications, 2008).

Lee, James G. 'Counterspace Operations for Information Dominance', in Bruce M. DeBlois, ed. *Beyond the Paths of Heaven: The Emergence of Space Power Thought* (Air University Press, 1999).

Lieven, Dominic, 'Russian Empires', in Sally N. Cummings and Raymond Hinnebusch, ed. *Sovereignty After Empire: Comparing the Middle East and Central Asia* (Edinburgh University Press, 2011).

McCarthy, Daniel R. 'Introduction: Technology in world politics', in Daniel R. McCarthy, ed, *Technology and World Politics: An Introduction* (Routledge, 2018).

McKinley, Cynthia A.S. 'When the Enemy Has Our Eyes', in Bruce M. DeBlois, ed. *Beyond the Paths of Heaven: The Emergence of Space Power Thought* (Air University Press, 1999).

Mueller, Karl P. 'Totem and Taboo: Depolarizing the Space Weaponization Debate', in John M. Logsdon and Gordon Adams (eds.) *Space Weapons: Are They Needed?* (Washington D.C.: Space Policy Institute, 2003).

Neufeld, Michael J. 'Cold War—But No War—in Space', in Alexander C.T. Geppert, Daniel Brandou, Tilmann Siebeneicher, ed. *Militarizing Outer Space: Astroculture, Dystopia, and the Cold War* (Palgrave, 2021).

Oikonomou, Iraklis, 'The EMWF: EU Armaments and Space Policy', in Thomas Hörber and Paul Stephenson, ed. *European Space Policy: European Integration and the Final Frontier* (Routledge, 2016).

Oldenziel, Ruth, 'Islands: The United States as a Networked Empire' in Gabrielle Hecht, ed. *Entangled Geographies: Empire and Technopolitics in the Global Cold War* (MIT Press, 2011).

Orsekes, Naomi, 'Introduction', in Naomi Orsekes and John Krige, ed. *Science and Technology in the Global Cold War* (MIT Press, 2014).

Peldszus, Regina, 'Architectures of Command: The Dual-Use Legacy of Mission Control Centers', in Alexander C.T. Geppert, Daniel Brandou, Tilmann Siebeneicher, ed. *Militarizing Outer Space: Astroculture, Dystopia, and the Cold War* (Palgrave, 2021).

Peoples, Columba, 'Haunted dreams: Critical theory, technology and the militarization of space', in Natalie Bormann and Michael Sheehan, ed. *Securing Outer Space* (Routledge, 2009).

Poole, Robert, 'The Myth of Progress: *2001—A Space Odyssey*' in Alexander C.T. Geppert, ed. *Limiting Outer Space: Astroculture After Apollo* (Palgrave, 2018).

Schmalzer, Sigrid, 'Self-reliant Science: The Impact of the Cold War on Science in Socialist China', in Naomi Orsekes and John Krige, ed. *Science and Technology in the Global Cold War* (MIT Press, 2014).

Sheehan, Michael, 'Profaning the Path to the Sacred: The Militarisation of the European Space Programme', in Natalie Bormann and Michael Sheehan, ed. *Securing Outer Space* (Routledge, 2009).

———, 'West European Integration and the Militarization of Space, 1945–1970', in Alexander C.T. Geppert, Daniel Brandou, Tilmann Siebeneicher, ed. *Militarizing Outer Space: Astroculture, Dystopia, and the Cold War* (Palgrave, 2021).

Siddiqi, Asif, 'Dispersed Sites: San Marco and the Launch from Kenya' in John Krige, ed. *How Knowledge Moves: Writing the Transnational History of Science and Technology* (University of Chicago Press, 2019).

———, 'Soviet Spacepower During the Cold War', in Paul G. Gillespie and Grant T. Weller, ed. *Harnessing the Heavens: National Defense Through Space* (Imprint Publications, 2008).

Wentz, Larry K. 'Communications support for the high technology battlefield', in Alan D. Campen, ed. *The First Information War* (AFCEA, 1992).

Xiaobing Li, *The Dragon in the Jungle: The Chinese Army in the Vietnam War* (Oxford University Press, 2020).

Xiaoming Zhang, *Deng Xiaoping's Long War: The Military Conflict Between China and Vietnam, 1979–1991* (University of North Carolina Press, 2015).

You Ji, 'The Indian Ocean: A Grand Sino-Indian Game of "Go"', in David Brewster, ed. *India and China at Sea* (Oxford University Press, 2018).

Zuoyue Wang, 'The Cold War and the Reshaping of Transnational Science in China', in Naomi Orsekes and John Krige, ed. *Science and Technology in the Global Cold War* (MIT Press, 2014).

JOURNAL ARTICLES

Acton, James M. 'Escalation through Entanglement: How the Vulnerability of Command-and-Control Systems Raises the Risks of Inadvertent Nuclear War,' *International Security*, 2018, 43:1.

Ahmed, Raja Qaismer, 'Emerging Trends of Space Weaponization: India's Quest for Space Weapons and Implications for Security in South Asia', *Astropolitics*, 2020, 18:2.

Alexis-Martin, Becky, 'The nuclear imperialism-necropolitics nexus: contextualizing Chinese-Uyghur oppression in our nuclear age', *Eurasian Geography and Economics*, 2019, 60:2.

Baldwin, David A. 'Interdependence and power: a conceptual analysis,' *International Organization*, 1980, 34:4.

Bateman, Aaron, 'Intelligence and alliance politics: America, Britain, and the Strategic Defense Initiative', *Intelligence and National Security*, 2021, 36:7.

———, 'Mutually assured surveillance at risk: Anti-satellite weapons and Cold War arms control', *Journal of Strategic Studies*, 2022, 45:1.

———, 'Technological Wonder and Strategic Vulnerability: Satellite Reconnaissance and American National Security during the Cold War', *Journal of Intelligence and Counterintelligence*, 2020, 33:2.

Baylis John, and Kristan Stoddart, 'The British Nuclear Experience: The Role of Ideas and Beliefs (Part One)', *Diplomacy and Statecraft*, 2012, 23:2.

Beidleman, Scott W. 'GPS Vs Galileo: Balancing for Position in Space', *Astropolitics*, 2005, 3:2.

Betts, Richard K. 'Analysis, War, and Decision: Why Intelligence Failures Are Inevitable', *World Politics*, 1978, 31:1.

Betz, David J. 'The More You Know, the Less You Understand', *Journal of Strategic Studies*, 2006, 29:3.

Beyerchen, Alan, 'Clausewitz, Nonlinearity, and the Unpredictability of War', *International Security*, 1992, 17:3.

Billaud Pierre, and Venance Journe, 'The Real Story Behind the Making of the French Hydrogen Bomb: Chaotic, Unsupported, but Successful', *Nonproliferation Review*, 2008, 15:2.

Blasko, Dennis J. 'The PLA army after "below the neck" reforms: contributing to China's joint warfighting, deterrence, and MOOTW Posture', *Journal of Strategic Studies*, 2021, 44:2.

Bowen, Bleddyn E. 'Cascading Crises: Orbital Debris and the Widening of Space Security', *Astropolitics*, 2014, 12:1.

———, 'From the sea to outer space: The command of space as the foundation of spacepower theory', *Journal of Strategic Studies*, 2019, 42:3–4.

Bronk, Justin, 'Britain's "Independent" V-Bomber Force and US Nuclear Weapons, 1957–1962', *Journal of Strategic Studies*, 2014, 37:6–7.

Brooks, Stephen G. and William C. Wohlforth, 'Power, Globalization, and the End of the Cold War,' *International Security*, 2000, 25:3.

Brun, Itai, '"While You're Busy Making Other Plans"—The "Other RMA"', *Journal of Strategic Studies*, 2010, 33:4.

Bruusgaard, Kristin Ven, 'Russian nuclear strategy and conventional inferiority', *Journal of Strategic Studies*, 2021, 44:1.

Bull, Hedley, 'The great irresponsibles? The United States, the Soviet Union, and world order', *International Journal*, 1980, 35:3,.

Burris, Matthew, 'Astroimpolitic: Organizing Outer Space by the Sword', *Strategic Studies Quarterly*, 2013, Autumn.

Byman, Daniel L. and Matthew C. Waxman, 'Kosovo and the Great Air Power Debate', *International Security*, 2000, 28:4.

Chandrashekar, S. 'Space, War, and Deterrence: A Strategy for India', *Astropolitics*, 2016, 14:2–3.

Cheng, Dean, 'China's Military Role in Space', *Strategic Studies Quarterly*, 2012, Spring.

Cohn, Carol, 'Sex and Death in the Rational World of Defense Intellectuals', *Signs*, 1987, 12:4.

Craig, Campbell, 'Solving the nuclear dilemma: Is a world state necessary?', *Journal of International Political Theory*, 2019, 15:3.

Downing, John, 'The Intersputnik System and Soviet Television', *Soviet Studies*, 1985, 37:4.

Duvall, Raymond, and Jonathan Havercroft, 'Taking sovereignty out of this world: space weapons and empire of the future', *Review of International Studies*, 2008, 34:4.

Finlan, Alastair, 'British Special Forces in the Falklands War of 1982', *Small Wars and Insurgencies*, 2002, 13:3.

Ford, Christopher, and David Rosenberg, 'The Naval Intelligence Underpinnings of Reagan's Maritime Strategy', *Journal of Strategic Studies*, 2005, 28:2.

Futter, Andrew '"Cyber" semantics: why we should retire the latest buzzword in security studies', *Journal of Cyber Policy*, 2018, 3:2.

Gaddis, John Lewis, 'The Long Peace: Elements of Stability in the Postwar International System', *International Security*, 1986, 10:4.

Gökalp, Iskender, 'On the Analysis of Large Technical Systems', *Science, Technology, and Human Values*, 1992, 17:1.

Gorman, Alice, 'La Terre et l'Espace: Rockets, Prisons, Protests and Heritage in Australia and French Guiana', *Archaeologies*, 2007, 3.

Gray, Colin S. and John B. Sheldon, 'Space Power and the Revolution in Military Affairs: A Glass Half Full?', *Airpower Journal*, 1999, Autumn.

Hagt, Eric, and Matthew Durnin, 'Space, China's Tactical Frontier,' *Journal of Strategic Studies*, 2011, 34:5.

Hecht, Gabrielle, 'Rupture-Talk in the Nuclear Age: Conjugating Colonial Power in Africa', *Social Studies of Science*, 2002, 32:5–6.

Herrera, Geoffrey L. 'Technology and International Systems', *Millennium*, 2003, 32:3.

Honig, Jan Willem, 'The Tyranny of Doctrine and Modern Strategy: Small (and Large) States in a Double Bind,' *Journal of Strategic Studies*, 2016, 39:2.

Hunter, Cameron, 'The forgotten first iteration of the "Chinese space threat" to US National Security', *Space Policy*, 2019, 27.

Immerwahr, Daniel, 'Twilight of Empire', *Modern American History*, 2018, 1:1.

Kania, Elsa B. and John Costella, 'Seizing the commanding heights: the PLA Strategic Support Force in Chinese military power', *Journal of Strategic Studies*, 2021, 44:2.

Kay, W.D. 'John F. Kennedy and the Two Faces of the U.S. Space Program', *Presidential Studies Quarterly*, 1998, 28:3.

Khan, Zulfqar, and Ahmad Khan, 'Space Security Trilemma in South Asia', *Astropolitics*, 2020, 17:1.

Kristensen, Hans M. and Matt Korda, 'Indian Nuclear Forces 2020', *Bulletin of the Atomic Scientists*, 2020, 76:4.

Lambakis, Stephen, 'Space Control in Desert Storm and Beyond', *Orbis*, 1995, Summer.

Launius, Roger D. 'Interpreting the Moon Landings: Project Apollo and the Historians', *History and Technology*, 2006, 22:3.

Lele, Ajey, 'Space Security Dilemma: India and China', *Astropolitics*, 2019, 17:1.

Lewis, John Wilson and Hua Di, 'China's Ballistic Missile Programs: Technologies, Strategies, Goals', *International Security*, 1992, 17:2.

Lewis, John Wilson and Xue Litai, 'Making China's nuclear war plan', *Bulletin of the Atomic Scientists*, 2012, 68:5.

Liemer, Ross, and Christopher F. Chyba, 'A Verifiable Limited Test Ban for Anti-Satellite Weapons', *Washington Quarterly*, 2010, 33:3.

Logan, David C. 'Are they reading Schelling in Beijing? The dimensions, drivers, and risks of nuclear-conventional entanglement in China', *Journal of Strategic Studies*, 2020, Online First.

Long, Austin, and Brendan Rittenhouse Green, 'Stalking the Secure Second Strike: Intelligence, Counterforce, and Nuclear Strategy', *Journal of Strategic Studies*, 2015, 38:1–2.

Lopez, Laura Delgado, 'Predicting and Arms Race in Space: Problematic Assumptions for Space Arms Control', *Astropolitics*, 2012, 10:1.

Maher, Neil M. 'Grounding the Space Race', *Modern American History*, 2018, 1:1.

Maile, David Uahikeaikalei'ohu, 'Resurgent Refusals: Protecting Mauna a Wākea and Kanaka Maoli Decolonization', *Hūlili: Multidisciplinary Research on Hawaiian Well-Being*, 2019, 11:1.

Mann, Edward, 'Desert Storm: The First Information War?', *Airpower Journal*, Winter 1994.

Mathieu, Charlotte, 'Assessing Russia's space cooperation with China and India—Opportunities and challenges for Europe', *Acta Astronautica*, 2010, 66.

McCarthy, Daniel R. 'Technology and "the International" or: How I Learned to Stop Worrying and Love Determinism', *Millennium*, 2013, 41:3.

McQuaid, Kim, 'Sputnik Reconsidered: Image and Reality in the Early Space Age', *Canadian Review of American Studies*, 2007, 37:3.

Moltz, James Clay, 'The Changing Dynamics of Twenty-First-Century Space Power', *Strategic Studies Quarterly*, 2019, 13:1.

Moore, Richard, 'Bad Strategy and Bomber Dreams: A New View of the Blue Streak Cancellation', *Contemporary British History*, 2013, 27:2.

Morgan-Owen, David, 'History and the Perils of Grand Strategy', *The Journal of Modern History*, 2020, 92.

Moss, Tristan, '"There Are Many Other Things More Important to Us Than Space Research": The Australian Government and the Dawn of the Space Age, 1956–62', *Australian Historical Studies*, 2020, 51:4.

Mowthorpe, Matthew, 'The Soviet/Russian approach to military space', *The Journal of Slavic Military Studies*, 2002, 15:3.

Murthia, K.R. Sridhara, and A. Bhaskaranarayanab, and H.N. Madhusudanab, 'New developments in Indian space policies and programmes—The next five years', *Acta Astronautica*, 2010, 66.

Neufeld, Michael J. 'The Nazi aerospace exodus: towards a global, transnational history', *History and Technology*, 2012, 28:1.

Pagedas, C.A. 'The afterlife of Blue Streak: Britain's American answer to Europe', *The Journal of Strategic Studies*, 1995, 18:2.

Pelopidas, Benoît, and Sébastien Philippe, 'Unfit for purpose: reassessing the development and deployment of French nuclear weapons (1956–1974)', *Cold War History*, 2020, online first.

Peoples, Columba, and Tim Stevens, 'At the Outer Limits of the International: Orbital infrastructures and the technopolitics of planetary (in)security', *European Journal of International Security*, 2020, 5:3.

Podvig, Pavel, 'Did Star Wars Help End the Cold War? Soviet Response to the SDI Program', *Science and Global Security*, 2017, 25:1.

———, 'History and the Current Status of the Russian Early-Warning System', *Science and Global Security*, 2002, 10.

Pollpeter, Kevin, 'Space, the New Domain: Space Operations and Chinese Military Reforms', *Journal of Strategic Studies*, 2016, 39:5–6.

Powel, Brieg, 'Blinkered Learning, Blinkered Theory: How Histories in Textbooks Parochialize IR', *International Studies Review*, 2020, 22:4.

Ramana, M.V., R. Rajaraman, and Zia Mian, 'Nuclear Early Warning in South Asia: Problems and Issues', *Economic and Political Weekly*, 2004, 39:3.

Rid, Thomas, 'Cyber War Will Not Take Place', *Journal of Strategic Studies*, 2012, 35:1.

Sankaran, Jaganath, 'Limits of the Chinese Antisatellite Threat to the United States', *Strategic Studies Quarterly*, 2014, 8:4.

Sheehan, Michael, 'Counterspace Operations and the Evolution of US Military Space Doctrine', Air Power Review, 2009, 12:2.

Siddiqi, Asif, 'Shaping the World: Soviet-African Technologies from the Sahel to the Cosmos', *Comparative Studies of South Asia, Africa and the Middle East*, 2021, 41:1.

———, 'The Soviet Fractional Orbital Bombardment System (FOBS): A Short Technical History', *Quest*, 2000, 7:4.

———, 'Staring at the Sea: The Soviet RORSAT and EORSAT Programmes', *Journal of the British Interplanetary Society*, 1999, 52:11–12.

———, 'Whose India? SITE and the origins of satellite television in India', *History and Technology*, 2020, 36, 3–4.

Snyder, Glenn H. 'Deterrence and power', *The Journal of Conflict Resolution*, 1960, 4:2.

Stout, Mark, and Michael Warner, 'Intelligence is as intelligence does', *Intelligence and National Security*, 2018, 33:4.

Sukhankin, Sergey, 'From "Bridge of Cooperation" to A2/AD "Bubble": The Dangerous Transformation of Kaliningrad Oblast', *Journal of Slavic Military Studies*, 2018, 31:1.

Suzuki Kazuto, 'The contest for leadership in East Asia: Japanese and Chinese approaches to outer space', *Space Policy*, 2013, 29.

Thierry Tardy, 'Does European defence really matter? Fortunes and misfortunes of the Common Security and Defence Policy', *European Security*, 2018, First Online.

Tracy, Cameron L. and David Wright, 'Modelling the Performance of Hypersonic Boost-Glide Missiles', *Science & Global Security*, 2020, 28: 3.

Twigg, Judyth L. 'Russia's space program: continued turmoil', *Space Policy*, 1999, 15.

Warden, John A. 'Success in Modern War: A Response to Robert Pape's *Bombing to Win*', *Security Studies*, 1997, 7:2.

Watling, Jack, 'More Sensors Than Sense', *Whitehall Papers*, 2021, 99:1.

Watts, Barry D. 'Ignoring reality: Problems of theory and evidence in security studies', *Security Studies*, 1997, 7:2.

Wicken, Olav, 'Space science and technology in the Cold War: The ionosphere, the military and politics in Norway', *History and Technology*, 1997, 13:3.

Wuthnow, Joel, 'A Brave New World for Chinese Joint Operations', *Journal of Strategic Studies*, 2017, 40:1–2.

Yonggang Fan, 'Latecomer's Strategy: An assessment of BDS industrialization policy', *Space Policy*, 2016, 38.

Zhang, Yongjin, 'China and liberal hierarchies in global international society: power and negotiation for normative change', *International Affairs*, 2016, 92:4.

REPORTS, DOCUMENTS, PRIMARY SOURCES, THESES, PRESENTATIONS

Academy of Military Science Military Strategy Department, People's Liberation Army, *The Science of Military Strategy* (China Aerospace Studies Institute, 2013).

Air Force Doctrine Document 3–14.1, *Counterspace Operations*, 28/07/2011.

Assemblée National, 'Compte rendu Commission de la défense nationale et des forces armées', briefing by Chief of the French Navy, 13/10/2021, https://www.assemblee-nationale.fr/dyn/15/comptes-rendus/cion_def/l15cion_def2122008_compte-rendu (accessed 03/11/2021).

Bateman, Aaron, 'US-UK Military and Intelligence Cooperation in Space, 1960–1990', presentation to the Intelligence, Security, Strategic Studies Research Cluster, School of History, Politics and International Relations, University of Leicester, 25/2/2021.

Bernard, Richard L. *The Foreign Missile and Space Telemetry Collection Story—The First 50 Years: Part One—The 1950s and 1960s* (U.S. National Security Agency, 1998).

Bowen, Bleddyn E., 'UK Spacepower and the Integrated Review: The Search of Strategy' (Freeman Air and Space Institute, 2021).

Bowen, Bleddyn E., and Cameron Hunter, 'Chinese Fractional Orbital Bombardment', Policy Brief No. 78, Asia Pacific Leadership Network, 2021.

Cheng, Dean, Testimony before the Committee on Commerce, Science, and Transportation United States Senate, 9 April 2014.

Dickey, Robin, 'The Rise and Fall of Space Sanctuary in U.S. Policy', September 2020, Aerospace Corporation.

Ellis, Thomas, *Reds in Space: American Perceptions of the Soviet Space Programme from Apollo to Mir 1967–1991*, PhD Thesis, University of Southampton, 2018.

Engvall, John, *Russia's Military R&D Infrastructure: A Primer* (Swedish Ministry of Defence, 2021).

European Commission, 'Space Strategy for Europe', 26 October 2016.

European Global Navigation Satellite Systems Agency, 'GSA GNSS Market Report', 2019, Issue 6.

Federici, Gary, *From the Sea to the Stars: A Chronicle of the U.S. Navy's Space and Space-related Activities, 1944–2009* (US Navy, 2010).

French Ministry of the Armed Forces, 'Space Defence Strategy', 2019, p. 4 (English version), https://www.abc.net.au/news/2021-03-31/raaf-looks-to-space-as-it-celebrates-100-years/100039914.

Grau, Lester W. and Charles K. Bartles, *The Russian Way of War: Force Structure, Tactics, and Modernization of the Russian Ground Forces* (Foreign Military Studies Office, US Army, 2016).

Harrison, Todd, Kaitlyn Johnson, and Makena Young, 'Defense Against the Dark Arts in Space: Protecting Space Systems from Counterspace Weapons' (Center for Strategic and International Studies, 2021).

Hays, Peter L. *Struggling Towards Space Doctrine: U.S. Military Space Plans, Programs, and Perspectives During the Cold War*, PhD Thesis, Tufts University, 1994.

Heginbotham, Eric, et al., *The U.S.-China Military Scorecard: Forces, Geography, and the Evolving Balance of Power 1996–2017* (RAND, 2015).

Karako, Tom, and Masao Dahlgren, 'Complex Air Defense: Countering the Hypersonic Missile Threat' (CSIS, 2022).

Kelso, T.S. Keynote address, presented at the 6th Space Traffic Management Conference, Austin, TX, 19 February 2020, https://celestrak.com/publications/STM/2020/.

Kennedy, John F. Speech at Rice University, Houston, Texas, 12 September 1962, transcript available at: https://er.jsc.nasa.gov/seh/ricetalk.htm.

Kofman Michael, et al., *Russian Military Strategy: Core Tenets and Operational Concepts* (Center for Naval Analyses, 2021).

Krepinevich, Andrew, 'The Military-Technical Revolution: A Preliminary Assessment', Center for Strategic and Budgetary Assessments, 1992.

Kulacki, Gregory, and Jeffrey G. Lewis, *A Place for One's Mat: China's Space Program 1956–2003* (American Academy of Arts & Sciences, 2009).

Lal, Bhavya, Asha Balakrishnan, Becaja M. Caldwell, Reina S. Buenconsejo, Sara A. Carioscia, 'Global Trends in Space Situational Awareness and Space Traffic Management', 01/04/2018, Institute for Defense Analyses.

Marshall, Andrew, Memorandum of 23 August 1993, Office of the Secretary of Defense, Washington DC.

McDermott, Roger N., 'Russia's Electronic Warfare Capabilities to 2025' (International Centre for Defence and Security, 2017).

Morgan, Forrest E., 'Deterrence and First-Strike Stability in Space' (RAND, 2010).

Mueller, Karl P., 'The Absolute Weapon and the Ultimate High Ground: Why Nuclear Deterrence and Space Deterrence are Strikingly Similar—Yet Profoundly Different', in Michael Krepon and Julia Thompson, ed. *Anti-Satellite Weapons, Deterrence and Sino-American Space Relations* (Stimson Centre, 2013).

Nelson, Bill, State of NASA Address, 02/06/2021. Available at: https://www.youtube.com/watch?v=nDem528iNF0 (accessed 22/01/2022).

Peldszus, Regina, 'EU Space Surveillance & Tracking Support Framework', presented at 12th ESPI Autumn Conference, 27/09/2018.

Penent, Guilhelm, 'Introduction', in Guilhelm Penent, ed., 'Governing the Geostationary Orbit Orbital Slots and Spectrum Use in an Era of Interference' (IFRI, 2014).

Pfrang, Kaila, and Brian Weeden, 'Russian Co-orbital Anti-Satellite Testing Fact Sheet', August 2020, Secure World Foundation.

———, 'Russian Direct Ascent Anti-Satellite Testing Fact Sheet', August 2020, Secure World Foundation.

———, 'US Direct Ascent Anti-Satellite Testing Factsheet', August 2020, Secure World Foundation.

Podvig, Pavel, 'Russian space systems and the risk of weaponizing space', in Samuel Bendett et al., 'Advanced Military Technology in Russia', Chatham House, September 2021.

RAND Corporation, 'Preliminary Design of an Experimental World-Circling Spaceship', May 1946, https://www.rand.org/pubs/special_memoranda/SM11827.html (accessed 27/10/2021).

Richelson, Jeffrey T., *Soldiers, Spies and the Moon: Secret U.S. and Soviet Plans from the 1950s and 1960s*, National Security Archive Electronic Briefing Book No. 479 (National Security Archive, 2014) https://nsarchive2.gwu.edu/NSAEBB/NSAEBB479/ (accessed 27/10/2021).

Roberts, Thomas G., 'Space-based Missile Defense: How Much is Enough?', 13/03/2019, https://aerospace.csis.org/data/space-based-missile-interceptors/ (accessed 11/12/2021).

Shaw, George, 'MarRINav Final Report', 25/03/2020, NLA International Limited.

Starling, Clementine G. et al., 'The Future of Security in Space: A Thirty-Year Strategy', April 2021, Atlantic Council, https://www.atlanticcouncil.org/wp-content/uploads/2021/04/TheFutureofSecurityinSpace.pdf.

Sterling, Michael J., 'Soviet Reactions to NATO's Emerging Technologies for Deep Attack' (RAND, 1985).

Stokes, Mark A. and Dean Cheng, 'China's evolving space capabilities: Implications for U.S. interests', Project 2049 Institute, 26/04/2012.

Stroikos, Dimitrios, 'China, India in Space and the Orbit of International Society: Power, Status, and Order on the High Frontier', PhD Thesis, London School of Economics, November 2016.

Susumu Yoshitomi, 'SSA Capabilities and Policies in Japan', presentation at the Space Situational Awareness Workshop, Seoul, 24–25 January

2019. Available at: https://swfound.org/media/206349/susumu-yoshitomi-ssa-workshop-in-seoul-20190124.pdf (accessed 25/04/2021).

Thomas, Timothy, 'Russian Lessons Learned in Syria: An Assessment', June 2020 (MITRE Center for Technology and National Security).

Trevino, Natalie B., 'The Cosmos Is Not Finished', PhD Thesis, University of Western Ontario, 2020.

Union of Concerned Scientists, Satellite Database. Available at: https://www.ucsusa.org/resources/satellite-database.

United Nations Office of Outer Space Affairs, The Outer Space Treaty, Article I. Available at: https://www.unoosa.org/pdf/gares/ARES_21_2222E.pdf (accessed 31/01/2022).

U.S. Air Force, *AU-18 Military Space Primer* (Air University Press, 2009).

U.S. Central Intelligence Agency, 'The Ariane Space Launch Vehicle: Europe's Answer to the US Space Shuttle', 1 July 1983, Doc. No. CIA-RDP84M00044R000200610001–1. Available at: https://www.cia.gov/readingroom/document/cia-rdp84m00044r000200610001–1.

———, Memorandum: 'Intelligence as a force multiplier', 25/04/1986, Document No. CIA-RDP89B01330R000400750014–7. Available at: https://www.cia.gov/readingroom/document/cia-rdp89b01330r000400750014–7 (accessed 03/11/2021).

———, *North Korea Military Power* (US Government Publishing Office, 2021).

U.S. Government Accountability Office, 'Surplus Missile Motors: Sale Price Drives Potential Effects on DOD and Commercial Launch Providers', August 2017, GAO-17–609.

U.S. House of Representatives Committee on Armed Services, 'Report of the Commission to Assess United States National Security Space Management and Organization', 11/01/2001.

U.S. Joint Chiefs of Staff, Joint Publication 31–4, 'Space Operations', 26/10/2020.

U.S. Missile Defense Agency, 'PTSS Fact Sheet', https://www.mda.mil/global/documents/pdf/ptss.pdf (accessed 29/06/2021).

U.S. National Security Council Memorandum 345.

U.S. Navy, 'From the Sea to the Stars: A Chronicle of the U.S. Navy's Space and Space-related Activities, 1944–2009' (Naval History and Heritage Command, 2010).

U.S. Space Force, Space Capstone Publication, 'Spacepower Doctrine for Space Forces', June 2020.

Vidal, Florian, 'Russia's Space Policy: The Path of Decline', 07/01/2021, IFRI.

Watling, Jack and Nick Reynolds, 'Operation Z: The Death Throes of an Imperial Delusion', Royal United Services Institute, 22/04/2022.

Weeden, Brian, 'Current and Future Trends in Chinese Counterspace Capabilities', IFRI, November 2020.

Weeden, Brian, and Vitoria Samson, ed. 'Global Counterspace Capabilities: An Open Source Assessment', Secure World Foundation, April 2021.

The White House, 'Memorandum on Space Policy Directive 7: Infrastructure and Technology', 15 January 2021, https://trump-whitehouse.archives.gov/presidential-actions/memorandum-space-policy-directive-7/ (accessed 02/07/2021).

Woolf, Amy F. 'Conventional Prompt Global Strike and Long-Range Ballistic Missiles: Background and Issues', Congressional Research Service Report, 14 February 2020.

NEWS / MAGAZINES / WEBSITES

Bateman, Aaron, 'Spooks and satellites: the role of intelligence in Cold War American space policy', *The Space Review*, 16/11/2020 https://www.thespacereview.com/article/4070/1 (accessed 29/11/2020).

BBC, 'Space Force: Trump officially launches new US military service', 19 December 2019, https://www.bbc.co.uk/news/world-us-canada-50876429 (accessed 31/12/2021).

Billings, Lee, 'Q&A: Plotting U.S. Space Policy with White House Adviser Scott Pace', *Scientific American*, 06/11/2017, https://www.scientificamerican.com/article/q-a-plotting-u-s-space-policy-with-white-house-adviser-scott-pace/?utm_term=space_news_text_free&sf147509331=1&utm_content=buffer350f5&utm_medium=social&utm_source=twitter.com&utm_campaign=buffer (accessed 06/11/2017).

Broad, William J. 'How Space Became the Next "Great Power" Contest Between the U.S. and China', *New York Times*, 24/01/2021, https://www.nytimes.com/2021/01/24/us/politics/trump-biden-pentagon-space-missiles-satellite.html (accessed 09/01/2022).

Carstens, Karen, 'EU and US reach agreement over Galileo satellite', *Politico*, 25 February 2004, https://www.politico.eu/article/eu-and-us-reach-agreement-over-galileo-satellite/ (accessed 02/07/2021).

Center for Strategic and International Studies, 'Missiles of China', 12 April 2021, https://missilethreat.csis.org/missile/dong-feng-26-df-26/ (accessed 11/10/2021).

———, 'Missiles of Iran', https://missilethreat.csis.org/country/iran/ (accessed 18/10/2021).

———, 'Missiles of South Korea', https://missilethreat.csis.org/country/south-korea/ (accessed 18/10/2021).

———, 'Missiles of Taiwan', https://missilethreat.csis.org/country/taiwan/ (accessed 18/10/2021).

Chow, Denise, 'Jeff Bezos says spaceflight reinforced his commitment to solving climate change', ABC News, 20/07/2021, https://www.nbcnews.com/science/science-news/jeff-bezos-says-spaceflight-reinforced-commitment-solving-climate-chan-rcna1467 (accessed 08/01/2022).

Clark, Colin, 'US, China, Russia Test New Space War Tactics: Sats Buzzing, Spoofing, Spying', *Breaking Defense*, 28/10/2021, https://breakingdefense.com/2021/10/us-china-russia-test-new-space-war-tactics-sats-buzzing-spoofing-spying/ (accessed 15/12/2021).

Davis, Scott, 'The Story of the Kettering Radio Group', https://spacecentre.co.uk/blog-post/kettering-radio-group/ (accessed 21/07/2021).

Day, Dwayne, 'Above the clouds: the White Cloud ocean surveillance satellites', *The Space Review*, 13 April 2009, https://www.thespacereview.com/article/1351/1 (accessed 19/06/2021).

———, 'Applied witchcraft: American communications intelligence satellites during the 1960s', *The Space Review*, 19/10/2020, https://www.thespacereview.com/article/4051/1 (accessed 29/11/2020).

———, 'Burning Frost, the view from the ground: shooting down a spy satellite in 2008', *The Space Review*, 21/06/2021, https://www.thespacereview.com/article/4198/1 (accessed 01/07/2021).

———, 'Intersections in real time: the decision to build the KH-11 KENNEN reconnaissance satellite (part 1)', *The Space Review*, 09/09/2019, https://www.thespacereview.com/article/3791/1 (accessed 09/07/2021).

———, 'Intersections in real time: the decision to build the KH-11 KENNEN reconnaissance satellite (part 2)', *The Space Review*, 16/09/2019, https://www.thespacereview.com/article/3795/1 (accessed 09/07/2021).

———, 'The NRO and the Space Shuttle', *The Space Review*, 31/01/2022, https://www.thespacereview.com/article/4324/1 (accessed 08/02/2022).

———, 'Nuking Moscow with a Space Shuttle', *The Space Review*, 23/09/2021, https://www.thespacereview.com/article/3855/1 (accessed 21/12/2021).

———, 'Red Moon Revisited', *The Space Review*, 11/03/2021, https://www.thespacereview.com/article/3674/1 (accessed 12/02/2022).

———, 'Shipkillers: From Satellite to Shooter at Sea', *The Space Review*, 28/06/2021, https://www.thespacereview.com/article/4204/1 (accessed 29/06/2021).

———, 'Stealing secrets from the ether: missile and satellite telemetry interception during the Cold War', *The Space Review*, 17/01/2022, https://www.thespacereview.com/article/4316/1 (accessed 24/01/2022).

———, 'War at Sea, Seen from Above', *The Space Review*, 25/04/2022, https://www.thespacereview.com/article/4375/1 (accessed 26/04/2022).

The Economic Times, 'ISRO sets up dedicated control centre for Space Situational Awareness', 16 December 2020, https://economictimes.indiatimes.com/news/science/isro-sets-up-dedicated-control-centre-for-space-situational-awareness/articleshow/79763690.cms?from=mdr (accessed 25/04/2021).

Eleazer, Wayne, 'Battle of the Titans (part 1)', *The Space Review*, 28/09/2020, https://www.thespacereview.com/article/4034/1 (accessed 04/12/2021).

Erwin, Sandra, 'GAO steps up criticism of Space Force's missile-warning satellite procurement', *Space News*, 22/09/2021, https://spacenews.com/gao-steps-up-criticism-of-space-forces-missile-warning-satellite-procurement/ (accessed 30/10/2021).

———, 'Raymond's progress report on Space Force: "All the pieces are coming together"', *SpaceNews*, 24/08/2021, https://spacenews.com/raymonds-progress-report-on-space-force-all-the-pieces-are-coming-together/ (accessed 31/12/2021).

———, 'U.S. Army signs deal with SpaceX to assess Starlink broadband', *SpaceNews*, 26 May 2020, https://spacenews.com/u-s-army-signs-deal-with-spacex-to-assess-starlink-broadband/ (accessed 27/06/2021).

European Space Agency, 'SAR-Lupe', Earth Observation Portal, https://earth.esa.int/web/eoportal/satellite-missions/s/sar-lupe (accessed 29/06/2021).

Flewelling, Brien, 'Securing Cislunar Space: A Vision for U.S. Leadership', *SpaceNews*, 9 November 2020, https://spacenews.com/op-ed-securing-cislunar-space-a-vision-for-u-s-leadership/ (accessed 04/12/2020).

Globalsecurity.org, 'Chinese Ballistic Missile Early Warning', https://www.globalsecurity.org/space/world/china/warning.htm (accessed 29/06/2020).

Goff, Stan, 'India: A High-Tech Partner for European GNSS', *Inside GNSS*, 27/07/2017, http://www.insidegnss.com/node/5570 (accessed 12/09/2017).

———, 'Successful Launch of Japan's Second Michibiki Satellite to Boost QZSS', *Inside GNSS*, 09/06/2017, http://www.insidegnss.com/node/5499 (accessed 12/09/2017).

Goswami, Namrata, 'Explaining China's space ambitions and goals through the lens of strategic culture', *The Space Review*, https://www.thespacereview.com/article/3944/1 (accessed 03/12/2020).

———, 'Why the Chang'e-4 Moon Landing is Unique', *The Space Review*, 14 January 2019, https://www.thespacereview.com/article/3639/1 (accessed 04/12/2020).

Graff, Garrett M. 'We're the only plane in the sky', *Politico*, https://www.politico.com/magazine/story/2016/09/were-the-only-plane-in-the-sky-214230/ (accessed 01/11/2021).

Greene, Andrew, 'RAAF planning for new military space command as it celebrates 100th anniversary', ABC News, 30/03/2021, https://www.abc.net.au/news/2021-03-31/raaf-looks-to-space-as-it-celebrates-100-years/100039914 (accessed 22/01/2022).

Gunter's Space Page, https://space.skyrocket.de/index.html.

Hendrickx, Bart, 'Burevestnik: a Russian air-launched anti-satellite system,' *The Space Review*, April 27 2020, https://www.thespacereview.com/article/3931/1.

———, 'EKS: Russia's Space-based Early Warning System', *The Space Review*, 8 February 2021, https://www.thespacereview.com/article/4121/1 (accessed 28/07/2021).

———, 'The status of Russia's signals intelligence satellites', *The Space Review*, 5 April 2021, https://www.thespacereview.com/article/4154/1 (accessed 29/06/2021).

Hunter, Cameron, 'China and the US are both shooting for the Moon, but don't call it a space race', *The Conversation*, 08/11/2017, https://theconversation.com/china-and-the-us-are-both-shooting-for-the-moon-but-dont-call-it-a-space-race-85228 (accessed 12/02/2022).

Indian Space Research Organisation, 'Polar Satellite Launch Vehicle', https://www.isro.gov.in/launchers/pslv (accessed 22/05/2021).

Jewet, Rachel, 'OneWeb Partners With TrustComm as US Military Distribution Partner', *Via Satellite*, 26 March 2021, https://www.satellitetoday.com/government-military/2021/03/26/oneweb-partners-with-trustcomm-as-us-military-distribution-partner/ (accessed 27/06/2021).

Jonathan's Space Report, 10 May 2021, https://www.planet4589.org/space/jsr/back/news.792.txt (accessed 24/01/2022).

Kim, Dana, 'The "Democratization of Space" and the Increasing Effects of Commercial Satellite Imagery on Foreign Policy', CSIS, https://www.csis.org/democratization-space-and-increasing-effects-commercial-satellite-imagery-foreign-policy (accessed 06/02/2022).

Krepon, Michael, 'What is a Space Weapon?' *Arms Control Wonk*, 18/03/2010, https://www.armscontrolwonk.com/archive/402665/what-is-a-space-weapon/ (accessed 28/01/2022).

Lee, Jeong-Ho and Lee Jihye, 'South Korea Seeks to Move Up Its Spot in Global Space Race', *Bloomberg*, 14/07/2021, https://www.bloomberg.com/news/articles/2021–07–14/south-korea-seeks-to-move-up-its-spot-in-global-space-race (accessed 18/10/2021).

Lee, Taylor A. and Peter W. Singer, 'China's Space Program Is More Military Than You Might Think', *Defense One*, 16/07/2021, https://www.defenseone.com/ideas/2021/07/chinas-space-program-more-military-you-might-think/183790/ (accessed 09/01/2022).

Lewis, Jeffrey and Gregory Kulacki, 'A Different View of China's ASAT Test', presentation at New America, 13/11/07. Video available at: https://www.youtube.com/watch?v=baMZKLq8zKk (accessed 03/02/2022).

Listner, Michael, 'A U.S. return to the Moon is about preserving the rule of law', *Space News*, 17 April 2020, https://spacenews.com/op-ed-a-u-s-return-to-the-moon-is-about-preserving-the-rule-of-law/ (accessed 04/12/2020).

Lory, Gregoire, '"Europe is already a great space power", says Thierry Breton', *EuroNews*,https://www.euronews.com/2022/01/28/europe-is-already-a-great-space-power (accessed 29/01/2022).

Luft, Alistair, 'The OODA Loop and the Half-Beat', *The Strategy Bridge*, 17 March 2020, https://thestrategybridge.org/the-bridge/2020/3/17/the-ooda-loop-and-the-half-beat (accessed 03/08/2021).

National Aeronautics and Space Administration, 'About the Space Station solar arrays', 04/08/2017, https://www.nasa.gov/mission_pages/station/structure/elements/solar_arrays-about.html (accessed 11/12/2021).

North Atlantic Treaty Organization, 'NATO's Approach to Space', 02/12/2021, https://www.nato.int/cps/en/natohq/topics_175419.htm (accessed 09/01/2022).

Nuclear Threat Initiative, 'The CNS India and Pakistan Missile Launch Databases', https://www.nti.org/analysis/articles/cns-india-and-pakistan-missile-launch-databases/ (accessed 18/10/2021).

Osborn, Kris, 'Missile Defense Agency Optimistic About Glide Phase Interceptor Program', *National Interest*, 12/12/2021, https://nationalinterest.org/blog/buzz/missile-defense-agency-optimistic-about-glide-phase-interceptor-program-197867 (accessed 06/02/2022).

Park Si-soo, '"We go together": US Space Force chief seeks deeper space cooperation with South Korea', *Space News*, 18/10/2022, https://spacenews.com/we-go-together-us-space-force-chief-seeks-deeper-space-cooperation-with-south-korea/ (accessed 11/01/2022).

Russian Ministry of Defence, 'Chief of the General Staff of the Russian Armed Forces—First Deputy Minister of Defence General of the Army Valery Gerasimov meets with representatives of the military diplomatic corps accredited in Russia', 17/02/2019, https://eng.mil.ru/en/news_page/country/more.htm?id=12267331@egNews (accessed 03/11/2021).

Salmeri, Antonio, 'Op-ed | No, Mars is not a free planet, no matter what SpaceX says', *SpaceNews*, 05/12/2020, https://spacenews.com/op-ed-no-mars-is-not-a-free-planet-no-matter-what-spacex-says/ (accessed 29/04/2022).

Sevastopulo, Demetri, 'China conducted two hypersonic weapons tests this summer', *Financial Times*, 21/10/2021, https://www.ft.com/content/c7139a23-1271-43ae-975b-9b632330130b (accessed 25/01/2022).

Sheetz, Michael, and Yun Li, '"Adjust your location quickly"—How China warns residents before rockets crash down from space', CNBC, 26/11/2019, https://www.cnbc.com/2019/11/26/chinese-rocket-crushes-houses-after-government-warning-to-residents.html (accessed 07/06/2021).

Siddiqi, Asif, 'Transcending Gravity: The View from Postcolonial Dhaka to Colonies in Space', *Los Angeles Review of Books*, 12/10/2020, https://lareviewofbooks.org/article/transcending-gravity-the-view-from-postcolonial-dhaka-to-colonies-in-space/ (accessed 05/06/2021).

Thorbecke, Catherine, 'What to know about Richard Branson's spaceflight, as billionaires race to the cosmos', ABC News, 10/07/2021, https://abcnews.go.com/Business/richard-bransons-spaceflight-billionaires-race-cosmos/story?id=78754546 (accessed 29/04/2022).

United Nations Office of Outer Space Affairs, 'Long-term Sustainability of Outer Space Activities', https://www.unoosa.org/oosa/en/ourwork/topics/long-term-sustainability-of-outer-space-activities.html (accessed 02/01/2022).

U.S. National Geospatial Intelligence Agency, 'National Imagery and Mapping Agency', https://www.nga.mil/defining-moments/National_Imagery_and_Mapping_Agency.html (accessed 18/06/2021).

U.S. Space Force, 'AEHF Factsheet', 22 March 2017, https://www.spaceforce.mil/About-Us/Fact-Sheets/Article/2197713/advanced-extremely-high-frequency-system/ (accessed 01/11/2021).

Vizard, Frank, 'Safeguarding GPS', *Scientific American*, 14 April 2003, http://www.scientificamerican.com/article.cfm?id=safeguarding-gps (accessed 22/02/2013).

Werner, Debra, 'Spacety shares first images from small C-band SAR satellite', *SpaceNews*, 04/01/2021, https://spacenews.com/spacety-releases-first-sar-images/ (accessed 19/06/2021).

West, Jessica, 'Outer space', *First Committee Monitor*, 2021, 19:5.

Withington, Thomas, 'Russian EW: underused or a Potemkin capability?', *Shephard Media*, 01/03/2022, https://www.shephardmedia.com/news/digital-battlespace/russian-ew-underused-or-a-potemkin-capability/ (accessed 28/04/2022).

Woodward, Bob, 'CIA Aiding Iraq in Gulf War', *Washington Post*, 15/12/1986. Available at: https://www.cia.gov/readingroom/docs/CIA-RDP90–00965R000807560006–3.pdf.

World Nuclear Association, 'Small Nuclear Power Reactors', http://www.world-nuclear.org/information-library/nuclear-fuel-cycle/nuclear-power-reactors/small-nuclear-power-reactors.aspx (accessed 04/02/2022).

Zak, Anatoly, 'IS', *Russian Space Web*, http://www.russianspaceweb.com/is.html (accessed 04/12/2020).

———, 'Russian Communications Satellites', http://www.russianspaceweb.com/spacecraft_comsats.html (accessed 26/06/2021).

———, 'Russia's Military Spacecraft', http://russianspaceweb.com/spacecraft_military.html (accessed 28/11/20).

———, 'Salyut-3 (Almaz OPS-2) space station', *Russian Space Web*, http://www.russianspaceweb.com/almaz_ops2.html (accessed 01/01/2021).

———, 'Spooky world of military satellites', *Russian Space Web*, http://russianspaceweb.com/spacecraft_military.html (accessed 28/11/2021).

———, 'The Tselina signal-intelligence satellite family', http://russianspaceweb.com/tselina.html (accessed 28/11/2021).

———, 'Yantar-1KFT', http://russianspaceweb.com/spacecraft_military.html (accessed 28/11/2021).

INDEX